T0141818

Studies in Fuzziness and Soft Computing

Volume 341

About this Series

The series “Studies in Fuzziness and Soft Computing” contains publications on various topics in the area of soft computing, which include fuzzy sets, rough sets, neural networks, evolutionary computation, probabilistic and evidential reasoning, multi-valued logic, and related fields. The publications within “Studies in Fuzziness and Soft Computing” are primarily monographs and edited volumes. They cover significant recent developments in the field, both of a foundational and applicable character. An important feature of the series is its short publication time and world-wide distribution. This permits a rapid and broad dissemination of research results.

More information about this series at http://www.springer.com/series/2941

Cengiz Kahraman · Uzay Kaymak
Adnan Yazici
Editors

Fuzzy Logic in Its 50th Year

New Developments, Directions and Challenges

Editors
Cengiz Kahraman
Department of Industrial Engineering, Management Faculty
Istanbul Technical University
Istanbul
Turkey

Adnan Yazici
Department of Computer Engineering
Middle East Technical University
Ankara
Turkey

Uzay Kaymak
Information Systems Group, School of Industrial Engineering
Eindhoven University of Technology
Eindhoven
The Netherlands

ISSN 1434-9922 ISSN 1860-0808 (electronic)
Studies in Fuzziness and Soft Computing
ISBN 978-3-319-80962-5 ISBN 978-3-319-31093-0 (eBook)
DOI 10.1007/978-3-319-31093-0

Printed on acid-free paper

This Springer imprint is published by Springer Nature
The registered company is Springer International Publishing AG Switzerland

This book is dedicated to a pioneer of the fuzzy set theory, Prof. I. Burhan Turksen.

Prof. I. Burhan Turksen

I. Burhan Turksen received the B.S. and M.S. degrees in Industrial Engineering and the Ph.D. degree in Systems Management and Operations Research all from the University of Pittsburgh, PA, USA. He joined the Faculty of Applied Science and Engineering at the University of Toronto, Canada, and became there Full Professor in 1983. In the 1984–1985 academic year, he was a Visiting Professor at the Middle East Technical

University in Ankara, Turkey, and Osaka Prefecture University in Japan. Since 1987, he has been Director of the Knowledge/Intelligence Systems Laboratory. During the 1991–1992 academic year, he was a Visiting Research Professor at LIFE, Laboratory for International Fuzzy Engineering, and the Chair of Fuzzy Theory at Tokyo Institute of Technology. During 1996 academic year, he was Visiting Research Professor at the University of South Florida, USA, and Bilkent University, Ankara, Turkey. Since December 2005, he is appointed as the Head of Department of Industrial Engineering at TOBB Economics and Technology University in Ankara.

He was and/or is a member of the Editorial Boards of the following publications: Fuzzy Sets and Systems, Approximate Reasoning, Decision Support Systems, Information Sciences, Fuzzy Economic Review, Expert Systems and its Applications, Journal of Advanced Computational Intelligence, Information Technology Management, Transactions on Operational Research, Fuzzy Logic Reports and Letters, Encyclopedia of Computer Science and Technology, Failures and Lessons Learned in Information Technology, Applied Soft Computing. He is the co-editor of NATO-ASI Proceedings on Soft Computing and Computational Intelligence, and Editor of NATO-ASI Proceedings on Computer Integrated Manufacturing as well co-editor of two special issues of Robotics and Autonomous Systems.

He is a Fellow of IFSA and IEEE, and a member of IIE, CSIE, CORS, IFSA, NAFIPS, APEO, APET, TORS, ACM, etc. He is the founding President of CSIE. He was Vice-President of IIE, General Conference Chairman for IIE International Conference, and for NAFIPS in 1990. He served as Co-Chairman of IFES'91 and Regional Chairman of World Congress on Expert Systems, WCES'91, WCES'94, WCES'96 and WCES'98, Director of NATO-ASI'87 on Computer Integrated Manufacturing and Co-Director of NATO-ASI'96 on Soft Computing and Computational Intelligence. He was General Conference Chairman for Intelligent Manufacturing Systems, IMS'1998, IMS'2001, IMS'2003. He was the President of IFSA during 1997–2001 and Past President of IFSA, International Fuzzy Systems Association during 2001–2003. Currently, he is the President, CEO and CSO, of IIC, Information Intelligence Corporation. He has published near 300 papers in scientific journals and conference proceedings.

Preface

The fuzzy set theory, founded by Lotfi A. Zadeh in 1965, had covered a lot of ground with excellent successes in almost all branches of science. Especially, fuzzy control technologies provided definite evidences to prove its power and usability in the solutions of real-life problems. These evidences caused the harsh criticisms to lose their value within the 50 years of the fuzzy set theory.

This book aims at presenting the latest position of the fuzzy set theory within the science branches and drawing the frame of its expansion. In our book, the new extensions of fuzzy sets are summarized by their pioneers: Intuitionistic fuzzy sets by Atanassov (Chapter "Mathematics of IntuitionisticFuzzy Sets"), hesitant fuzzy sets by Torra (Chapter "A Review of Hesitant Fuzzy Sets: Quantitative and Qualitative Extensions"), etc. Later, the position of the fuzzy set theory within earth and space sciences, human sciences, etc., is summarized. The last section of the book includes some applications of the fuzzy set theory.

The first chapter is on the emergence of fuzzy sets from a historical perspective. It tries to suggest some reasons why the fuzzy set theory came to life 50 years ago by pointing out the existence of streams of thought in the first half of the twentieth century in logic, linguistics, and philosophy, which paved way to the idea of moving away from the Boolean framework, through the proposal of many valued logics and the study of the vagueness phenomenon in natural languages.

The second chapter is on fuzzy decision-making. The chapter presents the pioneers of fuzzy decision-making: researchers, universities, and countries, and the media publishing fuzzy decision-making papers including the journals, books, and proceedings. It also includes numerical examples of decision-making problems with their solutions.

The third chapter is on mathematics of Intuitionistic Fuzzy Sets (IFSs). Short firsthand remarks on the history and theory of IFSs are given. Influences of other areas of mathematics for development of the IFSs theory are discussed. On the basis of results in IFSs theory, some ideas for development of other mathematical areas are offered.

The fourth chapter presents some comments on ordinary reasoning with fuzzy sets. It tries to serve as a theoretical support for computing with words, but, and specially, to motivate young researchers to go further in its extension towards the different types of commonsense reasoning.

The fifth chapter includes a review of hesitant fuzzy sets. Many researchers have paid attention to it and have proposed different extensions both in quantitative and qualitative contexts. Several concepts, basic operations, and its extensions are revised in this chapter.

The sixth chapter is on type-1 to type-n fuzzy logic and systems. The motivation for using fuzzy systems, the mathematical concepts of type-1 to type-n fuzzy sets, logic, and systems as well as their applications in solving real-world problems is presented.

The seventh chapter is on fuzzy sets in earth and space sciences. It reviews and analyzes the papers utilizing fuzzy logic in earth and space science problems from Scopus database. The graphical and tabular illustrations are presented for the subject areas, publication years, and sources of the papers on earth and space sciences.

The eighth chapter is on fuzzy sets and fuzzy logic in the human sciences. It surveys the history of fuzzy set applications in the human sciences, and then elaborates the possible reasons why fuzzy set concepts have been relatively under-utilized therein.

The ninth chapter is on fuzzy entropy used for predictive analytics. It develops models for predictive maintenance in a big data environment. The authors apply interval-valued fuzzy sets and various entropy measures defined on them to perform feature selection on process diagnostics. They show how these models can be utilized as the basis for decision support systems in process industries to aid predictive maintenance.

The 10th chapter is on fuzzy sets in agriculture. It aims at contributing to a refinement of review studies by applying fuzzy sets, fuzzy logic, and fuzzy cognitive mapping to the exploration of agriculture modeling and management.

The 11th chapter is on solving a multiobjective truck and trailer routing problem with fuzzy constraints. The fuzzy model is generalized in this work from a multiobjective approach by incorporating an objective to minimize the violation of constraints. We present and discuss the computational experiments carried out to solve the multiobjective truck and trailer routing problem with fuzzy constraint using benchmark instances with sizes ranging from 50 to 199 customers.

The 12th chapter is on health service network design under epistemic uncertainty. It provides a comprehensive review of the related literature to the HSND problem after explaining health, health system, and HSND problem briefly. The review is followed by a typical mathematical model for the concerned problem. Finally, a fuzzy programming approach is described in brief and different fuzzy measures, i.e., possibility, necessity, and credibility measures are applied to the compact form of the proposed mathematical programming model.

The 13th chapter is on robotics and control systems. It presents fuzzy control techniques as well as fuzzy mathematical scheduling model for an m machine robotic cell with one manipulator robot. Furthermore, it proposes an integrated fuzzy robotic control system, in which the fuzzy optimization model is solved at every predetermined period of time such as beginning of shifts or days, etc. Then based upon on the solutions obtained, input parameters, and unpredictable disturbances, the autonomous fuzzy control is executed at each unit of time. These two modules transfer information and feedback to each other via an intermediate collaborative module.

The 14th chapter is on the usage of fuzzy sets in the evaluation of socio-ecological systems through an interval-valued intuitionistic fuzzy multi-criteria approach. It demonstrates how to incorporate fuzzy sets theory into social sciences. An illustrative example which consists of a wide variety of social actors is used to evaluate sustainable management options based on interval-valued intuitionistic fuzzy TOPSIS method.

The 15th chapter is on the survey on models and methods for solving fuzzy linear programming problems. The solution approaches are divided into four areas: (1) Linear Programming (LP) problems with fuzzy inequalities and crisp objective function, (2) LP problems with crisp inequalities and fuzzy objective function, (3) LP problems with fuzzy inequalities and fuzzy objective function, and (4) LP problems with fuzzy parameters.

The 16th chapter is on applications of fuzzy mathematical programming approaches in supply chain planning problems. It aims to provide the useful and updated information about different sources and types of uncertainty in supply chain planning problems and the strategies used to confront with uncertainty in such problems. A hyper methodological framework is proposed to cope with uncertainty in supply chain planning problems.

We hope that this book will provide a useful resource of ideas, techniques, and methods for the development of the fuzzy set theory. We are grateful to the referees whose valuable and highly appreciated works contributed to select the high quality chapters published in this book. We would like to also thank Prof. Janusz Kacprzyk, the editor of Studies in Fuzziness and Soft Computing at Springer for his supportive role in this process, and to Assoc. Prof. Sezi Cevik Onar for her help to accelerate the processes to catch the deadlines.

Cengiz Kahraman
Uzay Kaymak
Adnan Yazıcı

Contents

Part I
Fuzzy Sets Theory: From the Past to the Future

The Emergence of Fuzzy Sets: A Historical Perspective

Didier Dubois and Henri Prade

Abstract This paper tries to suggest some reasons why fuzzy set theory came to life 50 years ago by pointing out the existence of streams of thought in the first half of the XXth century in logic, linguistics and philosophy, that paved the way to the idea of moving away from the Boolean framework, through the proposal of many-valued logics and the study of the vagueness phenomenon in natural languages. The founding paper in fuzzy set theory can be viewed as the crystallization of such ideas inside the engineering arena. Then we stress the point that this publication in 1965 was followed by several other seminal papers in the subsequent 15 years, regarding classification, ordering and similarity, systems science, decision-making, uncertainty management and approximate reasoning. The continued effort by Zadeh to apply fuzzy sets to the basic notions of a number of disciplines in computer and information sciences proved crucial in the diffusion of this concept from mathematical sciences to industrial applications.

Keywords Fuzzy sets · Many-valued logics · Vagueness · Possibility theory · Approximate reasoning

1 Introduction

The notion of a fuzzy set stems from the observation made by Zadeh [60] fifty years ago in his seminal paper that

> more often than not, the classes of objects encountered in the real physical world do not have precisely defined criteria of membership.

By "precisely defined", Zadeh means all-or-nothing, thus emphasizing the continuous nature of many categories used in natural language. This observation emphasizes the gap existing between mental representations of reality and usual

D. Dubois (✉) · H. Prade
IRIT-CNRS, Université Paul Sabatier, 31062 Toulouse Cedex 09, France
e-mail: Didier.Dubois@irit.fr

C. Kahraman et al. (eds.), *Fuzzy Logic in Its 50th Year*,
Studies in Fuzziness and Soft Computing 341,
DOI 10.1007/978-3-319-31093-0_1

mathematical representations thereof, which are traditionally based on binary logic, precise numbers, differential equations and the like. Classes of objects referred to in Zadeh's quotation exist only through such mental representations, e.g., through natural language terms such as *high* temperature, *young* man, *big* size, etc., and also with nouns such as *bird, chair*, etc. Classical logic is too rigid to account for such categories where it appears that membership is a gradual notion rather than an all-or-nothing matter.

The ambition of representing human knowledge in a human-friendly, yet rigorous way might have appeared like a futile exercise not worth spending time on, and even ridiculous from a scientific standpoint, only one hundred years ago. However in the meantime the emergence of computers has significantly affected the landscape of science, and we have now entered the era of information management. The development of sound theories and efficient technology for knowledge representation and automated reasoning has become a major challenge, now that many people possess computers and communicate with them in order to find information that helps them when making decisions. An important issue is to store and exploit human knowledge in various domains where objective and precise data are seldom available. Fuzzy set theory participates to this trend, and, as such, has close connection with Artificial Intelligence. This chapter is meant to account for the history of how the notion of fuzzy set could come to light, and what are the main landmark papers by its founder that stand as noticeable steps towards the construction of the fuzzy set approach to classification, decision, human knowledge representation and uncertainty. Besides, the reader is invited to consult a recent personal account, written by Zadeh [85], of the circumstances in which the founding paper on fuzzy sets was written.

2 A Prehistory of Fuzzy Sets

This section gives some hints to works what can be considered as forerunners of fuzzy sets. Some aspects of the early developments are described in more details by Gottwald [28] and Ostasiewicz [45, 46]. This section freely borrows from [17], previously written with the later author.

2.1 Graded Membership to Sets Before Zadeh

In spite of the considerable interest for multiple-valued logics raised in the early 1900s by Jan Łukasiewicz and his school who developed logics with intermediary truth value(s), it was the American philosopher Black [7] who first proposed so-called "consistency profiles" (the ancestors of fuzzy membership functions) in order to "characterize vague symbols."

As early as in 1946, the philosopher Abraham Kaplan argued in favor of the usefulness of the classical calculus of sets for practical applications. The essential novelty he introduces with respect to the Boolean calculus consists in entities which have a degree of vagueness characteristic of actual (empirical) classes (see Kaplan [33]). The generalization of the traditional characteristic function has been first considered by Weyl [55] in the same year; he explicitly replaces it by a continuous characteristic function. They both suggested calculi for generalized characteristic functions of vague predicates, and the basic fuzzy set connectives already appeared in these works.

Such calculus has been presented by Kaplan and Schott [34] in more detail, and has been called the calculus of empirical classes (CEC). Instead of notion of "property", Kaplan and Schott prefer to use the term "profile" defined as a type of quality. This means that a profile could refer to a simple property like *red, green*, etc. or to a complex property like *red and 20 cm long, green and 2 years old*, etc. They have replaced the classical characteristic function by an indicator which takes on values in the unit interval. These values are called the weight from a given profile to a specified class. In the work of Kaplan and Schott, the notion of "empirical class" corresponds to the actual notion of "fuzzy set", and a value in the range of the generalized characteristic function (indicator, in their terminology) is already called by Kaplan and Schott a "degree of membership" (Zadehian grade of membership). Indicators of profiles are now called membership functions of fuzzy sets. Strangely enough it is the mathematician of probabilistic metric spaces, Karl Menger, who, in 1951, was the first to use the term "ensemble flou" (the French counterpart of "fuzzy set") in the title of a paper [40] of his.

2.2 Many-Valued Logics

The Polish logician Jan Łukasiewicz (1878–1956) is considered as the main founder of multi-valued logic. This is an important point as multi-valued logic is to fuzzy set theory what classical logic is to set theory. The new system he proposed has been published for the first time in Polish in 1920. However, the meaning of truth-values other than "true" and "false" remained rather unclear until Zadeh introduced fuzzy sets. For instance, Łukasiewicz [38] interpreted the third truth-value of his 3-valued logic as "possible", which refers to a modality rather than a truth-value. Kleene [35] suggests that the third truth-value means "unknown" or "undefined". See Ciucci and Dubois [9] for a overview of such epistemic interpretations of three-valued logics. On the contrary, Zadeh [60] considered intermediate truth-degrees of fuzzy propositions as ontic, that is, being part of the definition of a gradual predicate. Zadeh observes that the case where the unit interval is used as a membership scale "corresponds to a multivalued logic with a continuum of truth values in the interval [0, 1]", acknowledging the link between fuzzy sets and many-valued logics. Clearly, for Zadeh, such degrees of truth do not refer to any kind of uncertainty, contrary to what is often found in more recent texts about fuzzy sets by various authors. Later on,

Zadeh [70] would not consider fuzzy logic to be another name for many-valued logic. He soon considered that fuzzy truth-values should be considered as fuzzy sets of the unit interval, and that fuzzy logic should be viewed as a theory of approximate reasoning whereby fuzzy truth-values act as modifiers of the fuzzy statement they apply to.

2.3 The Issue of Vagueness

More than one hundred years ago, the American philosopher Peirce [47] was one of the first scholars in the modern age to point out, and to regret, that

> Logicians have too much neglected the study of vagueness, not suspecting the important part it plays in mathematical thought.

Some time later, Russell pointed out that the law of excluded middle cannot be applied to vague predicates [49]. Even Wittgenstein [57] pointed out that concepts in natural language do not possess a clear collection of properties defining them, but have extendable boundaries, and that there are central and less central members in a category.

The claim that fuzzy sets are a basic tool for addressing vagueness of linguistic terms has been around for a long time. For instance, Novák [44] insists that fuzzy logic is tailored for vagueness and he opposes vagueness to uncertainty.

Nevertheless, in the last thirty years, the literature dealing with vagueness has grown significantly, and much of it is far from agreeing on the central role played by fuzzy sets in this phenomenon. Following Keefe and Smith [53], vague concepts in natural language display at least one among three features:

- **The existence of borderline cases**: That is, there are some objects such that neither a concept nor its negation can be applied to them. For a borderline object, it is difficult to make a firm decision as to the truth or the falsity of a proposition containing a vague predicate applied to this object, even if a precise description of the latter is available. The existence of borderline cases is sometimes seen as a violation of the law of excluded middle.
- **Unsharp boundaries**: The extent to which a vague concept applies to an object is supposed to be a matter of degree, not an all-or-nothing decision. It is relevant for predicates referring to continuous scales, like *tall, old*, etc. This idea can be viewed as a specialization of the former, if we regard as borderline cases objects for which a proposition is neither totally true nor totally false. In the following we shall speak of "gradualness" to describe such a feature. Using degrees of appropriateness of concepts to objects as truth degrees of statements involving these concepts goes against the Boolean tradition of classical logic.
- **Susceptibility to Sorites paradoxes**. This is the idea that the presence of vague propositions make long inference chains inappropriate, yielding debatable results. The well-known examples deal with heaps of sand (whereby, since adding a grain of sand to a small heap keeps its small, all heaps of sand should

be considered small), young persons getting older by one day, bald persons that are added one hair, etc.

Since their inception, fuzzy sets have been controversial for philosophers, many of whom are reluctant to consider the possibility of non-Boolean predicates, as it questions the usual view of truth as an absolute entity. A disagreement opposes those who, like Williamson, claim a vague predicate has a standard, though ill-known, extension [56], to those who, like Kit Fine, deny the existence of a decision threshold and just speak of a truth value gap [24]. However, the two latter views reject the concept of gradual truth, and concur on the point that fuzzy sets do not propose a good model for vague predicates. One of the reasons for the misunderstanding between fuzzy sets and the philosophy of vagueness may lie in the fact that Zadeh was trained in engineering mathematics, not in the area of philosophy. In particular, vagueness is often understood as a defect of natural language (since it is not appropriate for devising formal proofs, it questions usual rational forms of reasoning). Actually, vagueness of linguistic terms was considered as a logical nightmare for early 20th century philosophers. In contrast, for Zadeh, going from Boolean logic to fuzzy logic is viewed as a positive move: it captures tolerance to errors (softening blunt threshold effects in algorithms) and may account for the flexible use of words by people [73]. It also allows for information summarization: detailed descriptions are sometimes hard to make sense of, while summaries, even if imprecise, are easier to grasp [69].

However, the epistemological situation of fuzzy set theory itself may appear kind of unclear. Fuzzy sets and their extensions have been understood in various ways in the literature: there are several notions that are appealed to in connection with fuzzy sets, like similarity, uncertainty and preference [19]. The concept of similarity to prototypes has been central in the development of fuzzy sets as testified by numerous works on fuzzy clustering. It is also natural to represent incomplete knowledge by fuzzy sets (of possible models of a fuzzy knowledge base, or fuzzy error intervals, for instance), in connection to possibility theory [18, 74]. Utility functions in decision theory also appear as describing fuzzy sets of good options. These topics are not really related to the issue of vagueness.

Indeed, in his works, Zadeh insists that, even when applied to natural language, fuzziness is not vagueness. The term fuzzy is restricted to sets where the transition between membership and non-membership is gradual rather than abrupt, not when it is crisp but unknown. Zadeh [73] argues as follows:

> Although the terms fuzzy and vague are frequently used interchangeably in the literature, there is, in fact, a significant difference between them. Specifically, a proposition, p, is fuzzy if it contains words which are labels of fuzzy sets; and p is vague if it is both fuzzy and insufficiently specific for a particular purpose. For example, “Bob will be back in a few minutes” is fuzzy, while “Bob will be back sometime” is vague if it is insufficiently informative as a basis for a decision. Thus, the vagueness of a proposition is a decision-dependent characteristic whereas its fuzziness is not.

Of course, the distinction made by Zadeh may not be so strict as he claims. While “in a few minutes” is more specific than “sometime” and sounds less vague, one may argue that there is some residual vagueness in the former, and that the latter does not sound very crisp after all. Actually, one may argue that the notion of non-Boolean linguistic categories proposed by Zadeh from 1965 on is capturing the idea of gradualness, not vagueness in its philosophical understanding. Zadeh repetitively claims that gradualness is pervasive in the representation of information, especially human-originated.

The connection from gradualness to vagueness does exist in the sense that, insofar as vagueness refers to uncertainty about meaning of natural language categories, gradual predicates tend to be more often vague than Boolean ones: indeed, it is more difficult to precisely measure the membership function of a fuzzy set representing a gradual category than to define the characteristic function of a set representing the extension of a Boolean predicate [13]. In fact, the power of expressiveness of real numbers is far beyond the limited level of precision perceived by the human mind. Humans basically handle meaningful summaries. Analytical representations of physical phenomena can be faithful as models of reality, but remain esoteric to lay people; the same may hold for real-valued membership grades. Indeed, mental representations are tainted with vagueness, which encompasses at the same time the lack of specificity of linguistic terms, and the lack of well-defined boundaries of the class of objects they refer to, as much as the lack of precision of membership grades. So moving from binary membership to continuous is a bold step, and real-valued membership grades often used in fuzzy sets are just another kind of idealization of human perception, that leaves vagueness aside.

3 The Development of Fuzzy Sets and Systems

Having discussed the various streams of ideas that led to the invention of fuzzy sets, we now outline the basic building blocks of fuzzy set theory, as they emerged from 1965 all the way to the early 1980s, under the impulse of the founding father, via several landmark papers, with no pretense to exhaustiveness. Before discussing the landmark papers that founded the field, it is of interest to briefly summarize how L. A. Zadeh apparently came to the idea of developing fuzzy sets and more generally fuzzy logic. See also [50] for historical details. First, it is worth mentioning that, already in 1950, after commenting the first steps towards building thinking machines (a recently hot topic at the time), he indicated in his conclusion [58]:

> Through their association with mathematicians, the electronic engineers working on thinking machines have become familiar with such hitherto remote subjects as Boolean algebra, multivalued logic, and so forth.

which shows an early concern for logic and many-valued calculi. Twelve years later, when providing “a brief survey of the evolution of system theory [59] he wrote (p. 857)

> There are some who feel that this gap reflects the fundamental inadequacy of the conventional mathematics - the mathematics of precisely-defined points, functions, sets,

> probability measures, etc. - for coping with the analysis of biological systems, and that to deal effectively with such systems, which are generally orders of magnitude more complex than man-made systems, we need a radically different kind of mathematics, the mathematics of fuzzy or cloudy quantities which are not describable in terms of probability distributions.

This quotation shows that Zadeh was first motivated by an attempt at dealing with complex systems rather than with man-made systems, in relation with the current trends of interest in neuro-cybernetics in that time (in that respect, he pursued the idea of applying fuzzy sets to biological systems at least until 1969 [64]).

3.1 Fuzzy Sets: The Founding Paper and Its Motivations

The introduction of the notion of a fuzzy set by Zadeh [60] was motivated by the fact that, quoting the founding paper:

> imprecisely defined "classes" play an important role in human thinking, particularly in the domains of pattern recognition, communication of information, and abstraction.

This seems to have been a recurring concern in all of Zadeh's fuzzy set papers since the beginning, as well as the need to develop a sound mathematical framework for handling this kind of "classes". This purpose required an effort to go beyond classical binary-valued logic, the usual setting for classes. Although many-valued logics had been around for a while, what is really remarkable is that due to this concern, Zadeh started to think in terms of *sets* rather than only in terms of degrees of truth, in accordance with intuitions formalized by Kaplan but not pursued further. Since a set is a very basic notion, it was opening the road to the introduction of the fuzzification of any set-based notions such as relations, events, or intervals, while sticking to the many-valued logic point of view only does not lead you to consider such generalized notions. In other words, while Boolean algebras are underlying both propositional logic and naive set theory, the set point of view may be found richer in terms of mathematical modeling, and the same thing takes place when moving from many-valued logics to fuzzy sets.

A fuzzy set can be understood as a class equipped with an ordering of elements expressing that some objects are more inside the class than others. However, in order to extend the Boolean connectives, we need more than a mere relation in order to extend intersection, union and complement of sets, let alone implication. The set of possible membership grades has to be a complete lattice [27] so as to capture union and intersection, and either the concept of residuation or an order-reversing function in order to express some kind of negation and implication. The study of set operations on fuzzy sets has in return strongly contributed to a renewal of many-valued logics under the impulse of Hájek [29] (see [14] for an introductory overview).

Besides, from the beginning, it was made clear that fuzzy sets were not meant as probabilities in disguise, since one can read [60] that

> the notion of a fuzzy set is completely non-statistical in nature

and that it provides

> a natural way of dealing with problems where the source of imprecision is the absence of sharply defined criteria of membership rather than the presence of random variables.

Presented as such, fuzzy sets are prima facie not related to the notion of uncertainty. The point that typicality notions underlie the use of gradual membership functions of linguistic terms is more connected to similarity than to uncertainty. As a consequence,

- originally, fuzzy sets were designed to formalize the idea of soft classification, which is more in agreement with the way people use categories in natural language.
- fuzziness is just implementing the concept of gradation in all forms of reasoning and problem-solving, as for Zadeh, everything is a matter of degree.
- a degree of membership is an abstract notion to be interpreted in practice.

Important definitions appear in the founding paper such as cuts (a fuzzy set can be viewed as a family of nested crisp sets called its cuts), the basic fuzzy set-theoretic connectives (e.g. minimum and product as candidates for intersection, inclusion via inequality of membership functions) and the extension principle whereby the domain of a function is extended to fuzzy set-valued arguments. As pointed out earlier, according to the area of application, several interpretations can be found such as degree of similarity (to a prototype in a class), degree of plausibility, or degree of preference [19]. However, in the founding paper, membership functions are considered in connection with the representation of human categories only. The three kinds of interpretation would become patent in subsequent papers.

3.2 *Fuzzy Sets and Classification*

A popular part of the fuzzy set literature deals with fuzzy clustering where gradual transitions between classes and their use in interpolation are the basic contribution of fuzzy sets. The idea that fuzzy sets would be instrumental to avoid too rough classifications was provided very early by Zadeh, along with Bellman and Kalaba [3]. They outline how to construct membership functions of classes from examples thereof. Intuitively speaking, a cluster gathers elements that are rather close to each other (or close to some core element(s)), while they are well-separated from the elements in the other cluster(s). Thus, the notions of graded proximity, similarity (dissimilarity) are at work in fuzzy clustering. With gradual clusters, the key issue is to define fuzzy partitions. The most widely used definition of a fuzzy partition originally due to Ruspini [48], where the sum of membership grades of one element to the various classes is 1. This was enough to trigger the fuzzy clustering literature, that culminated with the numerous works by Bezdek and colleagues, with applications to image processing for instance [5].

3.3 Fuzzy Events

The idea of replacing sets by fuzzy sets was quickly applied by Zadeh to the notion of event in probability theory [63]. The probability of a fuzzy event is just the expectation of its membership function. Beyond the mathematical exercise, this definition has the merit of showing the complementarity between fuzzy set theory and probability theory: while the latter models uncertainty of events, the former modifies the notion of an event admitting it can occur to some degree when observing a precise outcome. This membership degree is ontic as it is part of the definition of the event, as opposed to probability that has an epistemic flavor, a point made very early by De Finetti [12], commenting Łukasiewicz logic. However, it took 35 years before a generalization of De Finetti's theory of subjective probability to fuzzy events was devised by Mundici [42]. Since then, there is an active mathematical area studying probability theory on algebras of fuzzy events.

3.4 Decision-Making with Fuzzy Sets

Fuzzy sets can be useful in decision sciences. This is not surprising since decision analysis is a field where human-originated information is pervasive. While the suggestion of modelling fuzzy optimization as the (product-based) aggregation of an objective function with a fuzzy constraint first appeared in the last section of [61], the full-fledged seminal paper in this area was written by Bellman and Zadeh [4] in 1970, highlighting the role of fuzzy set connectives in criteria aggregation. That pioneering paper makes three main points:

1. Membership functions can be viewed as a variant of utility functions or rescaled objective functions, and optimized as such.
2. Combining membership functions, especially using the minimum, can be one approach to criteria aggregation.
3. Multiple-stage decision-making problems based on the minimum aggregation connective can then be stated and solved by means of dynamic programming.

This view was taken over by Tanaka et al. [54] and Zimmermann [86] who developed popular multicriteria linear optimisation techniques in the seventies. The idea is that constraints are soft. Their satisfaction is thus a matter of degree. They can thus be viewed as criteria. The use of the minimum operation instead of the sum for aggregating partial degrees of satisfaction preserves the semantics of constraints since it enforces all of them to be satisfied to some degree. Then any multi-objective linear programming problem becomes a max-min fuzzy linear programming problem.

3.5 Fuzzy Relations

Relations being subsets of a Cartesian product of sets, it was natural to make them fuzzy. There are two landmark papers by Zadeh on this topic in the early 1970s. One published in 1971 [65] makes the notions of equivalence and ordering fuzzy. A similarity relation extends the notion of equivalence, preserving the properties of reflexivity and symmetry and turning transitivity into maxmin transitivity. Similarity relations come close to the notion of ultrametrics and correspond to nested equivalence relations. As to fuzzy counterparts of order relations, Zadeh introduces first definitions of what will later be mathematical models of a form of fuzzy preference relations [25]. In his attempt the difficult part is the extension of antisymmetry that will be shown to be problematic. The notion of fuzzy preorder, involving a similarity relation turns out to be more natural than the one of a fuzzy order [6].

The other paper on fuzzy relations dates back to 1975 [71] and consists in a fuzzy generalization of the relational algebra. This seminal paper paved the way to the application of fuzzy sets to databases (see [8] for a survey), and to flexible constraint satisfaction in artificial intelligence [16, 41].

3.6 Fuzzy Systems

The application of fuzzy sets to systems was not obvious at all, as traditionally systems were described by numerical equations. Zadeh seems to have tried several solutions to come up with a notion of fuzzy system. Fuzzy systems [61] were initially viewed as systems whose state equations involve fuzzy variables or parameters, giving birth to fuzzy classes of systems. Another idea, hinted in 1965 [61], was that a system is fuzzy if either its input, its output or its states would range over a family of fuzzy sets. Later in 1971 [67], he suggested that a fuzzy system could be a generalisation of a non-deterministic system, that moves from a state to a fuzzy set of states. So the transition function is a fuzzy mapping, and the transition equation can be captured by means of fuzzy relations, and the sup-min combination of fuzzy sets and fuzzy relations. These early attempts were outlined before the emergence of the idea of fuzzy control [68]. In 1973, though, there was what looks like a significant move, since for the first time it was suggested that fuzziness lies in the description of approximate rules to make the system work. That view, developed first in [69], was the result of a convergence between the idea of system with the ones of fuzzy algorithms introduced earlier [62], and his increased focus of attention on the representation of natural language statements via linguistic variables. In this very seminal paper, systems of fuzzy if-then rules were first described, which paved the way to fuzzy controllers, built from human information, with the tremendous success met by such line of research in the early 1980s. To-day, fuzzy rule-based systems are extracted from data and serve as models of systems more than as controllers. However the linguistic connection is often lost, and such fuzzy systems are rather standard precise systems using membership function for interpolation, than approaches to the handling of poor knowledge in system descriptions.

3.7 Linguistic Variables and Natural Language Issues

When introducing fuzzy sets, Zadeh seems to have been chiefly motivated by the representation of human information in natural language. This focus explains why many of Zadeh's papers concern fuzzy languages and linguistic variables. He tried to combine results on formal languages and the idea that the term sets that contain the atoms from which this language was built contain fuzzy sets representing the meaning of elementary concepts [66]. This is the topic where the most numerous and extended papers by Zadeh can be found, especially the large treatise devoted to linguistic variables in 1975 [72] and the papers on the PRUF language [73, 78]. Basically, these papers led to the "computing with words" paradigm, which takes the opposite view of say, logic-based artificial intelligence, by putting the main emphasis, including the calculation method, on the semantics rather than the syntax. Starting with natural language sentences, fuzzy words are precisely modelled by fuzzy sets, and inference comes down to some form of non-linear optimisation. In a later step, numerical results are translated into verbal terms, using the so-called linguistic approximation. While this way of reasoning seems to have been at the core of Zadeh's approach, it is clear that most applications of fuzzy sets only use terms sets and linguistic variables in rather elementary ways. For instance, quite a number of authors define linguistic terms in the form of trapezoidal fuzzy sets on an abstract universe which is not measurable and where addition makes no sense, nor linear membership functions. In [72], Zadeh makes it clear that linguistic variables refer to objective measurable scales: only the linguistic term has a subjective meaning, while the universe of discourse contains a measurable quantity like height, age, etc.

3.8 Fuzzy Intervals

In his longest 3-part paper in 1975 [72], Zadeh points out that a trapezoidal fuzzy set of the real line can model imprecise quantities. These trapezoidal fuzzy sets, called fuzzy numbers, generalize intervals on the real line and appear as the mathematical rendering of fuzzy terms that are the values of linguistic variables on numerical scales. The systematic use of the extension principle to such fuzzy numbers and the idea of applying it to basic operations of arithmetics is a key-idea that will also turn out to be seminal. Since then, numerous papers have developed methods for the calculation with fuzzy numbers. It has been shown that the extension principle enable a generalization of interval calculations, hence opening a whole area of fuzzy sensitivity analysis that can cope with incomplete information in a gradual way. The calculus of fuzzy intervals is instrumental in various areas including:

- systems of linear equations with fuzzy coefficients (a critical survey is [36]) and differential equations with fuzzy initial values, and fuzzy set functions [37];
- fuzzy random variables for the handling of linguistic or imprecise statistical data [10, 26];
- fuzzy regression methods [43];
- operations research and optimisation under uncertainty [15, 31, 32].

3.9 *Fuzzy Sets and Uncertainty: Possibility Theory*

Fuzzy sets can represent uncertainty not because they are fuzzy but because crisp sets are already often used to represent ill-known values or situations, albeit in a crisp way like in interval analysis or in propositional logic. Viewed as representing uncertainty, a set just distinguishes between values that are considered possible and values that are impossible, and fuzzy sets just introduce grades to soften boundaries of an uncertainty set. So in a fuzzy set it is the set that captures uncertainty [20, 21]. This point of view echoes an important distinction made by Zadeh himself [73, 83] between

- conjunctive (fuzzy) sets, where the set is viewed as the conjunction of its elements. This is the case for clusters discussed in the previous section. But also with set-valued attributes like the languages spoken more or less fluently by an individual.
- and disjunctive fuzzy sets which corresponds to mutually exclusive possible values of an ill-known single-valued attribute, like the ill-known birth nationality of an individual.

In the latter case, membership functions of fuzzy sets are called possibility distributions [74] and act as elastic constraints on a precise value. Possibility distributions have an epistemic flavor, since they represent the information we have at our disposal about what values remain more or less possible for the variable under consideration, and what values are (already) known as impossible. Associated with a possibility distribution, is a possibility measure [74], which is a max-decomposable set function. Thus, one can evaluate the possibility of a crisp, or fuzzy, statement of interest, given the available information supposed to be represented by a possibility distribution. It is also important to notice that the introduction of possibility theory by Zadeh was part of the modeling of fuzzy information expressed in natural language [77]. This view contrasts with the motivations of the English economist Shackle [51, 52], interested in a non-probabilistic view of expectation, who had already designed a formally similar theory in the 1940s, but rather based on the idea of degree of impossibility understood as a degree of surprise (using profiles of the form $1\text{-}\mu$, where μ is a membership function). Shackle can also appear as a forerunner of fuzzy sets, of possibility theory, actually.

Curiously, apart from a brief mention in [76], Zadeh does not explicitly use the notion of necessity (the natural dual of the modal notion of possibility) in his work on possibility theory and approximate reasoning. Still, it is important to distinguish between statements that are *necessarily* true (to some extent), i.e. whose negation is almost impossible, from the statements that are only *possibly* true (to some extent) depending on the way the fuzzy knowledge would be made precise. The simultaneous use of the two notions is often required in applications of possibility theory [23] and in possibilistic logic [22].

3.10 Approximate Reasoning

The first illustration of the power of possibility theory proposed by Zadeh was an original theory of approximate reasoning [75, 79, 80], later reworked for emphasizing new points [82, 84], where pieces of knowledge are represented by possibility distributions that fuzzily restrict the possible values of variables, or tuples of variables. These possibility distributions are combined conjunctively, and then projected in order to compute the fuzzy restrictions acting on the variables of interest. This view is the one at work in his calculus of fuzzy relations in 1975. One research direction, quite in the spirit of the objective of "computing with words" [82], would be to further explore the possibility of a syntactic (or symbolic) computation of the inference step (at least for some noticeable fragments of this general approximate reasoning theory), where the obtained results are parameterized by fuzzy set membership functions that would be used only for the final interpretation of the results. An illustration of this idea is at work in possibilistic logic [22], a very elementary formalism handling pairs made of a Boolean formula and a certainty weight, that captures a tractable form of non-monotonic reasoning. Such pairs syntactically encode simple possibility distributions, combined and reasoned from in agreement with Zadeh's theory of approximate reasoning, but more in the tradition of symbolic artificial intelligence than in conformity with the semantic-based methodology of computing with words.

4 Conclusion

The seminal paper on fuzzy sets by Lotfi Zadeh spewed out a large literature, despite its obvious marginality at the time it appeared. There exist many forgotten original papers without off-springs. Why has Zadeh's paper encountered an eventual dramatic success? One reason is certainly that Zadeh, at the time when the fuzzy set paper was released, was already a renowned scientist in systems engineering. So, his paper, published in a good journal, was visible. However another reason for success lies in the tremendous efforts made by Zadeh in the seventies and the eighties to develop his intuitions in various directions covering many topics from theoretical computer sciences to computational linguistics, from system sciences to decision sciences.

It is not clear that the major successes of fuzzy set theory in applications fully meet the expectations of its founder. Especially there was almost no enduring impact of fuzzy sets on natural language processing and computational linguistics, despite the original motivation and continued effort about the computing with words paradigm. The contribution of fuzzy sets and fuzzy logic was to be found elsewhere, in systems engineering, data analysis, multifactorial evaluation, uncertainty modeling, operations research and optimisation, and even mathematics. The notion of fuzzy rule-based system (Takagi-Sugeno form, not Mamdani's, nor even the view developed in [81]) has now been integrated in both the neural net literature and the non-linear control one. These fields, just like in clustering, only use the notion of fuzzy partition and possibly interpolation between subclasses, and bear almost no connection to the issue of fuzzy modeling of natural language. These fields have come of age, and almost no progress can be observed that concern their fuzzy set ingredients. In optimisation, the fuzzy linear programming method is now used in applications, with only minor variants (if we set apart the handling of uncertainty proper, via fuzzy intervals).

However there are some topics where basic research seems to still be active, with high potential. See [11] for a collection of position papers highlighting various perspectives for the future of fuzzy sets. Let us cite two such topics. The notion of fuzzy interval or fuzzy number, introduced in 1975 by Zadeh [72], and considered with the possibility theory lenses, seems to be more promising in terms of further developments because of its connection with non-Bayesian statistics and imprecise probability [39] and its potential for handling uncertainty in risk analysis [1]. Likewise, the study of fuzzy set connectives initiated by Zadeh in 1965, has given rise to a large literature, and significant developments bridging the gap between many-valued logics and multicriteria evaluation, with promising applications (for instance [2]).

Last but not least, we can emphasize the influence of fuzzy sets on some even more mathematically oriented areas, like the strong impact of fuzzy logic on many-valued logics (triggered by Hajek [29]) and topological and categorical studies of lattice-valued sets [30]. There are very few papers in the literature that could influence such various areas of scientific investigation to that extent.

References

1. Baudrit, C., Guyonnet, D., Dubois, D.: Joint propagation and exploitation of probabilistic and possibilistic information in risk assessment. IEEE Trans. Fuzzy Syst. **14**, 593–608 (2006)
2. Beliakov, G., Pradera, A., Calvo, T.: Aggregation Functions: A Guide for Practitioners. Studies in Fuzziness and Soft Computing, vol. 221. Springer (2007)

3. Bellman, R.E., Kalaba, R., Zadeh, L.A.: Abstraction and pattern classification. J. Math. Anal. Appl. **13**, 1–7 (1966)
4. Bellman, R.E., Zadeh, L.A.: Decision making in a fuzzy environment. Manage. Sci. **17**, B141–B164 (1970)
5. Bezdek, J., Keller, J., Krishnapuram, R., Pal, N.: Fuzzy Models for Pattern Recognition and Image processing. Kluwer (1999)
6. Bodenhofer, U., De Baets, B., Fodor, J.C.: A compendium of fuzzy weak orders: Representations and constructions. Fuzzy Sets Syst. **158**(8), 811–829 (2007)
7. Black, M., Vagueness, M.: Phil. Sci. **4**, 427–455 (1937) (Reprinted in Language and Philosophy: Studies in Method. Cornell University Press, pp. 23–58. Ithaca, London, 1949. Also in Int. J. Gen. Syst. **17**, 107–128, 1990)
8. Bosc, P., Pivert, O.: Modeling and querying uncertain relational databases: a survey of approaches based on the possible world semantics. Int. J. Uncertainty, Fuzziness Knowl. Based Syst. **18**(5), 565–603 (2010)
9. Ciucci, D., Dubois, D.: A map of dependencies among three-valued logics. Inf. Sci. **250**, 162–177 (2013)
10. Couso, I., Dubois, D., Sanchez, L.: Random Sets and Random Fuzzy Sets as Ill-Perceived Random Variables. Springer, Springer Briefs in Computational Intelligence, 2014
11. De Baets, B., Dubois, D., Hüllermeier, E.: Special issue celebrating the 50th anniversary of fuzzy sets. Fuzzy Sets Syst **281**, 1–308 (2015)
12. De Finetti, B.: La logique de la probabilité, Actes Congrés Int. de Philos. Scient., Paris (1935) (Hermann et Cie Editions. Paris, pp. IV1–IV9, 1936)
13. Dubois, D.: Have fuzzy sets anything to do with vagueness? (with discussion) In: Cintula, P., Fermüller, C. (eds.) Understanding Vagueness-Logical, Philosophical and Linguistic Perspectives, vol. 36 of Studies in Logic. College Publications, pp. 317–346 (2012)
14. Dubois, D., Esteva, F., Godo, L., Prade, H.: Fuzzy-set based logics—an history-oriented presentation of their main developments. In: Gabbay, D.M., Woods, J. (Eds.) Handbook of The History of Logic. The Many Valued and Nonmonotonic Turn in Logic, vol. 8, Elsevier, pp. 325–449 (2007)
15. Dubois, D., Fargier, H., Fortemps, P.: Fuzzy scheduling: modelling flexible constraints vs. coping with incomplete knowledge. Eur. J. Oper. Res. **147**(2), 231–252 (2003)
16. Dubois, D., Fargier, H., Prade, H.: Possibility theory in constraint satisfaction problems: handling priority, preference and uncertainty. Appl. Intell. **6**, 287–309 (1996)
17. Dubois, D., Ostasiewicz, W., Prade, H.: Fuzzy sets: history and basic notions. In: Dubois, D. Prade, H. (eds.) Fundamentals of Fuzzy Sets. The Handbooks of Fuzzy Sets Series, Kluwer, pp. 21–124 (2000)
18. Dubois, D., Prade, H.: Possibility Theory: An Approach to Computerized Processing of Uncertainty. Plenum Press, New York (1988)
19. Dubois, D., Prade, H.: The three semantics of fuzzy sets. Fuzzy Sets Syst. **90**, 141–150
20. Dubois, D., Prade, H.: Gradual elements in a fuzzy set. Soft. Comput. **12**, 165–175 (2008)
21. Dubois, D., Prade, H.: Gradualness, uncertainty and bipolarity: making sense of fuzzy sets. Fuzzy Sets Syst. **192**, 3–24 (2012)
22. Dubois, D., Prade, H.: Possibilistic logic—an overview. In: Siekmann, J., Gabbay, D.M., Woods, J. (eds.) Handbook of the History of Logic. Computational Logic, vol. 9, pp. 283–342 (2014)
23. Dubois, D., Prade, H.: Possibility theory and its applications: where do we stand? In: Kacprzyk, J., Pedrycz, W. (eds.) Handbook of Computational Intelligence, pp. 31–60. Springer (2015)
24. Fine, K.: Vagueness, truth and logic. Synthese **30**, 265–300 (1975)
25. Fodor, J., Roubens, M.: Fuzzy Preference Modelling and Multicriteria Decision Support. Kluwer Academic Publication (1994)
26. Gil, M.A., González-Rodriguez, G., Kruse, R. (eds.) Statistics with imperfect data. Spec. Issue Inf. Sci. **245**, 1–3 (2013)
27. Goguen, J.A.: L-fuzzy sets. J. Math. Anal. Appl. **8**, 145–174 (1967)

28. Gottwald, S.: Fuzzy sets theory: some aspects of the early development, In: Skala, H.J., Termini, S., Trillas, E., (eds.) Aspects of Vagueness, pp. 13–29. D. Reidel (1984)
29. Hájek, P.: Metamathematics of Fuzzy Logic, Trends in Logic, vol. 4. Kluwer, Dordercht (1998)
30. Hoehle, U., Rodabaugh, S.E. (eds.) Mathematics of Fuzzy Sets: Logic, Topology, and Measure Theory, The Handbooks of Fuzzy Sets Series, vol. 3. Kluwer Academic Publishers, Dordrecht (1999)
31. Inuiguchi, M., Ramik, J.: Possibilistic linear programming: a brief review of fuzzy mathematical programming and a comparison with stochastic programming in portfolio selection problem. Fuzzy Sets Syst. **111**(1), 3–28 (2000)
32. Inuiguchi, M., Watada, J., Dubois, D. (eds.) Special issue on fuzzy modeling for optimisation and decision support. Fuzzy Sets Syst. **274** (2015)
33. Kaplan, A.: Definition and specification of meanings. J. Phil. **43**, 281–288 (1946)
34. Kaplan, A., Schott, H.F.: A calculus for empirical classes. Methods **III**, 165–188 (1951)
35. Kleene, S.C.: Introduction to Metamathematics. North Holland Pub. Co., Amsterdam (1952)
36. Lodwick, W.A., Dubois, D.: Interval linear systems as a necessary step in fuzzy linear systems. Fuzzy Sets Syst. **281**, 227–251 (2015)
37. Lodwick, W.A., Oberguggenberger, M. (eds.) Differential equations over fuzzy spaces—theory, applications, and algorithms. Spec Issue Fuzzy Sets Syst. **230**, 1–162 (2013)
38. Łukasiewicz, J.: Philosophical remarks on many-valued systems of propositional logic (1930) (Reprinted in Selected works. In Borkowski (ed.) Studies in Logic and the Foundations of Mathematics, pp. 153–179. North-Holland, Amsterdam, 1970)
39. Mauris, G.: Possibility distributions: a unified representation of usual direct-probability-based parameter estimation methods. Int. J. Approx. Reasoning **52**(9), 1232–1242 (2011)
40. Menger, K.: Ensembles flous et fonctions aleatoires. Comptes Rendus de lAcadmie des Sciences de Paris **232**, 2001–2003 (1951)
41. Meseguer, P., Rossi, F., Schiex, T.: Soft Constraints, Chapter 9 in Foundations of Artificial Intelligence, vol. 2, Elsevier, pp. 281–328 (2006)
42. Mundici, D.: Bookmaking Over Infinite-Valued Events. Int. J. Approximate Reasoning **43**(3), 223–240 (2006)
43. Muzzioli, S., Ruggieri, A., De Baets, B.: A comparison of fuzzy regression methods for the estimation of the implied volatility smile function. Fuzzy Sets Syst. **266**, 131–143 (2015)
44. Novák, V.: Are fuzzy sets a reasonable tool for modeling vague phenomena? Fuzzy Sets Syst. **156**(3), 341–348 (2005)
45. Ostasiewicz, W.: Pioneers of fuzziness. Busefal **46**, 4–15 (1991)
46. Ostasiewicz, W.: Half a century of fuzzy sets. Supplement to Kybernetika 28:17–20, 1992 (Proc. of the Inter. Symp. on Fuzzy Approach to Reasoning and Decision Making, Bechyne, Czechoslovakia, June 25-29, 1990)
47. Peirce, C.S.: Collected papers of Charles Sanders Peirce. In: Hartshorne, C., Weiss, P. (eds.) Harvard University Press. Cambridge, MA (1931)
48. Ruspini, E.H.: A new approach to clustering. Inform. Control **15**, 22–32 (1969)
49. Russell, B., Vagueness, B.: Austr. J. Philos. **1**, 84–92 (1923)
50. Seising, R.: Not, or, and,—not an end and not no end! The "Enric-Trillas-path" in fuzzy logic. In: Seising, R., Argüelles Méndez, L. (eds.) Accuracy and fuzziness. A Life in Science and Politics. A festschrift book to Enric Trillas Ruiz, pp. 1–59, Springer (2015)
51. Shackle, G.L.S.: Expectation in Economics, 2nd edn. Cambridge University Press, UK (1949)
52. Shackle, G.L.S.: Decision, Order and Time in Human Affairs, 2nd edn. Cambridge University Press, UK (1961)
53. Smith, P., Keefe, R.: Vagueness: A Reader. MIT Press, Cambridge, MA (1997)
54. Tanaka, H., Okuda, T., Asai, K.: On fuzzy-mathematical programming. J. Cybern. **3**(4), 37–46 (1974)
55. Weyl, H.: Mathematic and logic. Am. Math. Mon. **53**, 2–13 (1946)
56. Williamson, T.: Vagueness. Routledge, London (1994)
57. Wittgenstein, L.: Philosophical Investigations. Macmillan, New York (1953)

58. Zadeh, L.A.: Thinking machines. A new field in electrical engineering. Columbia Eng. Q. **3**, 12–13, 30–31 (1950)
59. Zadeh, L.A.: From circuit theory to system theory. Proc. IRE **50**, 856–865 (1962)
60. Zadeh, L.A.: Fuzzy sets. Inf. Control **8**(3), 338–353 (1965)
61. Zadeh, L.A.: Fuzzy sets and systems. In: Fox, J. (Ed.) Systems Theory, pp .29-37. Polytechnic Press, Brooklyn, NY (1966) (Proceedings of Symposium on System Theory. New York, 20–22 Apr 1965, (reprinted in Int. J. General Syst., **17**, 129–138, 1990)
62. Zadeh, L.A.: Fuzzy algorithms. Inf. Control **12**, 94–102 (1968)
63. Zadeh, L.A.: Probability measures of fuzzy events. J. Math. Anal. Appl. **23**, 421–427 (1968)
64. Zadeh, L.A.: Biological applications of the theory of fuzzy sets and systems. In: Kilmer, W.L., Proctor, L.D., Brown, L., et al. (eds.) Biocybernetics of the Central Nervous System (with a discussion by Boston, pp. 199–206 (discussion pp. 207–212) (1969)
65. Zadeh, L.A.: Similarity relations and fuzzy orderings. Inf. Sci. **3**, 177–200 (1971)
66. Zadeh, L.A.: Quantitative fuzzy semantics. Inf. Sci. **3**, 159–176 (1971)
67. Zadeh, L.A.: Toward a theory of fuzzy systems. In: Kalman, R.E., De Claris, N. (eds.) Aspects of Network and System Theory, pp. 469–490. Holt, Rinehart and Winston, New York, (1971) (originally NASA Contractor Report 1432, Sept. 1969)
68. Zadeh, L.A.: A rationale for fuzzy control. J. of Dyn. Syst. Measur. Control Trans. ASME, 3–4 (1972) (March issue)
69. Zadeh, L.A.: Outline of a new approach to the analysis of complex systems and decision processes. IEEE Trans. Syst. Man Cybern. **3**(1), 28–44 (1973)
70. Zadeh, L.A.: Fuzzy logic and approximate reasoning. Synthese **30**, 407–428 (1975)
71. Zadeh, L.A.: Calculus of fuzzy restrictions. In: Zadeh, L.A., Fu, K.S., Tanaka, K., Shimura, M. (eds.) Fuzzy sets and Their Applications to Cognitive and Decision Processes. Proceedings of U.S.-Japan Seminar on Fuzzy Sets and Their Applications. Berkeley, 1–4 July, 1974 (Academic Press, pp. 1–39, 1975)
72. Zadeh, L.A.: The concept of a linguistic variable and its application to approximate reasoning. Inf. Sci., Part I, **8**(3), 199–249; Part II, **8**(4), 301–357; Part III, **9**(1), 43–80 (1975)
73. Zadeh, L.A.: PRUF—a meaning representation language for natural languages. Int. J. Man Mach. Stud. **10**, 395–460 (1978)
74. Zadeh, L.A.: Fuzzy sets as a basis for a theory of possibility. Fuzzy Sets Syst. **1**, 3–28 (1978)
75. Zadeh, L.A.: A theory of approximate reasoning. In: Hayes, J.E., Mitchie, D., Mikulich, L.I. (eds.) Machine Intelligence, vol. 9, pp. 149–194 (1979)
76. Zadeh, L.A.: Fuzzy sets and information granularity. In: Gupta, M.M., Ragade, R.K., Yager, R.R. (eds.) Advances in Fuzzy Set Theory and Applications, pp. 3–18. North-Holland (1979)
77. Zadeh, L.A.: Possibility theory and soft data analysis. In: Cobb, L., Thrall, R. (eds.) Mathematical Frontiers of Social and Policy Sciences, pp. 69–129. Westview Press, Boulder Co (1982)
78. Zadeh, L.A.: Precisiation of meaning via translation into PRUF. In: Vaina, L., Hintikka, J. (eds.) Cognitive Constraints on Communication, pp. 373–402. Reidel, Dordrecht (1984)
79. Zadeh, L.A.: Syllogistic reasoning in fuzzy logic and its application to usuality and reasoning with dispositions. IEEE Trans. Syst. Man Cybern. **15**(6), 754–763 (1985)
80. Zadeh, L.A.: Knowledge representation in fuzzy logic. IEEE Trans. Knowl. Data Eng. **1**(1), 89–100 (1989)
81. Zadeh, L.A.: The calculus of fuzzy if-then rules. AI Expert **7**(3), 23–27 (1992)
82. Zadeh, L.A.: Fuzzy logic = computing with words. IEEE Trans. Fuzzy Syst. **4**(2), 103–111 (1996)
83. Zadeh, L.A.: Toward a theory of fuzzy information granulation and its centrality in human reasoning and fuzzy logic. Fuzzy Sets Syst. **90**, 111–128 (1997)
84. Zadeh, L.A.: Generalized theory of uncertainty (GTU)—principal concepts and ideas. Comput. Stat. Data Anal. **51**, 15–46 (2006)
85. Zadeh, L.A.: Fuzzy logic—a personal perspective. Fuzzy Sets Syst. **281**, 4–20 (2015)
86. Zimmermann, H.-J.: Fuzzy programming and linear programming with several objective functions. Fuzzy Sets Syst. **1**, 45–55 (1978)

Fuzzy Decision Making: Its Pioneers and Supportive Environment

Cengiz Kahraman, Sezi Çevik Onar and Başar Öztayşi

Abstract Fuzzy decision making is the collection of single or multicriteria techniques aiming at selecting the best alternative in case of imprecise, incomplete, and vague data. This chapter reviews the fuzzy decision making literature and summarizes the review results by tabular and graphical illustrations. The classification is based on the new extensions of fuzzy sets: Intuitionistic, hesitant, and type-2 fuzzy sets. Later, the media publishing fuzzy decision making papers, journals, books, conferences, and societies are summarized. Finally, fuzzy decision making examples are given for ordinary, intuitionistic, hesitant, and type-2 fuzzy sets.

Keywords Decision-making · Ordinary fuzzy set · Intuitionistic fuzzy set · Hesitant fuzzy set · Type-2 fuzzy set

1 Introduction

Decision making is the study of identifying and choosing alternatives based on the values and preferences of the decision maker [16]. A decision making process can be divided into the following eight steps Baker et al. [5]:

Step 1. Definition of the problem: This step includes the identification of root causes, limiting assumptions, system and organizational boundaries and interfaces, and any stakeholder issues

Step 2. Determination of requirements: This step includes the constraints describing the set of the feasible solutions of the decision problem

Step 3. Establishment of goals: This step includes the definition of the goals which are broad statements of intent and desirable programmatic values

---

C. Kahraman (✉) · S.Ç. Onar · B. Öztayşi
Department of Industrial Engineering, Istanbul Technical University,
Macka, 34367 Istanbul, Turkey
e-mail: kahramanc@itu.edu.tr

C. Kahraman et al. (eds.), *Fuzzy Logic in Its 50th Year*,
Studies in Fuzziness and Soft Computing 341,
DOI 10.1007/978-3-319-31093-0_2

Step 4. Identification of alternatives: This step includes the determination of the alternatives which offer different approaches for changing the initial condition into the desired condition

Step 5. Definition of criteria: This step includes the determination of the criteria which will discriminate among alternatives. The criteria must be based on the goals

Step 6. Selection of a decision making technique: This step includes the determination of the analysis technique which will solve the problem

Step 7. Evaluation of alternatives with respect to criteria: This step includes the application of the determined technique in Step 6

Step 8. Validation of solutions: This step includes the check of the validity of the solution obtained by the evaluation techniques

Real-world decision-making problems are usually too complex and ill-structured to be considered by classical decision making techniques. In the classical decision making techniques, human's judgments are represented as exact numbers. However, in many practical cases, the data may be imprecise, or the decision makers might be unable to assign exact numerical values to the evaluation. Since some of the evaluation criteria are subjective and qualitative in nature, it is very difficult for the decision maker to express the preferences using exact numerical values [35]. Instead, decision makers usually prefer making linguistic evaluations in the decision matrix. Besides, the conventional decision making approaches tend to be less effective in dealing with the imprecise or vague nature of the linguistic assessments [20]. The real case problems including decision making problems are complex and involve vagueness and fuzziness. This has led to the development of fuzzy set theory by Zadeh [44], as a means of representing and manipulating data that were not precise, but rather fuzzy. The theory of fuzzy logic provides a mathematical strength to capture the uncertainties associated with human cognitive process, such as thinking and reasoning [32].

The term *fuzzy decision making* has first been introduced by Bellman and Zadeh [6]. They defined the fuzzy decision making as follows: The term *decision making in fuzzy environment* means a decision making process in which the goals and/or the constraints, but not necessarily the system under control, are fuzzy in nature. This means that the goals and/or the constraints constitute classes of alternatives whose boundaries are not sharply defined [6]. Later, Baas and Kwakernaak [4] applied the most classic work on the fuzzy decision making and it was used as a benchmark for other similar fuzzy decision models. Dubois and Prade [15], Zimmermann [47, 48], Chen and Hwang [14], and Ribeiro [31] differentiated the family of fuzzy decision making methods into two main phases. The first phase is generally known as the rating process, dealing with the measurement of performance ratings or the degree of satisfaction with respect to all attributes of each alternative under fuzziness. The second phase, the ranking of alternatives that is carried out by ordering the existing alternatives according to the resulted aggregated performance ratings obtained from the first phase. Yager [40] presented some ideas on the applications of fuzzy sets to multi-objective decision making with particular emphasis on a means of including differing degrees of importance to different objectives.

After the introduction of fuzzy decision making by these pioneers, the classical decision making techniques have been transformed to their fuzzy versions within few years. Classical TOPSIS method was transformed to fuzzy TOPSIS by Chen [13]; Classical AHP was transformed to fuzzy AHP by Van Laarhoven and Pedrycz [37], Buckley [7], and Chang [11]. Classical VIKOR was transformed to fuzzy VIKOR by Opricovic [28].

After the introduction of the extensions of fuzzy sets such as type-2 fuzzy sets, intuitionistic fuzzy sets, fuzzy multisets, and hesitant fuzzy sets, the ordinary fuzzy versions of the decision making methods have started to extend to their new versions such as type-2 fuzzy TOPSIS [12], type-2 fuzzy AHP [19], type-2 fuzzy VIKOR [42], intuitionistic fuzzy TOPSIS [2], hesitant fuzzy TOPSIS [46], etc.

The rest of this chapter is organized as follows. Section 2 includes a literature review on fuzzy decision making based on the extensions of fuzzy sets. Section 3 presents the publication media of fuzzy decision making. Section 4 gives some fuzzy decision making applications. Section 5 concludes the chapter.

2 Literature Review

In the following, a literature review on decision making based oordinary fuzzy sets, hesitant fuzzy sets, intuitionistic fuzzy sets, and type-2 fuzzy sets, respectively.

2.1 *Ordinary Fuzzy Decision-Making (OFDM)*

If X is a collection of objects denoted generically by x, then a fuzzy set $\tilde{A}$ in X is a set of ordered pairs:

$$\tilde{A} = \left\{ \left(x, \mu_{\tilde{A}}(x) | x \in X \right) \right\} \quad (1)$$

$\mu_{\tilde{A}}(x)$ is called the membership function which maps X to the membership space M. Its range is the subset of nonnegative real numbers whose supremum is finite [44].

The search for OFDM based on Scopus database in October 2015 yielded 2159 papers. 1281 of them are articles while 775 of them are conference papers. The rest are review, book chapters, editorial, erratum, article in press, etc.

Figure 1 presents the publication frequencies of the papers on OFDM with respect to years. As it is clearly seen from the graph, significant increases in the frequencies of OFDM papers exist after the year 2000.

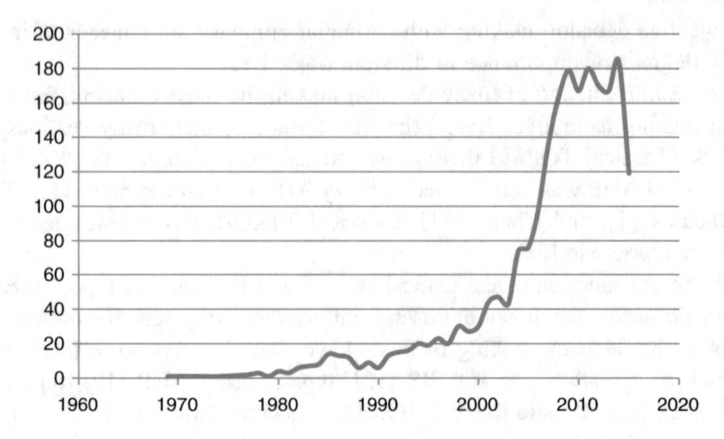

Fig. 1 Publication frequencies of papers on OFDM with respect to years

Table 1 shows the sources publishing OFDM papers and the frequencies of those papers. *Fuzzy Sets and Systems* and *Expert Systems with Applications* are by far the leading journals publishing OFDM papers.

Table 2 presents the authors most publishing OFDM papers.

Table 3 gives the affiliations of the authors publishing OFDM papers.

Figure 2 illustrates the source countries publishing OFDM papers. China is by far the first country publishing OFDM papers.

Figure 3 gives classifies the OFDM papers with respect to their subject areas. Engineering (55.1 %), computer sciences (52.3 %), mathematics (24.9 %), and decision sciences (13.7 %) are the first four areas that OFDM papers come from.

2.2 *Hesitant Fuzzy Decision-Making (HFDM)*

Hesitant fuzzy sets (HFSs), initially developed by Torra [34], are the extensions of regular fuzzy sets which handle the situations where a set of values are possible for the membership of a single element.

Torra [34] defines hesitant fuzzy sets (HFSs) as follow: Let X be a fixed set, a HFS on X is in terms of a function that when applied to X returns a subset of [0, 1]. Mathematical expression for HFS is as follows:

$$E = \{\langle x, h_E(x)\rangle | x \in X\}, \tag{2}$$

where $h_E(x)$ is a set of some values in [0, 1], denoting the possible membership degrees of the element $x \in X$ to the set E.

There are different types of HFSs that define membership degrees in various formats such as hesitant fuzzy linguistic term sets (HFLTS), interval valued hesitant

Table 1 Sources publishing OFDM papers and the frequencies of those papers

Source	Number of documents
Fuzzy Sets and Systems	79
Expert Systems with Applications	65
IEEE International Conference on Fuzzy Systems	53
Information Sciences	30
Xi Tong Gong Cheng Yu Dian Zi Ji Shu Systems Engineering and Electronics	28
Studies in Fuzziness and Soft Computing	24
European Journal of Operational Research	20
Journal of Intelligent and Fuzzy Systems	18
Computers and Industrial Engineering	17
International Journal of Production Research	17
Annual Conference of the North American Fuzzy Information Processing Society NAFIPS	16
Applied Mathematical Modelling	16
Cybernetics and Systems	16
Fuzzy Optimization and Decision Making	16
Knowledge Based Systems	16
Proceedings of the IEEE International Conference on Systems Man and Cybernetics	16
International Journal of Advanced Manufacturing Technology	15
Kongzhi Yu Juece Control and Decision	15
Xitong Gongcheng Lilun Yu Shijian System Engineering Theory and Practice	14
Advanced Materials Research	13
Applied Mechanics and Materials	13
Computers and Mathematics with Applications	12
International Journal of Information Technology and Decision Making	12
International Journal of Intelligent Systems	12
International Journal of Uncertainty Fuzziness and Knowledge Based Systems	12
Studies in Computational Intelligence	12
Soft Computing	10

fuzzy sets (IVHFS), triangular hesitant fuzzy sets (THFS), and trapezoidal hesitant fuzzy sets (TrHFS).

The search for HFDM based on Scopus database in October 2015 yielded 221 papers. 181 of them are articles while 25 of them are conference papers. The rest are review, editorial, erratum, and article in press.

Figure 4 presents the publication frequencies of the papers on HFDM with respect to years. As it is clearly seen from the graph, significant increases in the frequencies of HFDM papers exist after the year 2012.

Table 2 Authors most publishing OFDM papers

Authors	Number of OFDM papers he/she published
Herrera-Viedma, E.	22
Kahraman, C.	21
Yano, H.	20
Sakawa, M.	19
Kacprzyk, J.	18
Ekel, P.Y.	18
Lu, J.	17
Zhang, G.	15
Mousavi, S.M.	13
Chiclana, F.	12
Li, D.F.	12
Ruan, D.	11
Pedrycz, W.	11
Tavakkoli-Moghaddam, R.	11
Chen, S.Y.	11
Lee, H.S.	10
Liu, P.	10

Table 3 Affiliations of the authors publishing OFDM papers and frequencies

Affiliations	Number of papers
Islamic Azad University	44
University of Tehran	36
Istanbul Technical University	31
Huazhong University of Science and Technology	28
North China Electric Power University	27
Universidad de Granada	27
Galatasaray University	25
Southeast University	23
Dalian University of Technology	22
National Chiao Tung University Taiwan	21
National Taiwan Ocean University	21
National Taiwan University of Science and Technology	21
University of Technology Sydney	20

Table 4 shows the sources publishing hesitant fuzzy decision-making papers and the frequencies of those papers. *Journal of Intelligent and Fuzzy Systems*, *Information Sciences* and *Knowledge Based Systems* are by far the leading journals publishing HFDM papers.

Figure 5 presents the authors most publishing HFDM papers.

Table 5 gives the affiliations of the authors publishing HFDM papers.

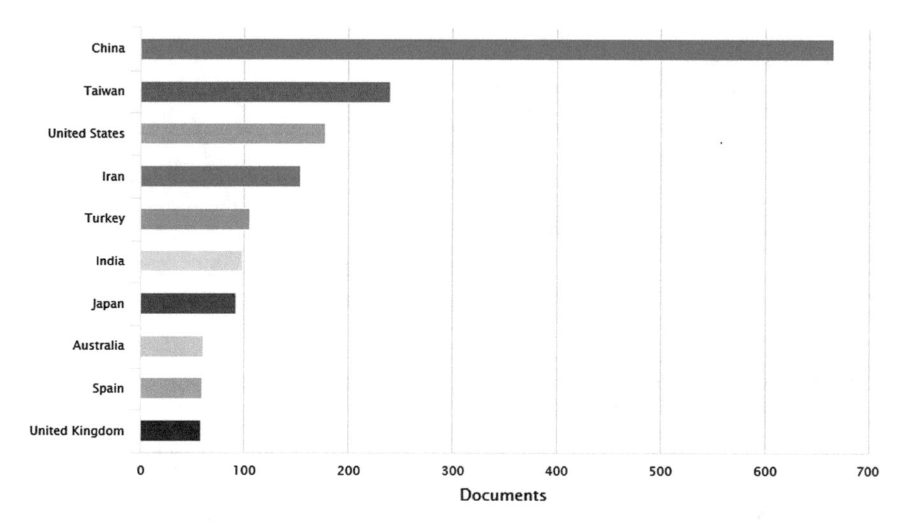

Fig. 2 Source countries publishing OFDM papers

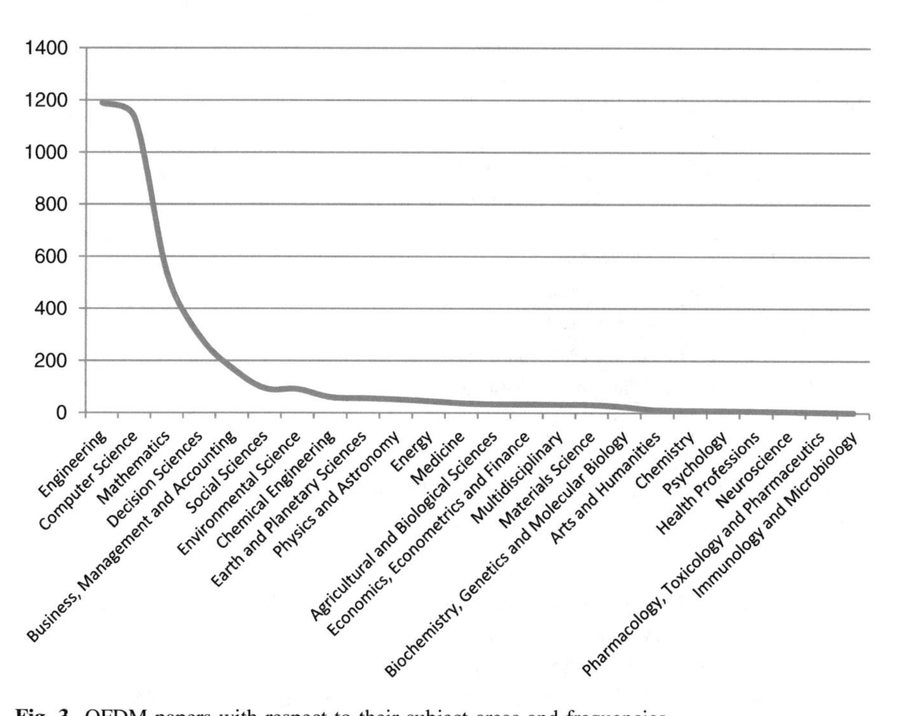

Fig. 3 OFDM papers with respect to their subject areas and frequencies

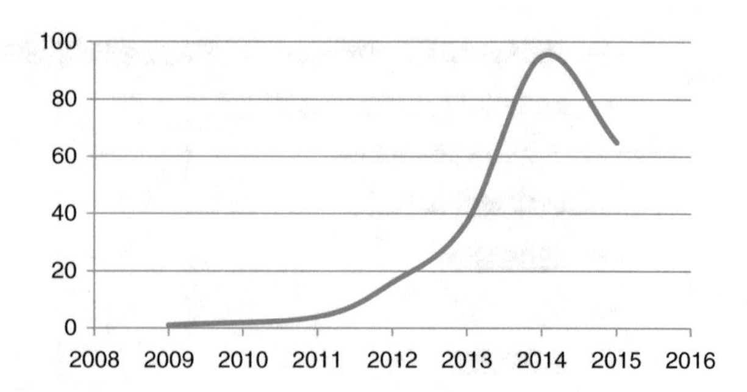

Fig. 4 Publication frequencies of the papers on HFDM with respect to years

Table 4 Sources publishing HFDM papers and the frequencies of those papers

Source	Number of documents
Journal of Intelligent and Fuzzy Systems	39
Information Sciences	16
Knowledge Based Systems	14
Mathematical Problems in Engineering	8
International Journal of Intelligent Systems	8
Journal of Applied Mathematics	6
Soft Computing	5
Computers and Industrial Engineering	4
Applied Mathematical Modelling	4
Journal of Information and Computational Science	4
Journal of Convergence Information Technology	4
International Journal of Uncertainty Fuzziness and Knowledge Based Systems	4
Expert Systems with Applications	4
International Journal of Computational Intelligence Systems	4
IEEE Transactions on Fuzzy Systems	4
Kongzhi Yu Juece Control and Decision	3
Journal of Computational Information Systems	3
International Journal of Fuzzy Systems	3
International Journal of Systems Science	3
International Journal of Digital Content Technology and Its Applications	3

Figure 6 illustrates the source countries publishing HFDM papers. China is again by far the first country publishing HFDM papers.

Figure 7 gives classifies the HFDM papers with respect to their subject areas. Computer sciences (77.2 %), mathematics (55.7 %), engineering (48.9 %), and decision sciences (17.8 %) are the first four areas that HFDM papers come from.

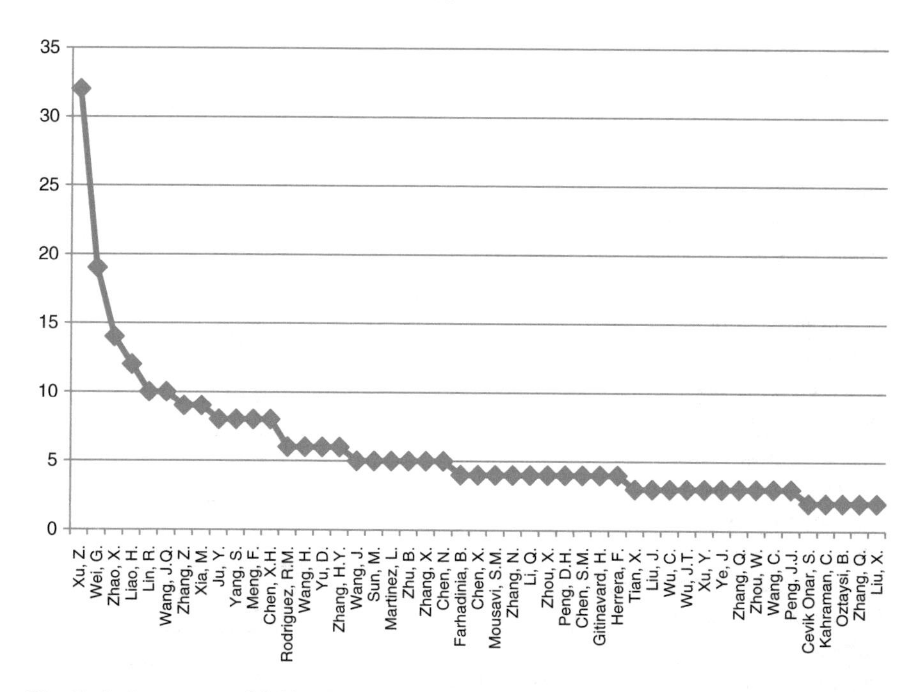

Fig. 5 Authors most publishing HFDM papers

Table 5 Affiliations of the authors publishing HFDM papers and frequencies

Affiliations	Number of papers
Central South University China	25
Southeast University	23
Chongqing University	21
Sichuan University	19
Beijing Institute of Technology	14
PLA University of Science and Technology	10
Hebei University	9
Shanghai Jiaotong University	9
Zhejiang University of Finance and Economics	8
Qingdao TechNological University	7
Southwest Jiaotong University	6
Tsinghua University	6
Universidad de Granada	6
Universidad de Jaen	5
Harbin Institute of Technology	5
National Taiwan University of Science and Technology	5
Zaozhuang University	5

(continued)

Table 5 (continued)

Affiliations	Number of papers
Iran University of Science and Technology	4
Nanjing University of Finance and Economics	4
Shahed University	4
Kunming University of Science and Technology	4
Harbin University of Science and Technology	4
Hunan University	4
Hebei University of Engineering	4
University of Manchester	3
Yangzhou University	3
Hohai University	3
Beijing Jiaotong Daxue	3
Beijing Normal University	3
Shaoxing University	3
Yunnan University	3
De Lin Institute of Technology	3
Hunan Institute of Science and Technology	3
Jining Medical University	2
Quchan Institute of Engineering and Technology	2
Quchan University of Advanced Technology	2
Hefei University of Technology	2
Zhejiang Normal University	2
Northwestern Polytechnical University	2
Central South University of Forestry and Technology	2
Beijing Forestry University	2
Anhui University of Technology	2
Universidad de Chile	2
Guangxi Normal University	2
Nanjing University of Aeronautics and Astronautics	2
Istanbul Technical University	2
South China University of Technology	2
Qufu Normal University	2

2.3 *Intuitionistic Fuzzy Decision-Making (IFDM)*

Atanassov [3] intuitionistic fuzzy sets (IFSs) include the membership value as well as the non-membership value for describing any x in X such that the sum of membership and non-membership is at most equal to 1.

Let $X \neq \emptyset$ be a given set. An intuitionistic fuzzy set (IFS) in X is an object A given by

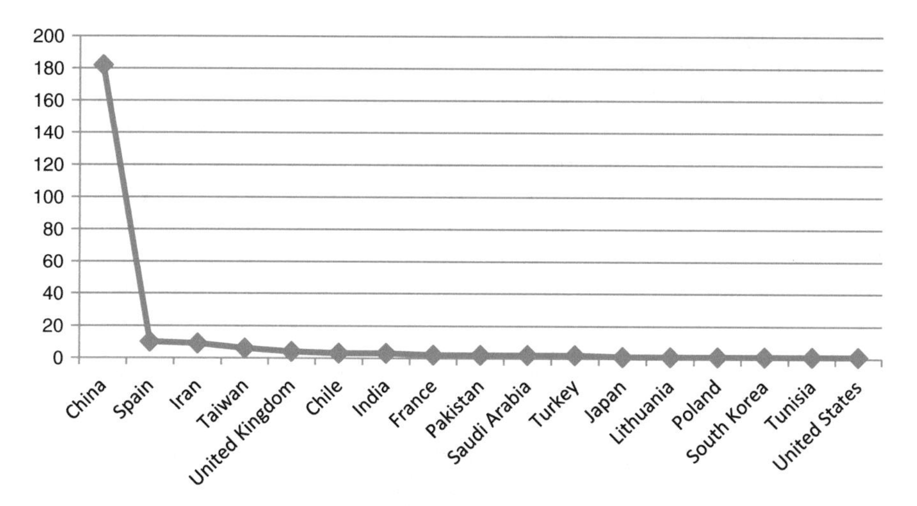

Fig. 6 Source countries publishing HFDM papers

$$\tilde{A} = \{x, \mu_{\tilde{A}}(x), v_{\tilde{A}}(x); x \in X\}, \tag{3}$$

where $\mu_{\tilde{A}}: X \to [0,1]$ and $v_{\tilde{A}}: X \to [0,1]$ satisfy the condition

$$0 \leq \mu_{\tilde{A}}(x) + v_{\tilde{A}}(x) \leq 1, \tag{4}$$

for every $x \in X$.

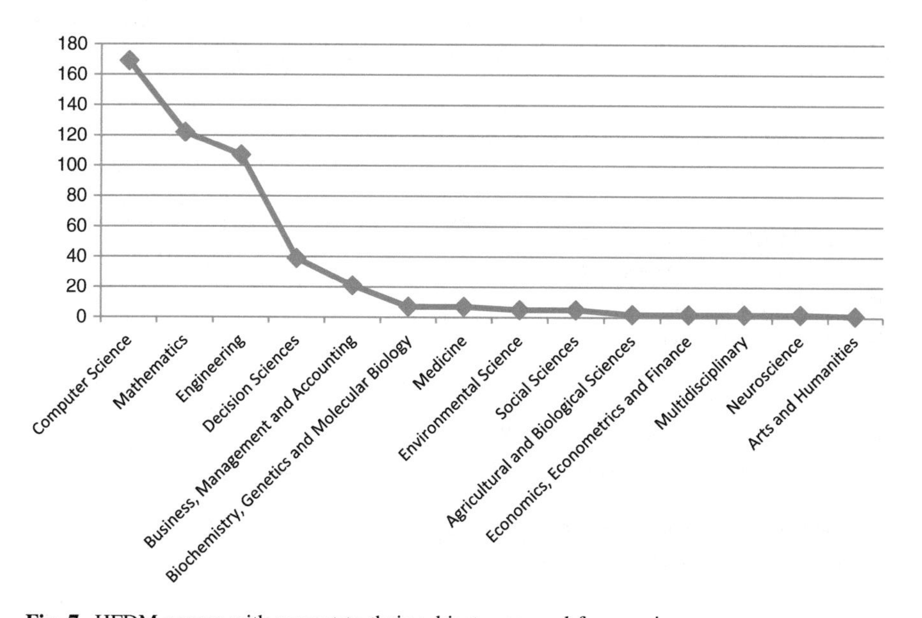

Fig. 7 HFDM papers with respect to their subject areas and frequencies

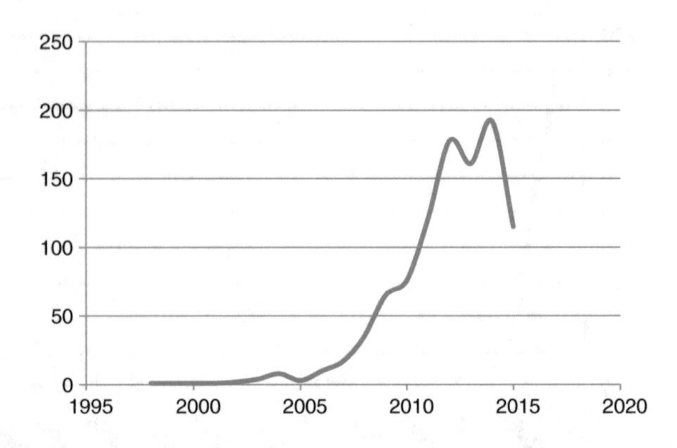

Fig. 8 Publication frequencies of the papers on IFDM with respect to years

There are different types of IFSs that define membership and nonmembership values in various formats such as interval valued intuitionistic fuzzy sets (IVIFS), triangular intuitionistic fuzzy sets (TIFS), and trapezoidal intuitionistic fuzzy sets (TrIFS).

The search for IFDM based on Scopus database in October 2015 yielded 994 papers. 719 of them are articles while 236 of them are conference papers. The rest are review, erratum, book and article in press.

Figure 8 presents the publication frequencies of the papers on IFDM with respect to years. As it is clearly seen from the graph, significant increases in the frequencies of IFDM papers exist after the year 2005.

Table 6 shows the sources publishing IFDM papers and the frequencies of those papers. *Journal of Intelligent and Fuzzy Systems*, *Expert Systems with Applications*, *Kongzhi Yu Juece Control and Decision*, and *Knowledge Based Systems*, *Applied Mathematical Modelling* are the leading journals publishing IFDM papers.

Table 6 Sources publishing IFDM papers and the frequencies of those papers

Source	Number of documents
Journal of Intelligent and Fuzzy Systems	47
Expert Systems with Applications	43
Kongzhi Yu Juece Control and Decision	33
Knowledge Based Systems	30
Applied Mathematical Modelling	27
Information Sciences	22
IEEE International Conference on Fuzzy Systems	21
International Journal of Uncertainty Fuzziness and Knowlege Based Systems	21
Advances in Information Sciences and Service Sciences	19
International Journal of Digital Content Technology and Its Applications	18

(continued)

Table 6 (continued)

Source	Number of documents
Mathematical Problems in Engineering	16
International Journal of Intelligent Systems	16
Fuzzy Optimization and Decision Making	16
IEEE Transactions on Fuzzy Systems	15
Journal of Applied Mathematics	13
Applied Mechanics and Materials	13
Journal of Computational Information Systems	12
Soft Computing	12
Journal of Convergence Information Technology	11
Xi Tong Gong Cheng Yu Dian Zi Ji Shu Systems Engineering and Electronics	11
Xitong Gongcheng Lilun Yu Shijian System Engineering Theory and Practice	11
International Journal of Computational Intelligence Systems	10
Journal of Information and Computational Science	10
International Journal of Advancements in Computing Technology	10
Communications in Computer and Information Science	10
Group Decision and Negotiation	9
Information Fusion	9
Journal of Systems Science and Systems Engineering	8
Technological and Economic Development of Economy	8

Figure 9 presents the authors most publishing IFDM papers.

Table 7 gives the affiliations of the authors publishing IFDM papers. Chinese universities again appear at the top of the table.

Figure 10 illustrates the source countries publishing IFDM papers. China is again by far the first country publishing IFDM papers.

Figure 11 gives classifies the IFDM papers with respect to their subject areas. Computer sciences (70.2 %), mathematics (40.3 %), engineering (44.4 %), and decision sciences (10.4 %) are the first four areas that IFDM papers come from.

2.4 *Type-2 Fuzzy Decision-Making (T2FDM)*

Zadeh [45] introduced type-2 fuzzy sets in 1975. A type-2 fuzzy set lets us incorporate uncertainty about the membership function into the fuzzy set theory. A type-2 fuzzy set $\tilde{A}$ in the universe of discourse X can be represented by a type-2 membership function $\mu_{\tilde{\tilde{A}}}$ shown as follows [27]:

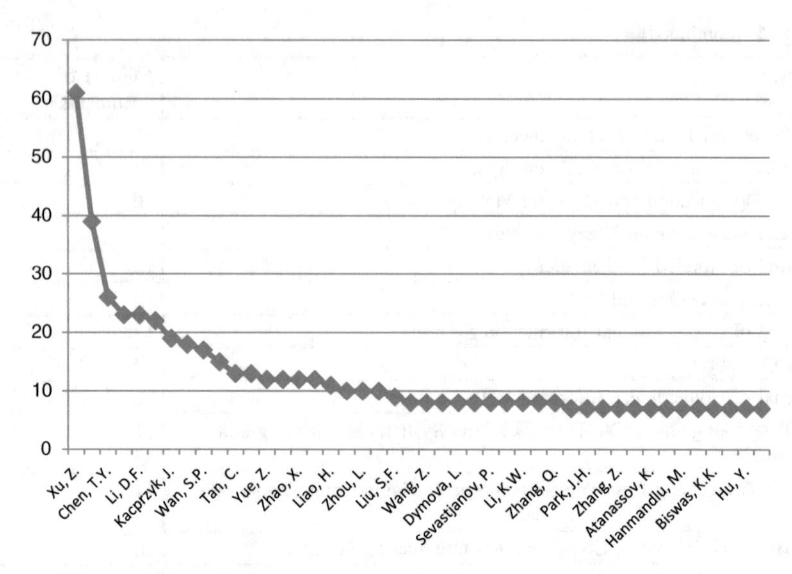

Fig. 9 Authors most publishing IFDM papers

Table 7 Affiliations of the authors publishing IFDM papers and frequencies

Affiliations	Number of papers
Central South University China	51
Chongqing University	51
Southeast University	44
PLA University of Science and Technology	32
Zhejiang University of Finance and Economics	28
Chang Gung University	25
Jiangxi University of Finance and Economics	24
Shanghai Jiaotong University	20
Systems Research Institute of the Polish Academy of Sciences	19
Hohai University	18
Sichuan University	18
Fuzhou University	18
Anhui University	17
Beijing Institute of Technology	17
Nanjing University of Aeronautics and Astronautics	14
Tsinghua University	14
Zhejiang Normal University	13
Beihang University	13
Guangdong University of Foreign Studies	12
Guangdong Ocean University	12
Hebei University	11

(continued)

Table 7 (continued)

Affiliations	Number of papers
Xiamen University	11
Harbin Institute of Technology	11
North China Electric Power University	11
Dalian University of Technology	10
Tianjin University	10
PLA Dalian Naval Academy	10
Zaozhuang University	10

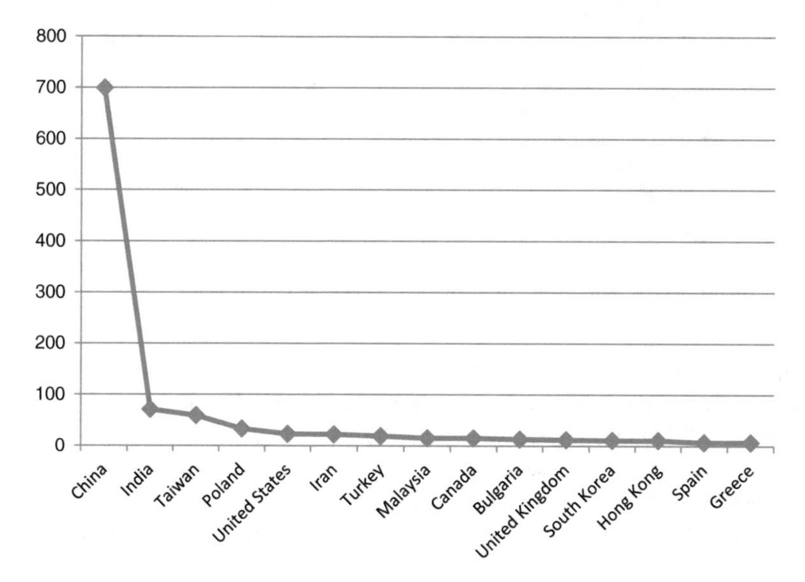

Fig. 10 Source countries publishing IFDM papers

$$\tilde{\tilde{A}} = \left\{ \left((x,u), \mu_{\tilde{\tilde{A}}}(x,u) \right) | \forall x \in X, \forall u \in J_x \subseteq [0,1], 0 \leq \mu_{\tilde{\tilde{A}}}(x,u) \leq 1 \right\} \quad (5)$$

where J_x denotes an interval [0, 1]. The type-2 fuzzy set $\tilde{\tilde{A}}$ also can be represented as follows [27]:

$$\tilde{\tilde{A}} = \int_{x \in X} \int_{u \in J_x} \mu_{\tilde{\tilde{A}}}(x,u)/(x,u), \quad (6)$$

where $J_x \subseteq [0,1]$ and $\int\int$ denote union over all admissible x and u.

The search for T2FDM based on Scopus database in October 2015 yielded 229 papers. 124 of them are articles while 95 of them are conference papers. The rest are conference review, erratum, book chapter, book and article in press.

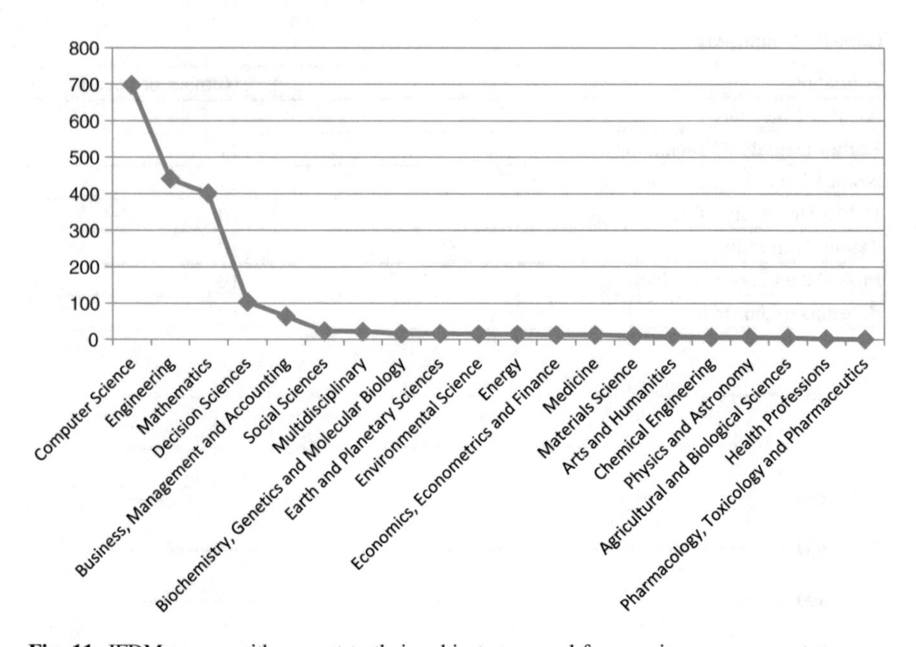

Fig. 11 IFDM papers with respect to their subject areas and frequencies

Figure 12 presents the publication frequencies of the papers on T2FDM with respect to years. As it is clearly seen from the graph, significant increases in the frequencies of T2FDM papers exist after the year 2007.

Table 8 shows the sources publishing T2FDM papers and the frequencies of those papers. *IEEE International Conference on Fuzzy Systems*, *Expert Systems with Applications*, *Information Sciences, Knowledge Based Systems*, *Soft Computing*, and *IEEE Transactions on Fuzzy Systems* are the leading journals publishing T2FDM papers.

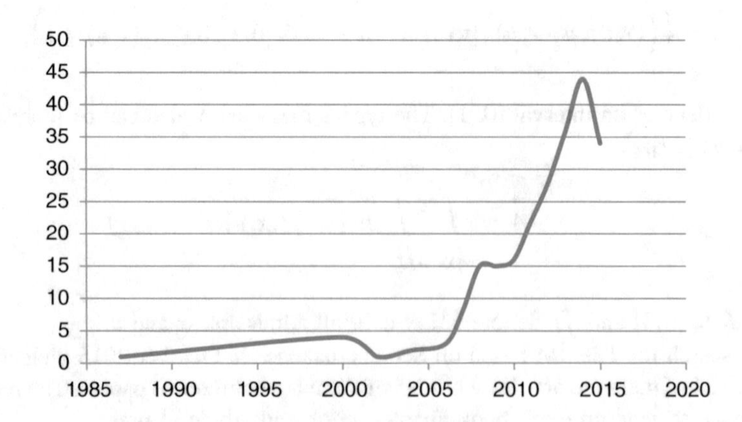

Fig. 12 Publication frequencies of the papers on T2FDM with respect to years

Table 8 Sources publishing T2FDM papers and the frequencies of those papers

Source	Number of documents
IEEE International Conference on Fuzzy Systems	21
Expert Systems with Applications	8
Information Sciences	7
Knowledge Based Systems	7
Soft Computing	7
IEEE Transactions on Fuzzy Systems	6
Annual Conference of the North American Fuzzy Information Processing Society NAFIPS	4
Computers and Industrial Engineering	3
International Journal of Computational Intelligence Systems	3
Studies in Computational Intelligence	3
Studies in Fuzziness and Soft Computing	3
Frontiers of Computer Science	2
International Journal of Electrical Power and Energy Systems	2
IEEE Transactions on Systems Man and Cybernetics Part B Cybernetics	2
International Journal of Information Technology and Decision Making	2
International Journal of Uncertainty Fuzziness and Knowledge Based Systems	2
Journal of Applied Mathematics	2
Journal of Industrial and Production Engineering	2
Journal of the Chinese Institute of Industrial Engineers	2
Advances in Soft Computing	2
Aip Conference Proceedings	2
International Journal of Advanced Manufacturing Technology	2
Engineering Applications of Artificial Intelligence	2
Communications in Computer and Information Science	2
Applied Mathematical Modelling	1
Applied Mathematical Sciences	1
Applied Mechanics and Materials	1
Arpn Journal of Engineering and Applied Sciences	1
Computers in Industry	1

Figure 13 presents the authors most publishing T2FDM papers.

Table 9 gives the affiliations of the authors publishing T2FDM papers.

Figure 14 illustrates the source countries publishing T2FDM papers. China and Taiwan are by far the first two countries publishing T2FDM papers.

Figure 15 gives classifies the T2FDM papers with respect to their subject areas. Computer sciences (78.7 %), engineering (41.7 %), mathematics (34.3 %), and decision sciences (8.3 %) are the first four areas that T2FDM papers come from.

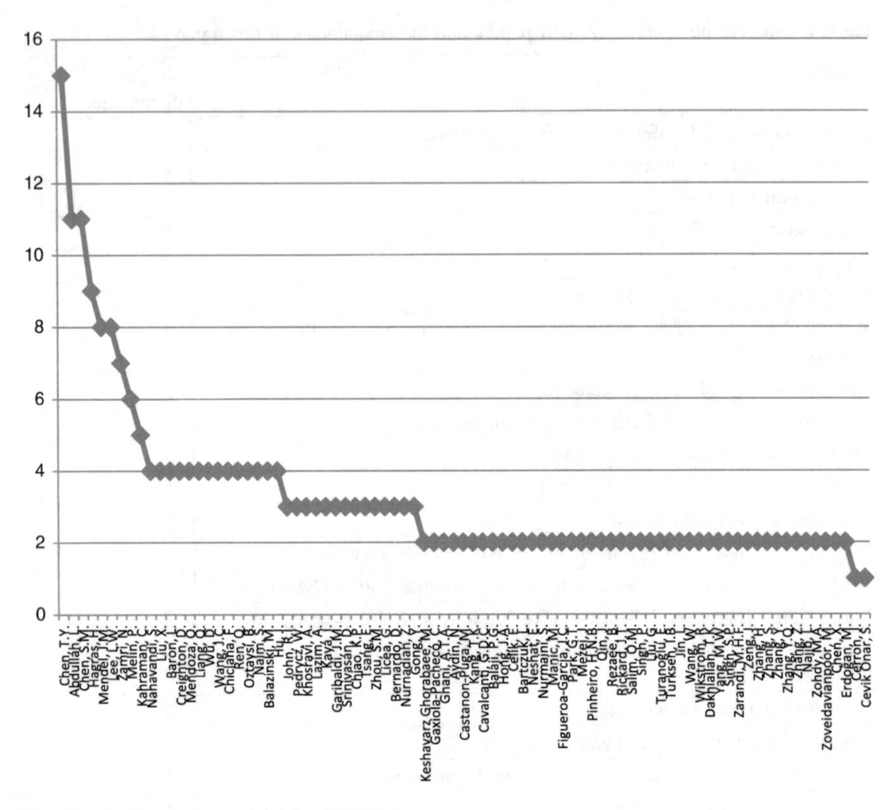

Fig. 13 Authors most publishing T2FDM papers

3 Fuzzy Decision Making Publishing Media

In this section, the journals, books, conferences, and societies, which are interested in fuzzy decision making are summarized.

In Table 10, you can find the list of journals publishing fuzzy decision making papers.

In Table 11, you can find the list of fuzzy decision making books.

In Table 12, you can find the list of conferences accepting fuzzy decision making papers.

4 Fuzzy Decision Making Applications

In this section, the following problem will be solved by ordinary fuzzy sets and three extensions in order to show how the above types of fuzzy sets can be used.

Table 9 Affiliations of the authors publishing T2FDM papers and frequencies

Affiliations	Number of papers
Chang Gung University	14
Universiti Malaysia Terengganu	13
National Taiwan University of Science and Technology	12
University of Essex	9
Istanbul Technical University	9
University of Southern California	9
Hebei University	7
Tijuana Institute of Technology	6
Yildiz Technical University	6
Southeast University	5
Universidad Autonoma de Baja California, Campus Tijuana	5
Central South University China	5
De Montfort University	5
Jinwen University of Science and Technology	5
Amirkabir University of Technology	4
Politechnika Czestochowska	4
Deakin University	4
Hohai University	3
Selcuk University	3

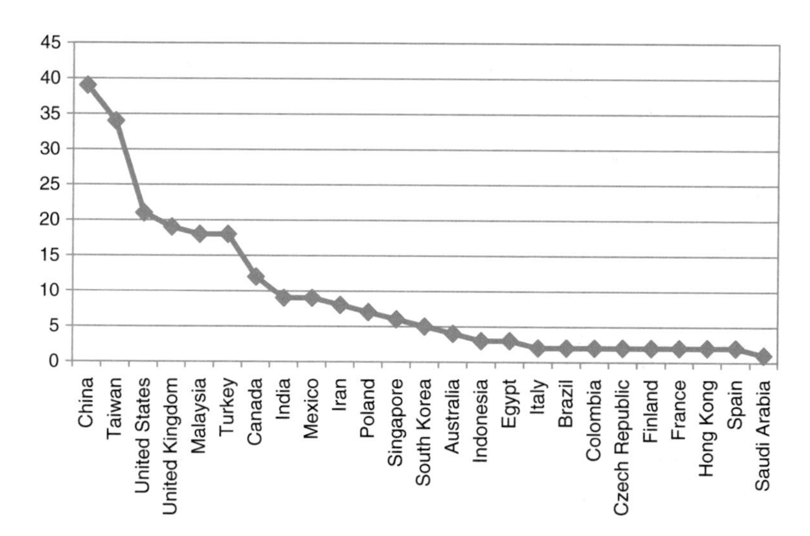

Fig. 14 Source countries publishing T2FDM papers

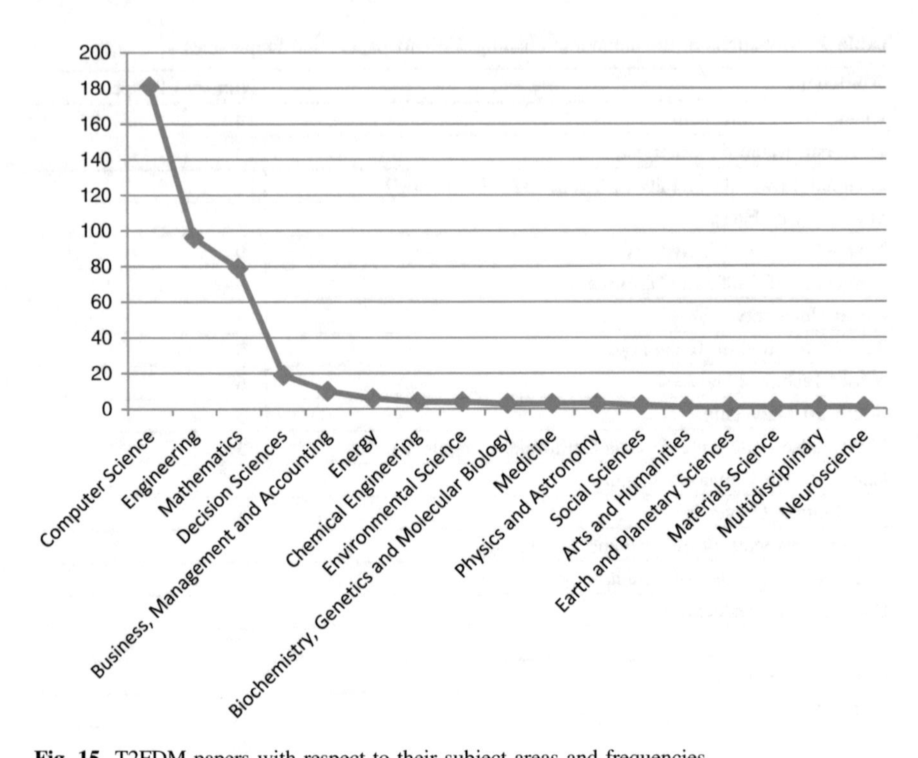

Fig. 15 T2FDM papers with respect to their subject areas and frequencies

Electric vehicles have been popular all around the world in recent years. Since one of the drawbacks of electric vehicles is the charge capacity of cars, locating charging stations has become a very important problem. Assume there are four possible locations and five criteria to be used for decision making. The criteria can be listed as follows:

- Destruction degree on vegetation and water (C1)
- Construction cost (C2)
- Annual operation and maintenance cost (C3)
- Traffic convenience (C4)
- Service capability (C5).

4.1 *Charging Station Location Selection Using Ordinary Fuzzy TOPSIS*

In Fuzzy TOPSIS approach the decision maker assigns importance weight for each criterion and score of each alternative with respect to the criteria. For the given problem the weights and scores are determined as shown in Table 13 [13].

Table 10 Journals publishing fuzzy decision making papers

Journal	Publisher	Number of fuzzy papers published
Fuzzy Sets and Systems	Elsevier	402
Expert Systems with Applications	Elsevier	381
IEEE International Conference on Fuzzy Systems	IEEE	314
Journal of Intelligent and Fuzzy Systems	IOS Press	177
Information Sciences	Elsevier	160
European Journal of Operational Research	Elsevier	130
Knowledge Based Systems	Elsevier	98
Applied Mathematical Modelling	Elsevier	83
Fuzzy Optimization and Decision Making	Springer	83
Computers and Industrial Engineering	Elsevier	74
International Journal of Uncertainty Fuzziness and Knowlege Based Systems	World Scientific	74
International Journal of Production Research	Taylor and Francis	71
International Journal of Advanced Manufacturing Technology	Springer	71
IEEE Transactions on Fuzzy Systems	IEEE	70
Soft Computing	Springer	64
International Journal of Intelligent Systems	John Wiley and Sons	58
Computers and Mathematics with Applications	Pergamon	52
Mathematical Problems in Engineering	Hindawi	51
International Journal of Computational Intelligence Systems	Atlantis	38
International Journal of Production Economics	Elsevier	38
Journal of Applied Mathematics	Hindawi	35
International Journal of Fuzzy Systems	Chinese Fuzzy Systems Association	35
Journal of Multiple Valued Logic and Soft Computing	Oldcity	34
International Journal of Approximate Reasoning	Elsevier	31
International Journal of Information Technology and Decision Making	World Scientific	30
International Journal of General Systems	Taylor and Francis	28
Engineering Applications of Artificial Intelligence	Elsevier	28
Group Decision and Negotiation	Springer	28
Technological and Economic Development of Economy	Taylor and Francis	27

Table 11 Fuzzy decision making books

Book	Editor and publisher	Tools and techniques
Fuzzy sets, decision making, and expert systems	Zimmermann [47, 48], Springer	Individual decision making in fuzzy environments, multi-person decision making in fuzzy environments, multi-criteria decision making in ıll-structured situations
Combining fuzzy imprecision with probabilistic uncertainty in decision making	Kacprzyk and Fedrizzi [17], Springer	Decision making in a probabilistic fuzzy environment, fuzzy statistical decision making, decision evaluation methods under uncertainty and imprecision
Multiperson decision making models using fuzzy sets and possibility theory	Kacprzyk and Fedrizzi [18], Springer	Multiperson decision making models using fuzzy sets and possibility theory
Fuzzy multiple objective decision making: methods and applications	Lai and Hwang [22], Springer	Multiple objective decision making, fuzzy multiple objective decision making, possibilistic multiple objective decision making
Fuzzy sets and fuzzy Decision-Making	Li and Yen [24], CRC Press	Multifactorial decision-making, additive standard multifactorial functions
Fuzzy logic for planning and decision making	Loostma [25], Springer	Fuzzy logic in operations research, fuzzy PERT, fuzzy SMART, fuzzy multi-objective optimization, fuzzy AHP
Fuzzy and multi-level decision making: an interactive computational approach	Lee and Shih [23], Springer	Fuzzy ınteractive multi-level decision making, possibility programming, aggregation of fuzzy systems in multi-level decisions
Dynamical aspects in fuzzy decision making	Yoshida [43], Physica-Verlag	Fuzzy dynamic programming, fuzzy portfolio model, interactive fuzzy programming, fuzzy optimization
Fuzzy reasoning in decision making and optimization	Carlsson and Fullér [9], Physica-Verlag	Fuzzy multicriteria decision making, fuzzy reasoning and decision making
Fuzzy decision making in modeling and control	Sousa and Kaymak [33], World Scientific	Model-based control with fuzzy decision functions, fuzzy decision functions, Fuzzy aggregated membership control
Multi-objective group decision making: methods, software and applications with fuzzy set techniques	Lu et al. [26], Imperial College Press	Fuzzy multi-objective decision making, fuzzy multi-objective DSS, fuzzy group decision making

(continued)

Table 11 (continued)

Book	Editor and publisher	Tools and techniques
Fuzzy sets and their extensions: representation, aggregation and models intelligent systems from decision making to data mining	Bustince et al. [8], Springer	Aggregation operators, voting systems, linguistic decision based model, group decision making
Fuzzy multi-criteria decision making: theory and applications with recent developments	Kahraman [21], Springer	Fuzzy AHP, fuzzy outranking methods, fuzzy TOPSIS, fuzzy multi-objective decision making, fuzzy geometric programming
Fuzzy multicriteria decision-making: models, methods and applications	Pedrycz et al. [29], John Wiley & Sons	Discrete models of multicriteria decision-making, fuzzy group decision making models
Fuzzy-like multiple objective decision making	Xu and Zhou [38], Springer	Fuzzy random multiple objective decision making, fuzzy-like multiple objective decision making, bifuzzy multiple objective decision making
Decision making in the manufacturing environment using graph theory and fuzzy multiple attribute decision making methods, vol. 2	Rao [30], Springer	Fuzzy ELECTRE, fuzzy PROMETHEE, fuzzy COPRAS, fuzzy VIKOR, fuzzy AHP
Intuitionistic fuzzy information aggregation: theory and applications	Xu and Cai [39], Springer	Intuitionistic preference relations, dynamic intuitionistic fuzzy multi-attribute decision making, nonlinear optimization models for multi-attribute group decision making
Fuzzy multiple objective decision making	Tzeng and Huang [36], Chapman and Hall/CRC	multi-objective evolutionary algorithms, network data envelopment analysis, fuzzy goal programming, multi-level multi-objective programming
Decision theory with imperfect information	Aliev and Huseynov [1], World Scientific	Hierarchical models for decision making, behavioral decision making with combined states, decision making under unprecisiated imperfect information

The scale used in this approach is given in Table 14.

The fuzzy values given in Table 14 are used to transform the linguistic evaluations into fuzzy numbers as given in Table 15.

The scores of the alternatives are then normalized to find the normalized scores (Table 16).

Table 12 Conferences accepting fuzzy decision making papers

Conference	Organizing institutes	Number of decision making papers presented
IEEE International Conference on Fuzzy Systems (FUZZIEEE)	IEEE	423
Annual Conference of the North American Fuzzy Information Processing Society (NAFIPS)	NAFIPS	136
IEEE International Conference on Systems Man and Cybernetics	IEEE	91
Conference of the European Society for Fuzzy Logic and Technology (EUSFLAT)	EUSFLAT	82
International Federation of Automatic Control Conference (IFAC)	IFAC	68
World Congress of the International Fuzzy Systems Association (IFSA)	IFSA	61
International Conference on Machine Learning and Cybernetics (ICMLC)	ICMLC	44
World Congress on Intelligent Control and Automation (WCICA)	WCICA	43
Computers and Industial Engineering (CIE) Conference	CIE	34
Information Processing and Management of Uncertainty in Knowledge-Based Systems (IPMU)	IPMU	33
International Federation for Information Processing (IFIP) Advances in Information and Communication Technology	IFIP	25
Joint Conference on Information Sciences	INFORMS	24
Fuzzy Logic in Nuclear Science (FLINS)	FLINS	22
International Joint Conference on Neural Networks, International Neural Network Society (INNS)	INNS	20
Industrial Electronics Conference	IEEE	15
Canadian Conference on Electrical and Computer Engineering	IEEE	14
ASME Design Engineering Technical Conference	ASME	13
IEEE International Symposium on Multiple-Valued Logic	IEEE	13
Annual Meeting of the Decision Sciences Institute (DSI)	DSI	12
IEEE Conference on Decision and Control	IEEE	12
IEEE International Symposium on Intelligent Control	IEEE	12
International IEEE Conference Intelligent Systems	IEEE	12
IEEE International Conference on Neural Networks Conference	IEEE	11

(continued)

Table 12 (continued)

Conference	Organizing institutes	Number of decision making papers presented
IEEE International Engineering Management Conference	IEEE	11
IEEE International Conference on Intelligent Robots and Systems	IEEE	10
IEEE Conference on Intelligent Transportation Systems Proceedings ITSC	IEEE	9
International Conference on Intelligent Systems and Knowledge Engineering (ISKE)	ISKE	9
Society of Instrument and Control Engineers (SICE) Annual Conference	SICE	9
Association for Computing Machinery (ACM) Symposium on Applied Computing	ACM	8
IEEE International Conference on Industrial Technology	IEEE	8

Table 13 Importance of each criterion and scores of each alternative with respect to the criteria

Criteria	Importance	Scores			
		A1	A2	A3	A4
C1	M	MG	MG	MP	G
C2	VH	MG	P	MG	G
C3	VH	G	P	VG	MP
C4	M	F	MP	MP	G
C5	H	F	MG	F	MG

Table 14 Linguistic scales used for fuzzy TOPSIS

Linguistic variables for importance	Fuzzy values	Linguistic variables for evaluations	Fuzzy values
Very low (VL)	(0, 0, 0.1)	Very poor (VP)	(0, 0, 1)
Low (L)	(0, 0.1, 0.3)	Poor (P)	(0, 1, 3)
Medium low (ML)	(0.1, 0.3, 0.5)	Medium poor (MP)	(1, 3, 5)
Medium (M)	(0.3, 0.5, 0.7)	Fair (F)	(3, 5, 7)
Medium high (MH)	(0.5, 0.7, 0.9)	Medium good (MG)	(5, 7, 9)
High (H)	(0.7, 0.9, 1)	Good (G)	(7, 9, 10)
Very high (VH)	(0.9, 1, 1)	Very good (VG)	(9, 10)

Table 15 Triangular fuzzy importance and scores

Criteria	Importance	Scores			
		A1	A2	A3	A4
C1	(0.3, 0.5, 0.7)	(7, 9, 10)	(1, 3, 5)	(7, 9, 10)	(3, 5, 7)
C2	(0.9, 1, 1)	(5, 7, 9)	(0,1,3)	(5, 7, 9)	(7, 9, 10)
C3	(0.9, 1, 1)	(7, 9, 10)	(3, 5, 7)	(9, 10)	(5, 7, 9)
C4	(0.3, 0.5, 0.7)	(7, 9, 10)	(5, 7, 9)	(7, 9, 10)	(7, 9, 10)
C5	(0.7, 0.9, 1)	(3, 5, 7)	(5, 7, 9)	(3, 5, 7)	(3, 5, 7)

Table 16 Normalized scores

	A1	A2	A3	A4
C1	(0.7, 0.9, 1)	(0.1, 0.3, 0.5)	(0.7, 0.9, 1)	(0.3, 0.5, 0.7)
C2	(0.5, 0.7, 0.9)	(0, 0.1, 0.3)	(0.5, 0.7, 0.9)	(0.7, 0.9, 1)
C3	(0.7, 0.9, 1)	(0.3, 0.5, 0.7)	(0.9, 1, 1)	(0.5, 0.7, 0.9)
C4	(0.7, 0.9, 1)	(0.5, 0.7, 0.9)	(0.7, 0.9, 1)	(0.7, 0.9, 1)
C5	(0.33, 0.56, 0.78)	(0.56, 0.78, 1)	(0.33, 0.56, 0.78)	(0.33, 0.56, 0.78)

Then the fuzzy importance weights are multiplied by the normalized scores in order to find the weighted normalized scores as in Table 17.

Using the values given in Table 17, the positive ideal solution (PIS) and negative ideal solution (NIS) are determined. Using these values, the distances to PIS and NIS and closeness coefficients are calculated as shown in Table 18.

Table 17 Weighted normalized scores

	A1	A2	A3	A4
C1	(0.21, 0.45, 0.7)	(0.03, 0.15, 0.35)	(0.21, 0.45, 0.7)	(0.09, 0.25, 0.49)
C2	(0.45, 0.7, 0.9)	(0, 0.1, 0.3)	(0.45, 0.7, 0.9)	(0.63, 0.9, 1)
C3	(0.63, 0.9, 1)	(0.27, 0.5, 0.7)	(0.81, 1, 1)	(0.45, 0.7, 0.9)
C4	(0.21, 0.45, 0.7)	(0.15, 0.35, 0.63)	(0.21, 0.45, 0.7)	(0.21, 0.45, 0.7)
C5	(0.23, 0.5, 0.78)	(0.39, 0.7,1)	(0.23, 0.5, 0.78)	(0.23, 0.5, 0.78)

Table 18 Distance measures and closeness coefficients

	d^-	d^+	CC_i	Ranking
A1	3.108	2.297	0.575	2
A2	2.089	3.295	0.387	4
A3	3.191	2.185	0.593	1
A4	2.934	2.457	0.544	3

4.2 Charging Station Location Selection Using HFLTS

When the same problem is handled using HFLTS, the input data contain linguistic operators such as lower than, between, higher than etc. For this problem, the pairwise comparison matrices are determined as given in Tables 19, 20, 21, 22, 23 and 24 [41].

Table 19 Pairwise comparison of the criteria using HFLTS

	C1	C2	C3	C4	C5
C1	–	Lower than very low	Is low	Is medium	Between low and medium
C2	Between high and absolute	–	Is high	Greater than high	Between medium and high
C3	Is vh	Is low	–	Between medium and high	Is high
C4	Is medium	Lower than low	Between medium and low	–	Is medium
C5	Between low and high	At most high	Is low	Is medium	–

Table 20 Pairwise comparison of the alternatives with respect to C1

	A1	A2	A3	A4
A1	–	Is medium	Between medium and vh	Greater than high
A2	Is medium	–	Between low and medium	Between high and vh
A3	Between very low and medium	Between medium and high	–	At most high
A4	Lower than low	Between low and very low	At most low	–

Table 21 Pairwise comparison of the alternatives with respect to C2

	A1	A2	A3	A4
A1	–	Between low and medium	Between low and medium	Is low
A2	Between very low and low	–	Between very low and low	Lower than very low
A3	Between medium and high	Between high and vh	–	Is low
A4	Is high	Greater than vh	Is medium	–

Table 22 Pairwise comparison of the alternatives with respect to C3

	A1	A2	A3	A4
A1	–	Between high and vh	Between low and medium	Greater than vh
A2	Between very low and low	–	At least vh	Is high
A3	Between medium and high	At most very low	–	Is absolute
A4	Lower than very low	Is low	Is none	–

Table 23 Pairwise comparison of the alternatives with respect to C4

	A1	A2	A3	A4
A1	–	Between high and vh	Is high	Between medium and low
A2	Between very low and low	–	Is medium	Is very low
A3	Is low	Is medium	–	Between very low and low
A4	Between medium and high	Is vh	Between high and vh	–

Table 24 Pairwise comparison of the alternatives with respect to C5

	A1	A2	A3	A4
A1	–	Between low and medium	Is medium	Is medium
A2	Between medium and high	–	Between medium and vh	Between medium and high
A3	Is medium	Between very low and medium	–	Between low and medium
A4	Is medium	Between low and medium	Between medium and high	–

Due to the page constraints only sample calculations are provided here. Using Table 19, the pessimistic and optimistic preference matrices are provided as given in Table 25.

Using the scale given in Table 26, the linguistic intervals are converted to numerical intervals.

In order to calculate the linguistic intervals given in Table T9, the mean value of each row in pessimistic and optimistic preference matrices is calculated. For example, in the interval value [(l, −0.25), (l, −0.0)], the lower bound is determined

Table 25 Pessimistic and optimistic preference matrices

Pessimistic collective preference					
	C1	C2	C3	C4	C5
C1	–	(n, −0.0)	(l, −0.0)	(m, −0.0)	(l, −0.0)
C2	(h, −0.0)	–	(h, −0.0)	(vh, −0.0)	(m, −0.0)
C3	(vh, −0.0)	(l, −0.0)	–	(m, −0.0)	(h, −0.0)
C4	(m, −0.0)	(n, −0.0)	(l, −0.0)	–	(m, −0.0)
C5	(m, −0.33)	(n, −0.0)	(l, −0.0)	(m, −0.0)	–
Optimistic collective preference					
	C1	C2	C3	C4	C5
C1	–	(n, −0.0)	(l, −0.0)	(m, −0.0)	(m, −0.0)
C2	(a, −0.0)	–	(h, −0.0)	(a, −0.0)	(h, −0.0)
C3	(vh, −0.0)	(l, −0.0)	–	(h, −0.0)	(h, −0.0)
C4	(m, −0.0)	(vl, −0.0)	(m, −0.0)	–	(m, −0.0)
C5	(h, −0.0)	(h, −0.0)	(l, −0.0)	(m, −0.0)	–

Table 26 The scale for the linguistic terms

Ni	vli	li	mi	hi	vhi	ai
0	1	2	3	4	5	6

by calculating the mean values of (n, −0.0), (l, −0.0), (m, −0.0), and (l, −0.0). In a similar way, the upper bound is the mean of the first line of optimistic preference matrix which is (n, −0.0), (l, −0.0), (m, −0.0), and (m, −0.0). Table 27 shows the interval utilities, midpoint values and weights of each criterion.

After linguistic intervals are calculated, they are transformed to interval utilities. By getting the mean of the upper and lower bounds of the interval, the midpoints are calculated. Finally, the midpoint values are normalized and the weights of each criterion are determined.

Table 27 Interval utilities, midpoint values and weights of each criterion

Criteria	Linguistic intervals	Interval utilities	Midpoints	Weights
C1	[(l, −0.25), (l, −0.0)]	[1.75, 2]	1.87	0.13
C2	[(h, −0.0), (vh, −0.0)]	[4, 5]	4.5	0.30
C3	[(h, −0.50), (h, −0.25)]	[3.5, 3.75]	3.62	0.24
C4	[(l, −0.0), (l, 0.50)]	[2, 2.5]	2.25	0.15
C5	[(l, −0.8), (m, 0.25)]	[1.92, 3.25]	2.58	0.17

Table 28 Linguistic intervals and weights of the criteria

	Wrt C1	Wrt C2	Wrt C3	Wrt C4	Wrt C5	Weighted score	Rank
Weights	0.13	0.30	0.24	0.15	0.17		
A1	0.38	0.24	0.36	0.31	0.24	0.297	1
A2	0.30	0.09	0.31	0.15	0.31	0.216	4
A3	0.23	0.29	0.28	0.18	0.21	0.249	2
A4	0.09	0.38	0.06	0.36	0.25	0.238	3

The process is repeated for the other matrices and the weights of the alternatives with respect each criterion are determined as shown in Table 28. In order to find the ranking of the alternatives, the weighted scores are calculated as in Table 28.

4.3 Charging Station Location Selection Using Pairwise Linguistic Evaluations Using IVIFS

This approach is based on pairwise linguistic comparison of the criteria and alternatives. As the evaluations are made using linguistic terms, they are transformed to IVIFS and the steps of the methodology are applied. The details of the methodology we apply can be obtained from Onar et al. [10]. The linguistic terms and related IVIFS numbers are listed in Table 29.

In accordance with the problem, the pairwise comparison matrices are formed and filled by the decision maker. Table 30 shows the pairwise comparison of the criteria and Table 31 shows the pairwise comparison of each alternative with respect to the criteria.

Due to the page constraints, we only give the details of the calculations for criteria weights. Table 32 gives the pairwise comparisons of the criteria expressed as IVIFS.

Table 29 Linguistic scale and its corresponding IVIFS

Linguistic terms		Membership
Absolutely low (AL)	AL	([0.10, 0.25], [0.65, 0.75])
Very low (VL)	VL	([0.15, 0.30], [0.60, 0.70])
Low (L)	L	([0.20, 0.35], [0.55, 0.65])
Medium low (ML)	ML	([0.25, 0.4]), [0.50, 0.60])
Equal (E)	E	([0.45, 0.55], [0.30, 0.45])
Medium high (MH)	MH	([0.50, 0.60], [0.25, 0.40])
High (H)	H	([0.55,0.65], [0.20, 0.35])
Very high (VH)	VH	([0.60,0.70], [0.15,0.30])
Absolutely high (AH)	AH	([0.65,0.75], [0.10,0.25])
Exactly equal	EE	([0.5,0.5], [0.5,0.5])

Table 30 Pairwise comparison of the criteria

Wrt goal	C1	C2	C3	C4	C5
C1	EE	AL	L	EE	ML
C2		EE	H	VH	H
C3			EE	H	H
C4				EE	EE
C5					EE

Table 31 Pairwise comparison of the alternatives with respect to the criteria

Wrt C1	A1	A2	A3	A4
A1	EE	E	VH	ML
A2		EE	VH	ML
A3			EE	AL
A4				EE
Wrt C2	A1	A2	A3	A4
A1	EE	H	EE	ML
A2		EE	VL	L
A3			EE	MH
A4				EE
Wrt C3	A1	A2	A3	A4
A1	EE	VH	ML	AH
A2		EE	AL	ML
A3			EE	AH
A4				EE
Wrt C4	A1	A2	A3	A4
A1	EE	MH	MH	ML
A2		EE	EE	VL
A3			EE	MH
A4				EE
Wrt C5	A1	A2	A3	A4
A1	EE	ML	EE	ML
A2		EE	MH	EE
A3			EE	MH
A4				EE

According to the steps given in Onar et al. [10], the scores of the IVIFS values are calculated using score function, as shown in Table 33.

Then, the interval multiplicative matrix $\tilde{A} = \left(\tilde{a}_{ij}\right)_{n \times n} = \left[10^{\left(\mu_{g_{ij}}^{-} - v_{g_{ij}}^{+}\right)}, 10^{\left(\mu_{g_{ij}}^{+} - v_{g_{ij}}^{-}\right)}\right]$ is calculated as given in Table 34.

Table 32 Pairwise comparisons of the criteria expressed as IVIFS

	C1	C2	C3	C4	C5
C1	([0, 5, 0, 5], [0, 5, 0, 5])	([0, 1, 0, 25], [0, 65, 0, 75])	([0, 2, 0, 35], [0, 55, 0, 65])	([0, 5, 0, 5], [0, 5, 0, 5])	([0, 25, 0, 4], [0, 5, 0, 6])
C2	([0, 65, 0, 75], [0, 1, 0, 25])	([0, 5, 0, 5], [0, 5, 0, 5])	([0, 55, 0, 65], [0, 2, 0, 35])	([0, 6, 0, 7], [0, 15, 0, 3])	([0, 55, 0, 65], [0, 2, 0, 35])
C3	([0, 55, 0, 65], [0, 2, 0, 35])	([0, 2, 0, 35], [0, 55, 0, 65])	([0, 5, 0, 5], [0, 5, 0, 5])	([0, 55, 0, 65], [0, 2, 0, 35])	([0, 55, 0, 65], [0, 2, 0, 35])
C4	([0, 5, 0, 5], [0, 5, 0, 5])	([0, 15, 0, 3], [0, 6, 0, 7])	([0, 2, 0, 35], [0, 55, 0, 65])	([0, 5, 0, 5], [0, 5, 0, 5])	([0, 5, 0, 5], [0, 5, 0, 5])
C5	([0, 5, 0, 6], [0, 25, 0, 4])	([0, 2, 0, 35], [0, 55, 0, 65])	([0, 2, 0, 35], [0, 55, 0, 65])	([0, 5, 0, 5], [0, 5, 0, 5])	([0, 5, 0, 5], [0, 5, 0, 5])

Table 33 Score judgement matrix

	C1	C2	C3	C4	C5
C1	[0, 0]	[−0.65, −0.4]	[−0.45, −0.2]	[0, 0]	[−0.35, −0.1]
C2	[0.4, 0.65]	[0, 0]	[0.2, 0.45]	[0.3, 0.55]	[0.2, 0.45]
C3	[0.2, 0.45]	[−0.45, −0.2]	[0, 0]	[0.2, 0.45]	[0.2, 0.45]
C4	[0, 0]	[−0.55, −0.3]	[−0.45, −0.2]	[0, 0]	[0, 0]
C5	[0.1, 0.35]	[−0.45, −0.2]	[−0.45, −0.2]	[0, 0]	[0, 0]

Table 34 Interval multiplicative matrix

	C1	C2	C3	C4	C5
C1	[1, 1]	[0.22, 0.4]	[0.35, 0.63]	[1, 1]	[0.45, 0.79]
C2	[2.51, 4.47]	[1, 1]	[1.58, 2.82]	[2, 3.55]	[1.58, 2.82]
C3	[1.58, 2.82]	[0.35, 0.63]	[1, 1]	[1.58, 2.82]	[1.58, 2.82]
C4	[1, 1]	[0.28, 0.5]	[0.35, 0.63]	[1, 1]	[1, 1]
C5	[1.26, 2.24]	[0.35, 0.63]	[0.35, 0.63]	[1, 1]	[1, 1]

In the next step, the priority vector of each criterion is calculated by using interval multiplicative matrix as in Table 35.

Possibility degree matrix is formed as given in Table Y8 and the weights of the criteria are determined.

The same steps are applied to the pairwise comparison matrices given in Table 36 and weights of the alternatives with respect to the criteria are determined. In order to find the overall rankings the weighted score of each alternative is calculated as in Table 37.

Table 35 Priority vectors

Criteria	Priority vector
C1	[0.079, 0.15]
C2	[0.227, 0.576]
C3	[0.16, 0.397]
C4	[0.095, 0.163]
C5	[0.104, 0.216]

Table 36 Possibility degree matrix and weights of the criteria

	C1	C2	C3	C4	C5	Weight
C1	0.50	0.00	0.00	0.40	0.25	0.133
C2	1.00	0.50	0.71	1.00	1.00	0.286
C3	1.00	0.29	0.50	0.99	0.84	0.256
C4	0.60	0.00	0.01	0.50	0.33	0.147
C5	0.75	0.00	0.16	0.67	0.50	0.179

Table 37 Weighted scores of the alternatives

Alternatives	wrt C1	wrt C2	wrt C3	wrt C4	wrt C5	Weighted Score	Ranking
Weights	0.13	0.29	0.26	0.15	0.18		
A1	0.28	0.26	0.32	0.28	0.17	0.266	3
A2	0.28	0.13	0.15	0.15	0.32	0.190	4
A3	0.13	0.32	0.34	0.25	0.26	0.277	1
A4	0.32	0.29	0.19	0.32	0.26	0.268	2

4.4 Charging Station Location Selection Using Interval Type-2 Fuzzy AHP

In this section, the same charging station location selection problem is solved by Interval Type-2 Fuzzy AHP method proposed by Kahraman et al. [19]. The linguistic scale used in this method is given in Table 38.

The evaluations given in Sect. 4.3 can be converted to the scale given in Table 38. The pairwise comparison matrix for the criteria is given in Table 39. The other comparison matrices are not given because of space constraints.

According to the steps provided by Kahraman et al. [19], the geometric mean for each row in Table 39 is calculated. The calculated mean values are given in Table 40.

Fuzzy weight of each criterion is determined by dividing the means given in Table 40 to the sum of all means. Interval Type-2 fuzzy weights of each criterion is given in Table 41. These values are later defuzzified and normalized in order to find the final crisp weights of the criteria.

Table 38 Linguistic scale used in interval type-2 fuzzy method

Linguistic variables	Trapezoidal interval type-2 fuzzy sets
Absolute (A)	[(7, 8, 9, 9; 1,1), (7.2, 8.2, 8.8, 8.9; 0.8, 0.8)]
Very strong (VS)	[(5, 6, 8, 9; 1, 1), (5.2, 6.2, 7.8, 8.8; 0.8, 0.8)]
Fairly strong (FS)	[(3, 4, 6, 7; 1, 1), (3.2, 4.2, 5.8, 6.8; 0.8, 0.8)]
Weak (W)	[(1, 2, 4, 5; 1, 1), (1.2, 2.2, 3.8, 4.8; 0.8, 0.8)]
Equal (E)	[(1, 1, 1, 1; 1, 1), (1, 1, 1, 1; 1, 1)]
If a factor i has one of the above linguistic variables assigned to it when compared with factor j. The j has the reciprocal value when compared with i	Reciprocals of above

Table 39 Pairwise comparisons expressed as type-2 fuzzy sets

	C1	C2	C3	C4	C5
C1	[(1, 1, 1, 1; 1, 1), (1, 1, 1, 1; 1, 1)]	[(0.14, 0.16, 0.25, 0.33; 1, 1), (0.14, 0.17, 0.23, 0.31; 0.8, 0.8)]	[(0.2, 0.25, 0.5, 1; 1, 1), (0.20, 0.26, 0.45, 0.83; 0.8, 0.8)]	[(1, 1, 1, 1; 1, 1), (1, 1, 1, 1; 1, 1)]	[(0.2, 0.25, 0.5, 1; 1, 1), (0.20, 0.26, 0.45, 0.83; 0.8, 0.8)]
C2	[(3, 4, 6, 7; 1, 1), (3.2, 4.2, 5.8, 6.8; 0.8, 0.8)]	[(1, 1, 1, 1; 1, 1), (1, 1, 1, 1; 1, 1)]	[(3, 4, 6, 7; 1, 1), (3.2, 4.2, 5.8, 6.8; 0.8, 0.8)]	[(3, 4, 6, 7; 1, 1), (3.2, 4.2, 5.8, 6.8; 0.8, 0.8)]	[(1, 2, 4, 5; 1, 1), (1.2, 2.2, 3.8, 4.8; 0.8, 0.8)]
C3	[(1, 2, 4, 5; 1, 1), (1.2, 2.2, 3.8, 4.8; 0.8, 0.8)]	[(0.14, 0.16, 0.25, 0.33; 1, 1), (0.14, 0.17, 0.23, 0.31; 0.8, 0.8)]	[(1, 1, 1, 1; 1, 1), (1, 1, 1, 1; 1, 1)]	[(1, 2, 4, 5; 1, 1), (1.2, 2.2, 3.8, 4.8; 0.8, 0.8)]	[(1, 2, 4, 5; 1, 1), (1.2, 2.2, 3.8, 4.8; 0.8, 0.8)9
C4	[(1, 1, 1, 1; 1, 1), (1, 1, 1, 1; 1, 1)]	[(0.14, 0.16, 0.25, 0.33; 1, 1), (0.14, 0.17, 0.23, 0.31; 0.8, 0.8)]	[(0.2, 0.25, 0.5, 1; 1, 1), (0.20, 0.26, 0.45, 0.83; 0.8, 0.8)]	[(1, 1, 1, 1; 1, 1), (1, 1, 1, 1; 1, 1)]	[(1, 1, 1, 1; 1, 1), (1, 1, 1, 1; 1, 1)]
C5	[(1, 2, 4, 5; 1, 1), (1.2, 2.2, 3.8, 4.8; 0.8, 0.8)]	[(0.2, 0.25, 0.5, 1; 1, 1), (0.20, 0.26, 0.45, 0.83; 0.8, 0.8)]	[(0.2, 0.25, 0.5, 1; 1, 1), (0.20, 0.26, 0.45, 0.83; 0.8, 0.8)]	[(1, 1, 1, 1; 1, 1), (1, 1, 1, 1; 1, 1)]	[(1, 1, 1, 1; 1, 1), (1, 1, 1, 1; 1, 1)]

The same steps are applied to the other matrices for comparing the alternatives with respect to each criterion. The resulting weights of the alternatives are given in Table 42. In order to rank the alternatives, the weighted scores are calculated.

Table 40 Geometric mean for each criterion row

Criteria	Means
C1	[(0.35, 0.40, 0.57, 0.80; 1, 1), (0.36, 0.41, 0.54, 0.73; 0.8, 0.8)]
C2	[(1.93, 2.63, 3.86, 4.43; 1, 1), (2.08, 2.76, 3.74, 4.32; 0.8, 0.8)]
C3	[(0.67, 1.05, 1.74, 2.10; 1, 1), (0.76, 1.12, 1.67, 2.03; 0.8, 0.8)]
C4	[(0.49, 0.52, 0.65, 0.80; 1, 1), (0.49, 0.53, 0.64, 0.76; 0.8, 0.8)]
C5	[(0.52, 0.65, 1, 1.37; 1, 1), (0.55, 0.68, 0.95, 1.27; 0.8, 0.8)]
Total	[(3.98, 5.28, 7.84, 9.52; 1, 1), (4.26, 5.53, 7.56, 9.12; 0.8, 0.8)]

Table 41 Fuzzy, defuzzified and normalized weights of the criteria

	Interval type-2 fuzzy weights	Defuzzified weights	Normalized weights
C1	[(0.037, 0.051, 0.10, 0.20; 1, 1), (0.039, 0.054, 0.098, 0.17; 0.8, 0.8)]	0.0918	0.079
C2	[(0.20, 0.33, 0.73, 1.11; 1,1), (0.22, 0.36, 0.67, 1.01; 0.8, 0.8)]	0.5577	0.482
C3	[(0.07, 0.13, 0.32, 0.52; 1, 1), (0.083, 0.14, 0.30, 0.47; 0.8, 0.8)]	0.2482	0.215
C4	[(0.051, 0.067, 0.12, 0.20; 1, 1), (0.054, 0.07, 0.11, 0.17; 0.8, 0.8)]	0.1036	0.090
C5	[(0.055, 0.084, 0.18, 0.34; 1, 1), (0.060, 0.0907, 0.17, 0.29; 0.8, 0.8)]	0.1555	0.134

Table 42 Weights and ranking of the alternatives using Interval Type-2 Fuzzy sets

	wrt C1	wrt C2	wrt C3	wrt C4	wrt C5	Weighted score	Ranking
Weights	0.08	0.48	0.21	0.09	0.13		
A1	0.23	0.23	0.34	0.31	0.15	0.249	3
A2	0.23	0.06	0.04	0.10	0.38	0.115	4
A3	0.04	0.40	0.55	0.22	0.24	0.366	1
A4	0.51	0.31	0.07	0.37	0.24	0.271	2

4.5 *Comparison of the Four Fuzzy Methods*

Table 43 shows the rankings of alternatives with respect to four methods given in Sects. 4.1–4.4. The alternative A3 is the best alternative with respect to the other three methods except HFLTS based method. The IVIFS based and Type-2 Fuzzy based methods yield the same rankings since these two methods use pairwise comparison among criteria and alternatives. The alternative A2 is the worst alternative with respect to all methods.

Since each method is based on a different definition of fuzzy sets, the observation of differences in the rankings is quite expected. The methods based on the

Table 43 Rankings of alternatives with respect to four methods

Methods	Alternative ranking
Location selection using ordinary fuzzy TOPSIS	A3 > A1 > A4 > A2
Location selection using HFLTS	A1 > A3 > A4 > A2
Location selection using pairwise evaluations based IVIFS	A3 > A4 > A1 > A2
Location selection using interval type-2 fuzzy AHP	A3 > A4 > A1 > A2

extensions of ordinary fuzzy sets should be considered as more reliable since they enable a better and detailed representation of the membership functions.

5 Conclusion

Decision making is an art and a science aiming at maximizing the expected value of our decision. Almost all of real world decision making problems emerge under risk and uncertainty conditions. If sufficient data exist to be able to obtain the probability distribution of decision making parameters, probability based risk models should be preferred. If uncertainty conditions exist without sufficient data, or if the past data are not reliable, or if the expert evaluations in the past are linguistic rather than exact numerical values, then the classical risk and uncertainty models cannot process these types of data. The fuzzy set theory can capture these uncertainties since it can work with vague, incomplete and imprecise data.

Fuzzy decision making has covered ground since Bellman and Zadeh [6] first work. This chapter showed which researchers, universities, and countries most contributed to it in this period. Besides, we examined the position of decision making within the extensions of ordinary fuzzy sets. Even the extensions of ordinary fuzzy sets can be divided into interval-valued fuzzy sets, hesitant fuzzy sets, intuitionistic fuzzy sets, type-2 fuzzy sets, fuzzy multisets, and stationary fuzzy sets, we illustrated its position in hesitant fuzzy sets, intuitionistic fuzzy sets, and type-2 fuzzy sets since these extensions have been much more used with respect to the other extensions.

For further research, this work can be expanded by examining the position of fuzzy decision making in the other extensions of fuzzy sets.

References

1. Aliev, R.A., Huseynov, O.R.: Decision Theory with Imperfect Information. World Scientific (2014)
2. Ashtiani, B., Haghighirad, F., Makui, A., Montazer, G.A.: Extension of fuzzy TOPSIS method based on interval-valued fuzzy sets. Appl. Soft Comput. J. **9**(2), 457–461 (2009)
3. Atanassov, K.T.: Intuitionistic fuzzy sets. Fuzzy Sets Syst. **20**, 87–96 (1986)

4. Baas, S.M., Kwakernaak, H.: Rating and ranking of multiple aspect alternatives using fuzzy sets. Automatica **13**(1), 47–58 (1977)
5. Baker, D., Bridges, D., Hunter, R., Johnson, G., Krupa, J., Murphy, J., Sorenson, K.: Guidebook to Decision-Making Methods, WSRC-IM-2002-00002. Department of Energy, USA (2002)
6. Bellman, R.E., Zadeh, L.A.: Decision Making is a Fuzzy Environment, managmenet. Science **17**(4), 141–164 (1970)
7. Buckley, J.J.: Fuzzy hierarchical analysis. Fuzzy Sets Syst. **17**(1), 233–247 (1985)
8. Bustince, H., Herrera, F., Montero, J.: Fuzzy Sets and Their Extensions: Representation, Aggregation and Models Intelligent Systems from Decision Making to Data Mining, Web Intelligence and Computer Vision. Springer, Berlin (2008)
9. Carlsson, C., Fullér, R.: Fuzzy Reasoning in Decision Making and Optimization. Physica-Verlag, Heidelberg (2002)
10. Onar, S.C., Oztaysi, B., Otay, I., Kahraman, C.: Multi-expert wind energy technology selection using interval-valued intuitionistic fuzzy sets. Energy **90**(1), 274–285 (2015)
11. Chang, D.-Y.: Applications of the extent analysis method on fuzzy AHP. Eur. J. Oper. Res. **95** (3), 649–655 (1996)
12. Chen, S.M., Lee, L.W.: Fuzzy multiple attributes group decision-making based on the interval type-2 TOPSIS method. Expert Syst. Appl. **37**, 2790–2798 (2010)
13. Chen, C.-T.: Extensions of the TOPSIS for group decision-making under fuzzy environment. Fuzzy Sets Syst. **114**, 1–9 (2000)
14. Chen, S.J., Hwang, C.L.: Fuzzy multiple attribute decision-making, methods and applications. Lecture Notes in Economics and Mathematical Systems, p 375. Springer, Heidelberg (19920
15. Dubois, D., Prade, H.: Fuzzy Sets and Systems: Theory and Applications. AcademicPress, New York (1980)
16. Harris, R.: Introduction to Decision Making. VirtualSalt (1998). http://www.virtualsalt.com/crebook5.htm
17. Kacprzyk, J., Fedrizzi, M.: Combining Fuzzy Imprecision with Probabilistic Uncertainty in Decision Making. Springer (1988)
18. Kacprzyk, J., Fedrizzi, M.: Multiperson Decision Making Models Using Fuzzy Sets and Possibility Theory. Springer, Netherlands (1990)
19. Kahraman, C., Öztayşi, B., Uçal, Sarı I., Turanoğlu, E.: Fuzzy analytic hierarchy process with interval type-2 fuzzy sets. Knowl. Based Syst. **59**, 48–57 (2014)
20. Kahraman, C., Ruan, D., Doğan, I.: Fuzzy group decision-making for facility location selection. Inf. Sci. **157**, 135–153 (2003)
21. Kahraman, C.: Fuzzy Multi-Criteria Decision Making Theory and Applications wit Recent Developments. Springer Optimization and Its Applications, vol. 16 (2008)
22. Lai Y.J., Hwang C.L.: Fuzzy Multiple objective decision making: methods and applications. Lecture Notes in Economics and Mathematical Systems, vol. 404. Springer, Berlin Heidelberg (1994)
23. Lee, E.S., Shih, H.S.: Fuzzy and Multi-Level Decision Making: An Interactive Computational Approach. Springer (2001)
24. Li H., Yen V.C.: Fuzzy Sets and Fuzzy Decision-Making. CRC Press (1995)
25. Loostma, F.A.: Fuzzy Logic for Planning and Decision Making. Springer, Dordrecht (1997)
26. Lu, J., Zhang, G., Ruan, D., Wu, F.: Multi-Objective Group Decision Making Methods, Software and Applications with Fuzzy Set Techniques. Imperial College Press, London (2007)
27. Mendel, J.M.: Interval type-2 fuzzy logic systems made simple. IEEE Trans. Fuzzy Syst. **14** (6), 808–821 (2006)
28. Opricovic, S.: Fuzzy VIKOR with an application to water resources planning. Expert Syst. Appl. **38**, 12983–12990 (2011)
29. Pedrycz, W., Ekel, P., Parreiras, R.: Fuzzy Multicriteria Decision-Making: Models, Methods and Applications. Wiley (2011)

30. Rao, R.V.: Decision Making in Manufacturing Environment Using Graph Theory and Fuzzy Multiple Attribute Decision Making Methods, vol. 2. Springer Series in Advanced Manufacturing (2013)
31. Ribeiro, R.A.: Fuzzy multiple attribute decision making: a review and new preference elicitation techniques. Fuzzy Sets Syst. **78**(2), 155–181 (1996)
32. Slowinski, R.: Fuzzy sets in decision analysis, operation research and statistics. Kluwer Academic Publishers, Boston (1998)
33. Sousa, J.M.C., Kaymak, U.: Fuzzy Decision Making in Modeling and Control, World Scientific Series in Robotics and Intelligent Systems, vol. 27 (2002)
34. Torra, V.: Hesitant fuzzy sets. Int. J. Intell. Syst. **25**(6), 529–539 (2010)
35. Tseng, M.L., Lin, Y.H., Chiu, A.S., Chen, C.Y.: Fuzzy AHP-approach to TQM strategy evaluation. Ind. Eng. Manage. Syst. Int. J. **7**(1), 34–43 (2008)
36. Tzeng, G.H., Huang J.J.: Fuzzy Multiple Objective Decision Making. Chapman and Hall/CRC (2013)
37. Van Laarhoven, P.J.M., Pedrycz, W.: A fuzzy extension of Saaty's priority theory. Fuzzy Sets Syst. **11**(1–3), 199–227 (1983)
38. Xu, J., Zhou, X.: Fuzzy-Like Multiple Objective Decision Making, Studies in Fuzziness and Soft Computing, vol. 263. Springer, Heidelberg (2011)
39. Xu, Z., Cai, X.: Intuitionistic Fuzzy Information Aggregation: Theory and Applications, Intuitionistic Fuzzy Information Aggregation: Theory and Applications, pp. 1–309 (2013)
40. Yager, R.R.: Fuzzy decision making using unequal objectives. Fuzzy Sets Syst. **1**, 87–95 (1978)
41. Yavuz, M., Oztaysi, B., Cevik, Onar S., Kahraman, C.: Multi-criteria evaluation of alternative-fuel vehicles via a hierarchical hesitant fuzzy linguistic model. Expert Syst. Appl. **42**(5), 2835–2848 (2015)
42. Yazici, I., Kahraman, C.: VIKOR method using interval type two fuzzy sets. J. Intell. Fuzzy Syst. **29**(1), 411–422 (2015)
43. Yoshida, Y.: Dynamical Aspects in Fuzzy Decision Making, Studies in Fuzziness and Soft Computing, vol. 73. Physica-Verlag, Heidelberg (2001)
44. Zadeh, L.A.: Fuzzy sets. Inf. Control **8**(3), 338–353 (1965)
45. Zadeh, L.A.: The concept of a linguistic variable and its application to approximate reasoning–1. Inf. Sci. **8**, 199–249 (1975)
46. Xu, Z., Zhang, X.: Hesitant Fuzzy Multi-Attribute Decision Making Based on TOPSIS with Incomplete Weight Information, Knowledge-Based Systems, vol. 52, pp. 53–64, Nov 2013
47. Zimmermann, H.J.: Fuzzy Sets, Decision Making, and Expert Systems. Springer, Netherlands (1987)
48. Zimmermann, H.J.: Fuzzy Sets, Decision Making, and Expert Systems. Kluwer, Boston (1987)

Part II
Mathematics of Fuzzy Sets

Mathematics of Intuitionistic Fuzzy Sets

Krassimir Atanassov

Abstract Short firsthand remarks on the history and theory of Intuitionistic Fuzzy Sets (IFSs) are given. Influences of other areas of mathematics for development of the IFSs theory are discussed. On the basis of results in IFSs theory, some ideas for development of other mathematical areas are offered.

Keywords Intuitionistic fuzzy set · Algebra · Analysis · Artificial intelligence · Geometry · Number theory · Topology

1 Introduction

In February 1983, to the concept of an Intuitionistic Fuzzy Set (IFS) originated and in June and August it was presented in two conferences—in Bulgaria, [1], presented by me, and in Poland [2], where my colleague St. Stoeva presented our joint paper. The new concept was defined as an extension of Zadeh's fuzzy sets [3]. The idea for IFSs occurred to me, while I was reading the first books for fuzzy sets by Kaufmann [4] and Dubois and Prade [5], that were on sale in Bulgarian bookshops via their Russian translations. In general, in my research, I have adhered to their notations. In the next year, it was an object of a lot research from the point of view of different mathematical areas. Nowadays, it is clear that some of the ideas, related to IFSs, can be transfered, on their turn, onto other mathematical areas. And this is the theme of the present paper.

K. Atanassov (✉)
Department of Bioinformatics and Mathematical Modelling, Institute of Biophysics and Biomedical Engineering—Bulgarian Academy of Sciences, Acad. G. Bonchev Street, Block 105, 1113 Sofia, Bulgaria
e-mail: krat@bas.bg

K. Atanassov
Intelligent Systems Laboratory, Professor Asen Zlatarov University, 8000 Bourgas, Bulgaria

C. Kahraman et al. (eds.), *Fuzzy Logic in Its 50th Year*,
Studies in Fuzziness and Soft Computing 341,
DOI 10.1007/978-3-319-31093-0_3

Let us have a fixed universe E and its subset A. The set

$$A^* = \{\langle x, \mu_A(x), \nu_A(x)\rangle | x \in E\},$$

where

$$0 \leq \mu_A(x) + \nu_A(x) \leq 1$$

is called IFS and the functions $\mu_A : E \to [0,1]$ and $\nu_A : E \to [0,1]$ represent *degree of membership (validity, etc.)* and *degree of non-membership (non-validity, etc.)*. Now, we can define also function $\pi_A : E \to [0,1]$ as

$$\pi(x) = 1 - \mu(x) - \nu(x)$$

which corresponds to *degree of indeterminacy (uncertainty, etc.)*.

For brevity, we shall write below A instead of A^*, whenever this is possible. Obviously, for every ordinary fuzzy set A: $\pi_A(x) = 0$ and for each $x \in E$ these sets have the form $\{\langle x, \mu_A(x), 1 - \mu_A(x)\rangle | x \in E\}$. Therefore, all results in the area of fuzzy sets, can be transformed without changes (only with adding this additional component) to the case of IFSs theory.

Fuzzy sets theory is related to different mathematical areas, as algebra, analysis, geometry, etc. Hence, IFSs theory is also related to these areas. In the next section, we discuss these relationships.

I will discuss here only the ideas realized with my participation and thus, I hope, to stimulate the rest colleagues to do the same about their results in the area of IFS theory.

2 Mathematics in Intuitionistic Fuzzy Sets

It is very difficult to estimate how many separate areas of mathematics have contributed to the development of IFSs theory. That is way here we will adopt the chronological approach when listing and describing the mathematical areas that have most powerfully influenced the area of IFSs.

2.1 Algebra

Historically, the research over IFSs started with defining algebraic operations, defined over these sets. Now, there are a lot of different operations:

- first—standard set theoretical operations as "$\cup$" and "$\cap$",
- second—algebraic and probabilistic operations "+" and "." (practically, my work over the IFSs theory started with definitions of these four operations);

- third—the average operation "@" (introduced independently by Buhaescu in [6] and me),
- fourth—algebraic operations "multiplication of a set with n" and "power of n" (introduced by…) and their dual operations "multiplication with $\frac{1}{n}$" and "$\sqrt[n]{}$" (introduced by B. Riečan and me),
- fifth—two types of the operation "subtraction" (algebraic and set-theoretical; introduced again by Riečan and me);
- sixth—the logical operations "negation" (more than 50 different negations), "implication" (more than 180 different implications) and "subtraction" (about 70 different subtractions).

Some relations have been introduced. The two most important (and standard) are:

$$A \subseteq B \quad \text{if and only if (iff)} \; (\forall x \in E)(\mu_A(x) \le \mu_B(x) \; \& \; \nu_A(x) \ge \nu_B(x)),$$

$$A \supseteq B \quad \text{iff } B \subseteq A,$$

$$A = B \quad \text{iff } (\forall x \in E)(\mu_A(x) = \mu_B(x) \; \& \; \nu_A(x) = \nu_B(x)).$$

During the years, the basic properties of all these operations have been researched.

Let us give some well-known definitions. Let us have a given set M, operation $*$ and a special unit element $e_* \in M$. Then, we can formulate the following well-known definitions:

- $\langle M, * \rangle$ is a groupoid if and only if (iff) $(\forall a, b \in M)(a * b \in M)$.
- $\langle M, * \rangle$ is a semigroup iff $\langle M, * \rangle$ is a groupoid and $(\forall a, b, c \in M)((a * b) * c = \; a * (b * c))$.
- $\langle M, *, e_* \rangle$ is a monoid iff $\langle M, * \rangle$ is a semigroup and $(\forall a \in M)(a * e_* = a = e_* * a)$.
- $\langle M, *, e_* \rangle$ is a commutative monoid iff $\langle M, *, e_* \rangle$ is a monoid and $(\forall a, b \in M)$ $(a * b = b * a)$.
- $\langle M, *, e_* \rangle$ is a group iff $\langle M, *, e_* \rangle$ is a monoid and $(\forall a \in M)(\exists b \in M)(a * b = e_* = b * a)$.
- $\langle M, *, e_* \rangle$ is a commutative group iff $\langle M, *, e_* \rangle$ is a group and $(\forall a, b \in M)$ $(a * b = b * a)$.

Let us define the following special IFSs

$$E^* = \{\langle x, 1, 0 \rangle | x \in E\},$$

$$U^* = \{\langle x, 0, 0 \rangle | x \in E\},$$

$$O^* = \{\langle x, 0, 1\rangle | x \in E\}.$$

Operation $\mathcal{P}(X)$—the set of all subsets of a given set X, now must be modified so that:

$$\mathcal{P}(E^*) = \{A | A = \{\langle x, \mu_A(x), \nu_A(x)\rangle | x \in E\}\},$$

$$\mathcal{P}(U^*) = \{B | B = \{\langle x, 0, \nu_B(x)\rangle | x \in E\}\},$$

$$\mathcal{P}(O^*) = \{O^*\}.$$

Therefore,

$$\emptyset \notin \mathcal{P}(O^*) \cup \mathcal{P}(E^*) \cup \mathcal{P}(U^*)$$

and for a fixed universe E, following [7]:

1. $\langle \mathcal{P}(E^*), \cap, E^*\rangle$ is a commutative monoid;
2. $\langle \mathcal{P}(E^*), \cup, O^*\rangle$ is a commutative monoid;
3. $\langle \mathcal{P}(E^*), +, O^*\rangle$ is a commutative monoid;
4. $\langle \mathcal{P}(E^*), ., E^*\rangle$ is a commutative monoid;
5. $\langle \mathcal{P}(E^*), @\rangle$ is a groupoid;
6. $\langle \mathcal{P}(U^*), \cap, U^*\rangle$ is a commutative monoid;
7. $\langle \mathcal{P}(U^*), \cup, O^*\rangle$ is a commutative monoid;
8. $\langle \mathcal{P}(U^*), +, O^*\rangle$ is a commutative monoid;
9. $\langle \mathcal{P}(U^*), ., U^*\rangle$ is a commutative monoid;
10. $\langle \mathcal{P}(U^*), @\rangle$ is a groupoid;
11. None of these objects is a (commutative) group.

2.2 *Set Theory and First Order Logic*

The most interesting is the case with the sixth group of operations, discussed in the previous section. As I mentioned in [8], historically, one of my mistakes was that for a long time I have been using only the simplest form of negation:

$$\neg_1 A = \{\langle x, \nu_A(x), \mu_A(x)\rangle | x \in E\}.$$

In this case the equality

$$\neg\neg A = A$$

holds, which resembles classical logic. Currently, a series of more than 40 new negations have been constructed by Atanassova, Dworniczak, Dimitrov and the author (see, e.g., [7]). The first four of them are:

$$\begin{aligned}
\neg_2 A &= \{\langle x, 1 - \mathrm{sg}\,(\mu_A(x)), \mathrm{sg}\,(\mu_A(x))\rangle | x \in E\}, \\
\neg_3 A &= \{\langle x, \nu_A(x), \mu_A(x)\nu_A(x) + \mu_A(x)\rangle | x \in E\}, \\
\neg_4 A &= \{\langle x, \nu_A(x), 1 - \nu_A(x)\rangle | x \in E\}, \\
\neg_5 A &= \{\langle x, \overline{\mathrm{sg}}(1 - \nu_A(x)), \mathrm{sg}\,(1 - \nu_A(x))\rangle | x \in E\},
\end{aligned}$$

where

$$\mathrm{sg}\,(x) = \begin{cases} 1 & \text{if } x > 0 \\ 0 & \text{if } x \leq 0 \end{cases} \qquad \overline{\mathrm{sg}}\,(x) = \begin{cases} 0 & \text{if } x > 0 \\ 1 & \text{if } x \leq 0 \end{cases}$$

The negations $\neg_2, \neg_3, \neg_4, \neg_5$ satisfy strongly intuitionistic properties.

In a series of papers co-authored by Trifonov, D. Dimitrov, N. Angelova, I have studied the properties of these negations. For example, it has been checked that $\neg_1$ satisfies all three properties below, while the rest negations satisfy only properties P1 and P3, where

Property P1: $A \to \neg\neg A$,
Property P2: $\neg\neg A \to A$,
Property P3: $\neg\neg\neg A = \neg A$.

By analogy with [9], we can show that negations $\neg_2, \ldots, \neg_5$ do not satisfy the Law for Excluded Middle ($P \vee \neg P$, where P is a propositional form), while they satisfy some of its modifications (e.g., $\neg\neg P \vee \neg P$). In [10], it is shown that the same negations do not satisfy De Morgan's Laws, but they satisfy some of their modifications.

With respect to some of the negations, or independently, more than 180 different implications have been introduced. Initially, for basis of this research, the author used Klir and Bo Yuan's book [11].

It is well-known that if we have some implication $\to$, then for a variable x we can define $\neg x = x \to \langle 0, 1\rangle$, and after this, we can define for two variables x and y: $x \vee_{\to,\neg,1} y = \neg x \to y$, $x \vee_{\to,\neg,2} y = \neg x \to \neg\neg y$, $x \wedge_{\to,\neg,1} y = \neg(x \to \neg y)$ and $x \wedge_{\to,\neg,1} y = \neg(\neg\neg x \to \neg y)$. Therefore, each of the implications generate one or two disjunctions and one or two conjunctions. An **open problem** is to construct all conjunctions and disjunctions, that can be generated by the existing already 185 implications. Having in mind that for predicate $P(x)$ with variable x, which can obtain values $a_1, a_2, \ldots$, it is valid: $\forall x P(x) = P(a_1) \wedge P(a_2) \wedge \ldots$ and $\forall x Q(x) = P(a_1) \vee P(a_2) \vee \ldots$. Hence, we can define as many quantifiers as many are the conjunctions and disjunctions. Having solved the previous problem, the new **open problem** arises: to construct all possible quantifiers on the basis of the conjunctions and disjunctions.

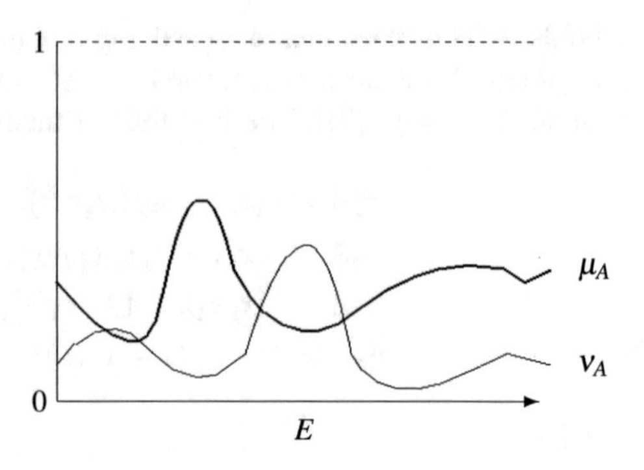

Fig. 1 First (standard) geometrical interpretation of an intuitionistic fuzzy set

2.3 Geometry

The ordinary fuzzy sets have only one geometrical interpretation, while in the IFS-case, this interpretation has two forms. The standard geometrical interpretation of the IFSs, proposed as an analogue of the geometrical (or analytical) interpretation of fuzzy sets, is shown in Fig. 1.

Its modification is given in Fig. 2.

The most important and useful (at least by the moment) geometrical interpretation of the IFSs is given on Fig. 3. Let a universe E be given and let us consider the figure F in the Euclidean plane with a Cartesian coordinate system. Triangle F was called "*IFS-interpretational triangle*" (see, e.g., [7, 12]).

Let $A \subseteq E$ be a fixed set. Then we can construct the function $f_A : E \to F$ such that if $x \in E$, then

$$p = f_A(x) \in F.$$

We can also construct the point p with coordinates $\langle a, b \rangle$, for which, $0 \leq a + b \leq 1$ and there hold the equalities $a = \mu_A(x), b = \nu_A(x)$.

Note that if there exist two different elements $x, y \in E$, $x \neq y$, for which $\mu_A(x) = \mu_A(y)$ and $\nu_A(x) = \nu_A(y)$ with respect to some set $A \subseteq E$, then $f_A(x) = f_A(y)$.

Each of the operations, mentioned in Sect. 2.1, has a geometrical interpretation. For example, if A and B are two IFSs over E, then the function $f_{A \cap B}$ assigns

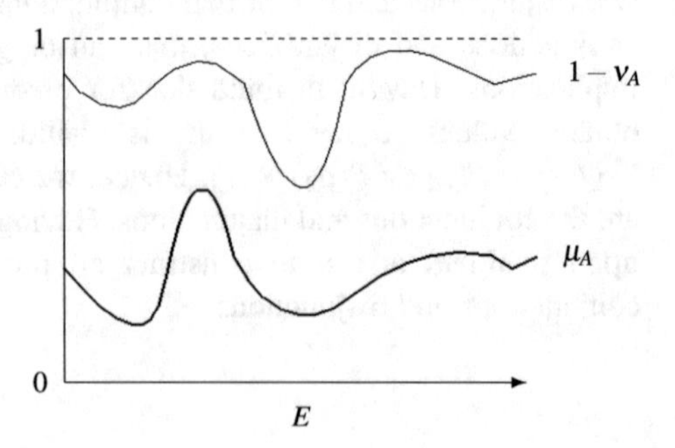

Fig. 2 First (basic) geometrical interpretation of an intuitionistic fuzzy set

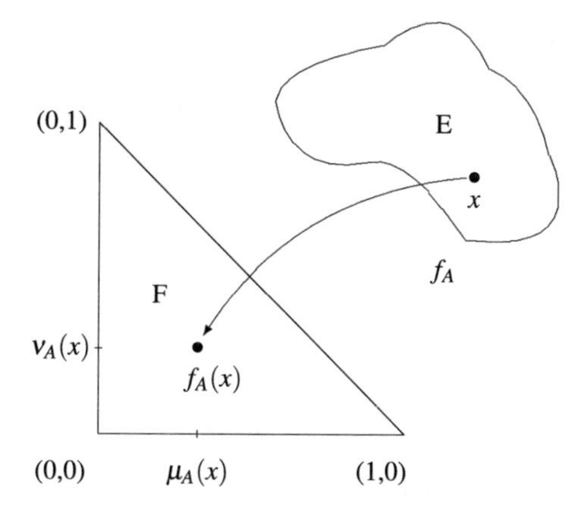

Fig. 3 Second geometrical interpretation of an intuitionistic fuzzy set

to $x \in E$ a point $f_{A \cap B}(x) \in F$ with coordinates $\langle \min(\mu_A(x), \mu_B(x)), \max(\nu_A(x), \nu_B(y)) \rangle$.

There exist three geometrical interpretations (see Fig. 4a–c) from which one is the general case (Fig. 4a) and two are particular cases (Fig. 4b, c).

If A and B are two IFSs over E, then the function f_{A+B} assigns to $x \in E$ a point $f_{A+B}(x) \in F$ with coordinates $\langle \mu_A(x) + \mu_B(x) - \mu_A(x).\mu_B(x), \nu_A(x).\nu_B(x) \rangle$.

There exists only one form of geometrical interpretation of this operation (see Fig. 5a).

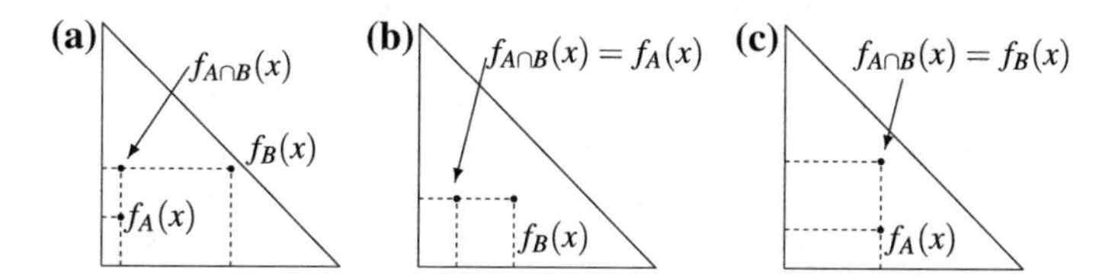

Fig. 4 Geometrical interpretations of the operation intersection of intuitionistic fuzzy sets (three cases)

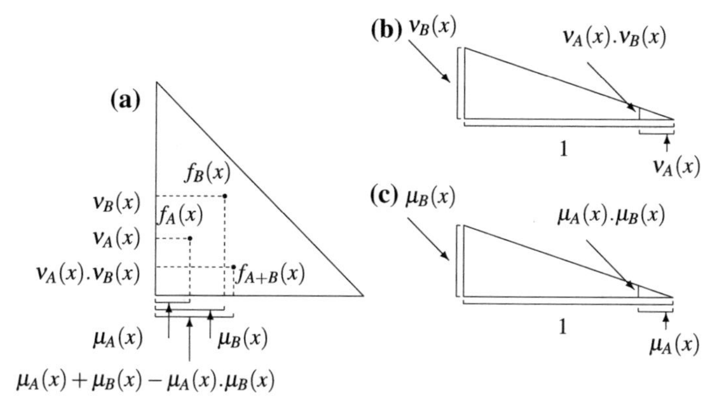

Fig. 5 Geometrical interpretations of the operation sum of intuitionistic fuzzy sets (three cases)

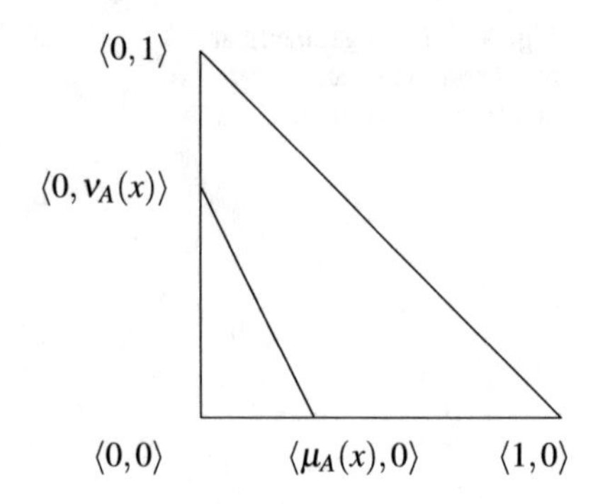

Fig. 6 Third geometrical interpretations of an intuitionistic fuzzy set

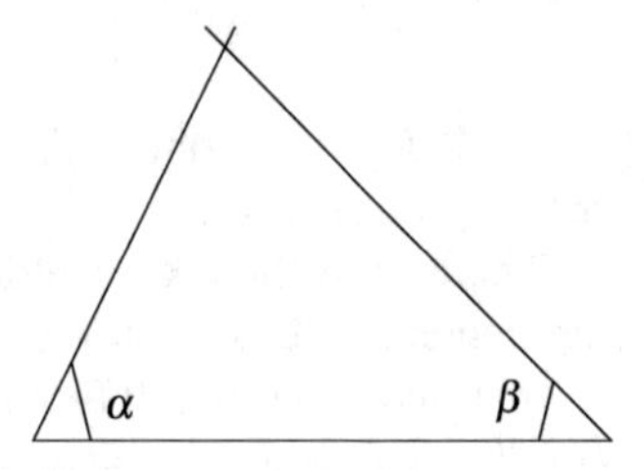

Fig. 7 Fourth geometrical interpretations of an intuitionistic fuzzy set

The construction of $\mu_A(x).\mu_B(x)$ and $\nu_A(x).\nu_B(x)$ are shown in Fig. 5b and c, respectively.

The author constructed two other geometrical interpretations (see Figs. 6 and 7, where $\alpha = \pi.\mu_A(x), \beta = \pi.\nu_A(x)$ and here $\pi = 3.14\ldots$). Two more geometrical interpretations were proposed by I. Antonov in [13] and by Danchev in [14]: the first one over a sphere and the second one on the equilateral triangle (see Fig. 8). In [15], Yang and Chiclana constructed another spherical interpretation.

E. Szmidt and J. Kacprzyk constructed another three-dimensional geometrical interpretation [16–18] (see Fig. 9).

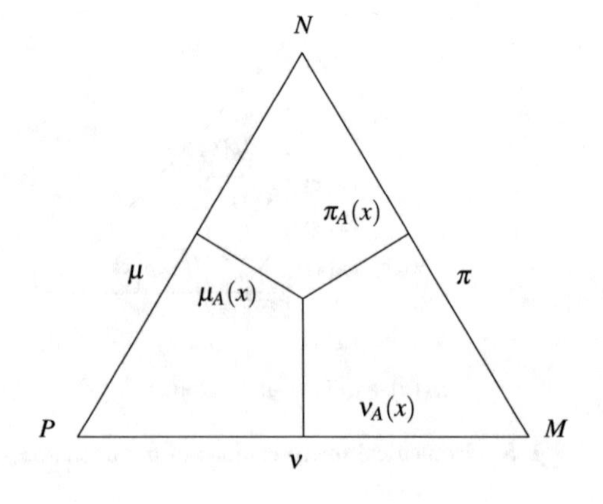

Fig. 8 Fifth geometrical interpretations of an intuitionistic fuzzy set

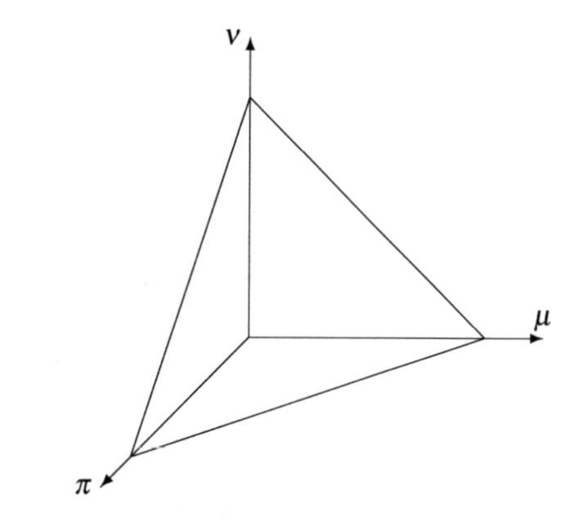

Fig. 9 Sixth (three-dimensional) geometrical interpretations of an intuitionistic fuzzy set

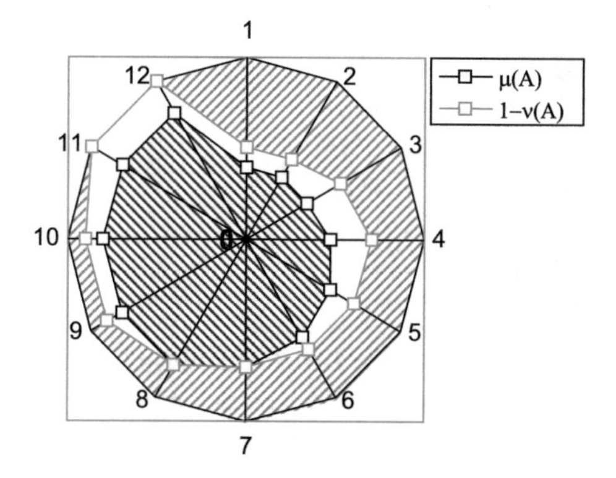

Fig. 10 Seventh (radar chart based) geometrical interpretations of an intuitionistic fuzzy set

In [19], a geometrical interpretation based on radar charts was proposed by V. Atanassova. In Fig. 10, the innermost zone corresponds to the membership degree, the outermost zone to the non-membership degree and the region between both zones denotes the degree of uncertainty. This IFS-interpretation can be especially useful for data in time series, multivaried data sets and other data with cyclic trait.

2.4 Modal Logic

In the beginning of March 1983, the idea for introducing of modal type of operators over IFSs originated. At that moment, I showed the results of my research to my colleague (and former teacher in Mathematical Faculty of Sofia University) George Gargov (1947–1996) who named the new sets "Intuitionistic Fuzzy Sets".

The first two modal operators over IFSs transform an IFS into a fuzzy set (as it was shown, a fuzzy sets is a particular case of an IFS). These two modal operators are similar to the operators "*necessity*" and "*possibility*" defined in some modal logics. Their properties resemble these of the modal logic (see e.g. [20]).

Let, for every IFS A,

$$\Box A = \{\langle x, \mu_A(x), 1 - \mu_A(x)\rangle | x \in E\},$$

$$\Diamond A = \{\langle x, 1 - \nu_A(x), \nu_A(x)\rangle | x \in E\}.$$

Obviously, if A is an ordinary fuzzy set, then

$$\Box A = A = \Diamond A. \tag{3}$$

These equalities show that the modal operators do not have analogues in the case of fuzzy sets; and therefore, this is a new demonstration of the fact that IFSs are proper extensions of the ordinary fuzzy sets.

It was checked that for every IFS A and for $\neg = \neg_1$, i.e., the standard negation,

(a) $\neg\Box\neg A = \Diamond A$,
(b) $\neg\Diamond\neg A = \Box A$,
(c) $\Box A \subseteq A \subseteq \Diamond A$,
(d) $\Box\,\Box A = \Box A$,
(e) $\Box\Diamond A = \Diamond A$,
(f) $\Diamond\Box A = \Box A$,
(g) $\Diamond\Diamond A = \Diamond A$,

where $\neg$ is the first (standard, classical) negation.

It is important to note that for $\neg = \neg_2$ equalities (a) and (b) are not valid. Instead of them, the equalities

$$\neg\Box\neg A = \Diamond\neg\neg A \text{ and } \neg\Diamond\neg A = \Box\neg\neg A$$

hold.

Two different geometrical interpretations of both operators are given in Figs. 11, 12, 13 and 14, respectively.

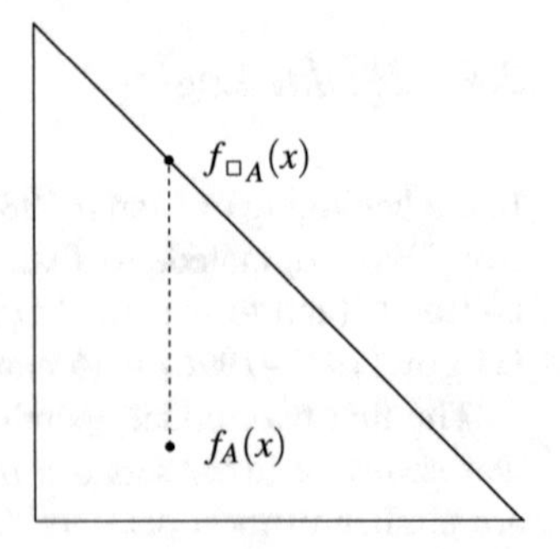

Fig. 11 First geometrical interpretations of the operator necessity

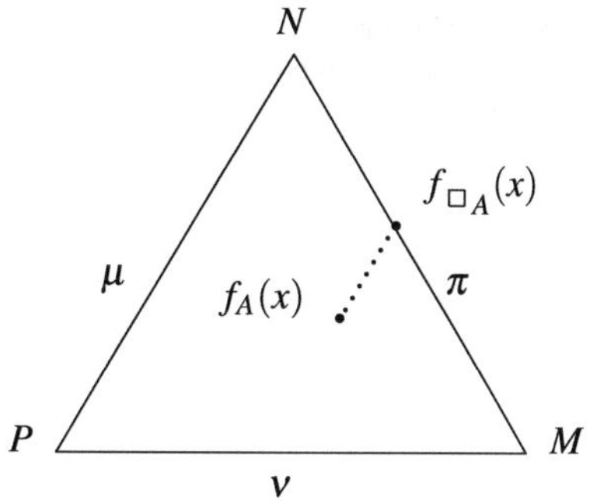

Fig. 12 Second geometrical interpretations of the operator necessity

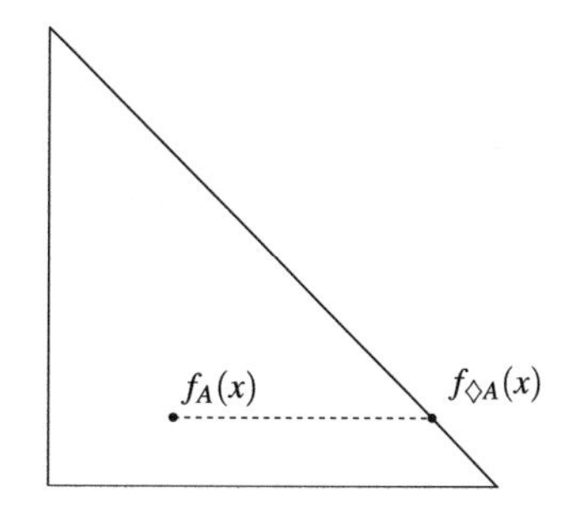

Fig. 13 First geometrical interpretations of the operator possibility

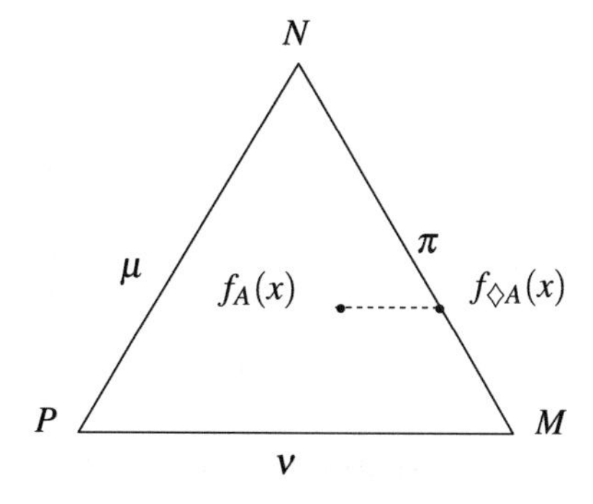

Fig. 14 Second geometrical interpretations of the operator possibility

2.5 *Analysis and Topology*

Next idea for development of the IFSs theory is related to the topological operators "closure" and "interior". They were defined in October 1983 by the author and their basic properties were studied. Three years later, the relations between the modal and the topological operators over IFSs were studied, too (see [12]). These first topological operators are defined for every IFS A by

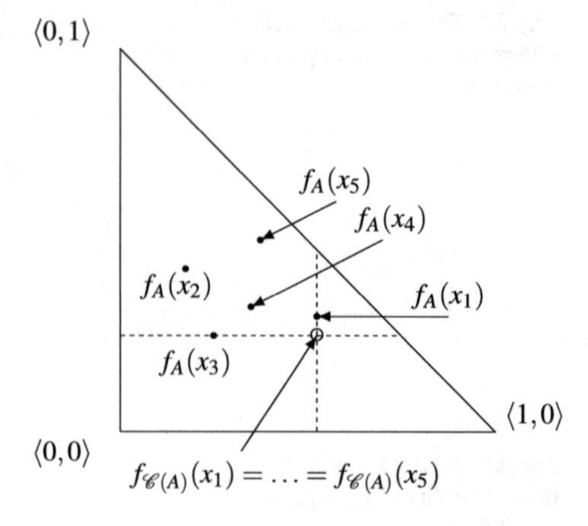

Fig. 15 Geometrical interpretations of the operator closure

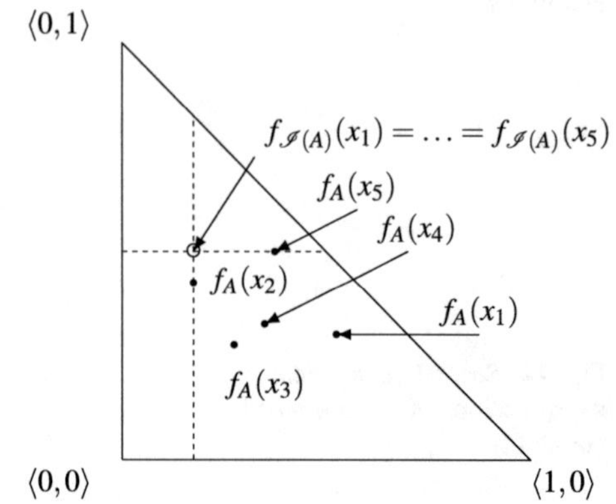

Fig. 16 Geometrical interpretations of the operator interior

$$\mathscr{C}(A) = \{\langle x, \sup_{y \in E} \mu_A(y), \inf_{y \in E} \nu_A(y)\rangle | x \in E\},$$

$$\mathscr{I}(A) = \{\langle x, \inf_{y \in E} \mu_A(y), \sup_{y \in E} \nu_A(y)\rangle | x \in E\}.$$

The geometrical interpretations of both operators are given in Figs. 15 and 16, respectively.

It is important to mention, that the two operators $\mathscr{C}$ and $\mathscr{I}$ are analogous to the intuitionistic fuzzy quantifiers $\exists$ and $\forall$, respectively, from intuitionistic fuzzy predicative calculus. In [12], a lot of properties of these two operators were discussed.

2.6 Temporal Logic

Let E be a universe, and T be a non-empty set. We call the elements of T "time-moments". Based on the definition of IFS, in [7, 21–23] we defined another type of an IFSs, called *Temporal IFSs (TIFSs)* as the following:

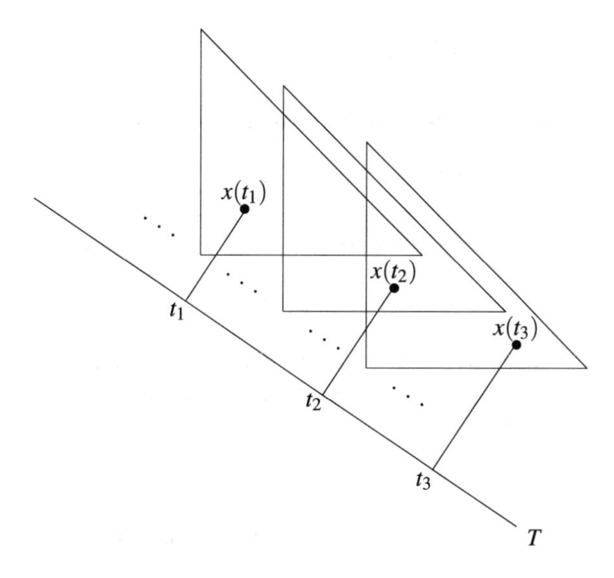

Fig. 17 The geometrical interpretation of a temporal intuitionistic fuzzy set

$$A(T) = \{\langle x, \mu_A(x,t), \nu_A(x,t)\rangle | \langle x,t\rangle \in E \times T\},$$

where

(a) $A \subseteq E$ is a fixed set,
(b) $\mu_A(x,t) + \nu_A(x,t) \leq 1$ for every $\langle x,t\rangle \in E \times T$,
(c) $\mu_A(x,t)$ and $\nu_A(x,t)$ are the degrees of membership and non-membership, respectively, of the element $x \in E$ at the time-moment $t \in T$.

The geometrical interpretation of the TIFSs is shown in Fig. 17.

Obviously, the list of mathematical tools, used for developing the IFS theory is longer, but above we discussed at least the basic areas. In the next Section, we discuss the opposite situation: elements of IFS theory that give rise to problems in different areas of mathematics and can help for their development.

3 Intuitionistic Fuzzy Sets as Tools for Development of Some Mathematical Areas

Here, we continue to follow the chronological order of events. Probably, the first case, when I realized that there was a possibility for introducing a new mathematical concept on the basis of an already defined object in IFS theory, was the operator D_α introduced in the next section.

3.1 Modal Logic

Historically, the first extension of the modal operators $\Box$ and $\Diamond$ is the operator

$$D_\alpha(A) = \{\langle x, \mu_A(x) + \alpha.\pi_A(x), \nu_A(x) + (1-\alpha).\pi_A(x)\rangle | x \in E\},$$

where $\alpha \in [0,1]$ is a fixed number (see, e.g. [7, 12]). Obviously,

$$D_0(A) = \Box A \text{ and } D_1(A) = \Diamond A.$$

Therefore, at least in IFSs theory there is a modal operator that has as boundary points both standard (IFS) modal operators. Now, the following questions provoke our interest:

1. Does the operator D_α have analogues in ordinary modal logic?
2. What properties would such an operator have?
3. If that operator exists, can the extended modal operator be extended again?

While the reason for the first two questions is seen in the above definition, the reason for the third question is in the next extended modal operator in IFSs theory:

$$F_{\alpha,\beta}(A) = \{\langle x, \mu_A(x) + \alpha.\pi_A(x), \nu_A(x) + \beta.\pi_A(x)\rangle | x \in E\},$$

where $\alpha, \beta \in [0,1]$ and $\alpha + \beta \leq 1$ (see, e.g. [7, 12]). Obviously,

$$F_{\alpha,1-\alpha}(A) = D_\alpha(A).$$

Operator $F_{\alpha,\beta}$ is the first IFS operator that transforms an IFS to an IFS (rather than to a fuzzy set). The geometrical interpretations of both new operators are given in Figs. 18 and 19.

Other five operators were introduced before 1990 (see [7, 12]). Thus, these seven operators were a basis of the definition of one operator that generalizes all of them. It has the form

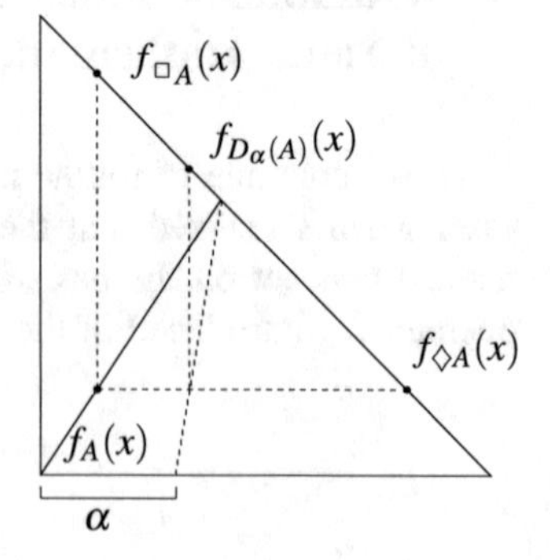

Fig. 18 Geometrical interpretations of the operator D_α

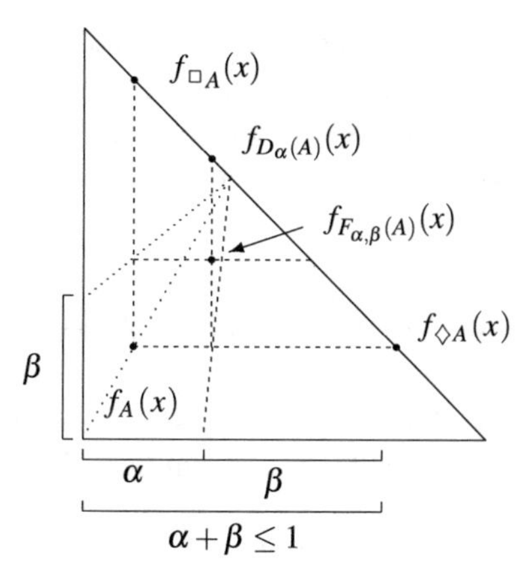

Fig. 19 Geometrical interpretations of the operator $F_{\alpha,\beta}$

$$X_{a,b,c,d,e,f}(A) = \{\langle x, a.\mu_A(x) + b.(1 - \mu_A(x) - c.\nu_A(x)), d.\nu_A(x) + e.(1 - f.\mu_A(x) - \nu_A(x))\rangle | x \in E\},$$

where the constants $a, b, c, d, e, f \in [0, 1]$ and

$$a + e - e.f \leq 1,$$

$$b + d - b.c \leq 1$$

$$b + e \leq 1.$$

As it was mentioned in [24], in [7, 12] the third condition was omitted through my fault.

Since the beginning of the new century, the group of seven operators has been firstly modified to another analogical group, each operator from which exhibits dual behaviour to its corresponding operator from the first group. For instance, operator that corresponds to operator $F_{\alpha,\beta}$ from the first group, now has dual operator $f_{\alpha,\beta}$ in the form

$$f_{\alpha,\beta}(A) = \{\langle x, \nu_A(x) + \alpha \cdot \pi_A(x), \mu_A(x) + \beta \cdot \pi_A(x)\rangle | x \in E\},$$

where $\alpha, \beta \in [0, 1]$ and $\alpha + \beta \leq 1$ (see, e.g. [7]). We can notice that $f_{0,0}$ coincides with the first negation $\neg_1$.

The first group with seven operators has also been extended by changing parameters α and β with elements of whole IFS. For example, operator $F_{\alpha,\beta}$ can obtain the form

$$F_B(A) = \{\langle x, \mu_A(x) + \mu_B(x).\pi_A(x), \nu_A(x) + \nu_B(x).\pi_A(x)\rangle \ x \in E\},$$

where A and B are IFSs.

A completely different group of operators was introduced by the author (in the 1990s and during the last ten years, see [12]), Dencheva [25], Cuvalcioglu [26–29]. Perhaps, these operators do not have analogues in ordinary modal logic, but it will be interesting to research if they can obtain such ones.

3.2 Analysis and Topology

The following operators are defined in [7, 30] as extensions of the two topological operators $\mathcal{C}$ and $\mathcal{I}$ from Sect. 2.5:

$$\mathcal{C}_\mu(A) = \{\langle x, \sup_{y\in E} \mu_A(y), \min(1 - \sup_{y\in E} \mu_A(y), \nu_A(x))\rangle | x \in E\},$$

$$\mathcal{C}_\nu(A) = \{\langle x, \mu_A(x), \inf_{y\in E} \nu_A(y)\rangle | x \in E\},$$

$$\mathcal{I}_\mu(A) = \{\langle x, \inf_{y\in E} \mu_A(y), \nu_A(x)\rangle | x \in E\},$$

$$\mathcal{I}_\nu(A) = \{\langle x, \min(1 - \sup_{y\in E} \nu_A(y), \mu_A(x)), \sup_{y\in E} \nu_A(y)\rangle | x \in E\}.$$

In [7, 31], the two new topological operators $\mathcal{C}^*_\mu$ and $\mathcal{I}^*_\nu$ were introduced as

$$\mathcal{C}^*_\mu(A) = \{\langle x, \min(\sup_{y\in E} \mu_A(y), 1 - \nu_A(x)), \min(1 - \sup_{y\in E} \mu_A(y), \nu_A(x))\rangle | x \in E\},$$

$$\mathcal{I}^*_\nu(A) = \{\langle x, \min(1 - \sup_{y\in E} \nu_A(y), \mu_A(x)), \min(\sup_{y\in E} \nu_A(y), 1 - \mu_A(x))\rangle | x \in E\}.$$

The geometrical interpretations of these operators, applied to the IFS A in Fig. 20 are shown in Figs. 21, 22, 23, 24, 25, and 26.

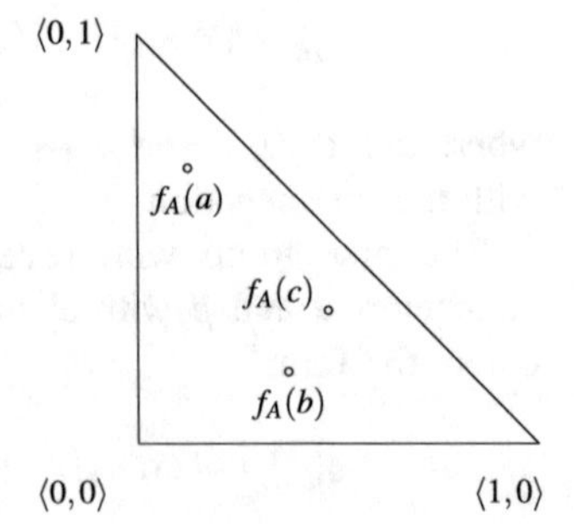

Fig. 20 Geometrical interpretations of three elements a, b and c of an intuitionistic fuzzy set

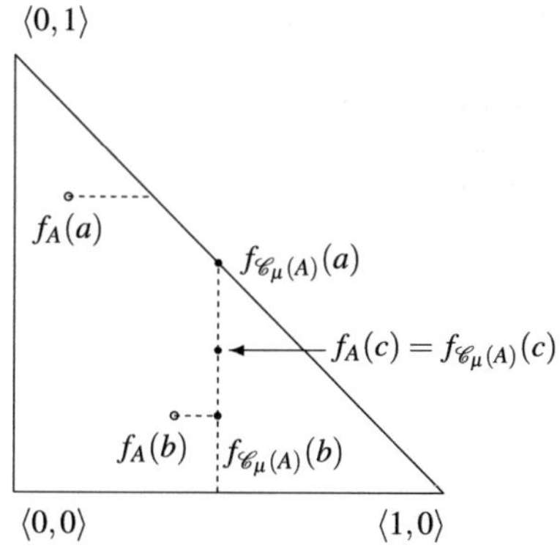

Fig. 21 Geometrical interpretations of the application of operator C_μ over the three elements a, b and c and of an intuitionistic fuzzy set

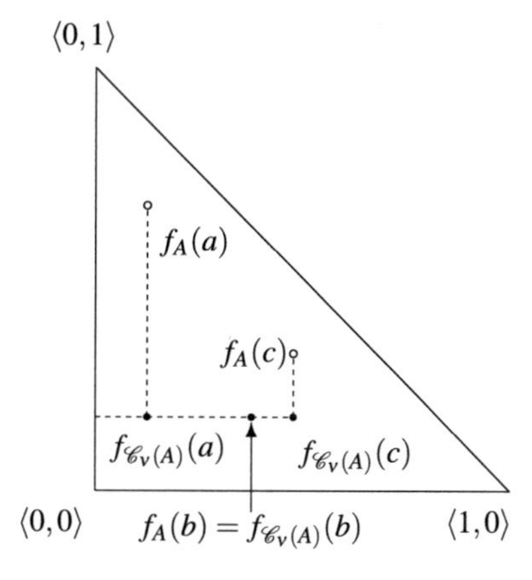

Fig. 22 Geometrical interpretations of the application of operator C_ν over the three elements a, b and c of an intuitionistic fuzzy set

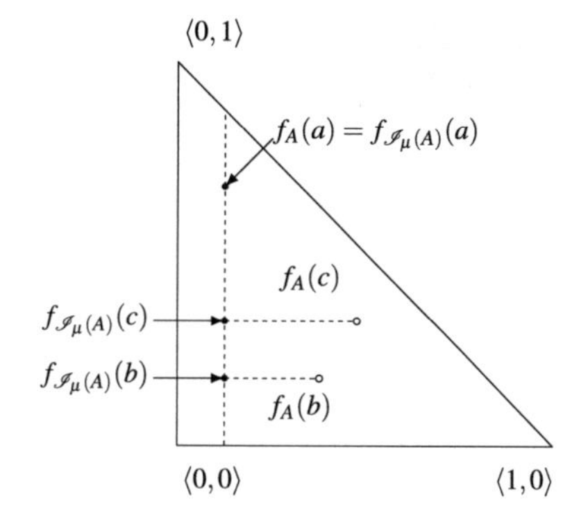

Fig. 23 Geometrical interpretations of the application of operator I_μ over the three elements a, b and c of an intuitionistic fuzzy set

Fig. 24 Geometrical interpretations of the application of operator I_ν over the three elements a, b and c of an intuitionistic fuzzy set

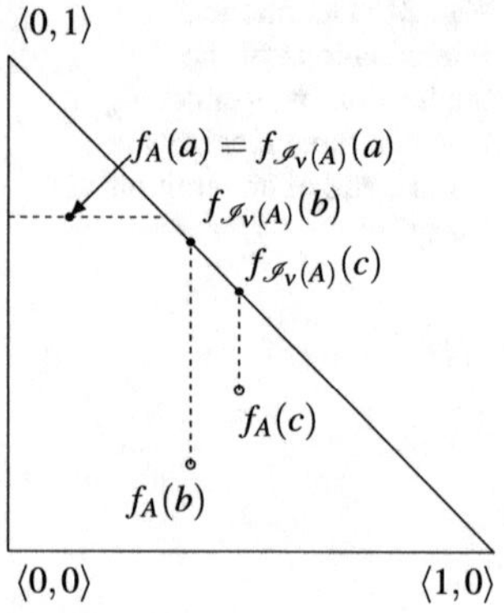

⟨0,1⟩
$f_A(a) = f_{\mathscr{I}_\nu(A)}(a)$
$f_{\mathscr{I}_\nu(A)}(b)$
$f_{\mathscr{I}_\nu(A)}(c)$
$f_A(c)$
$f_A(b)$
⟨0,0⟩
⟨1,0⟩

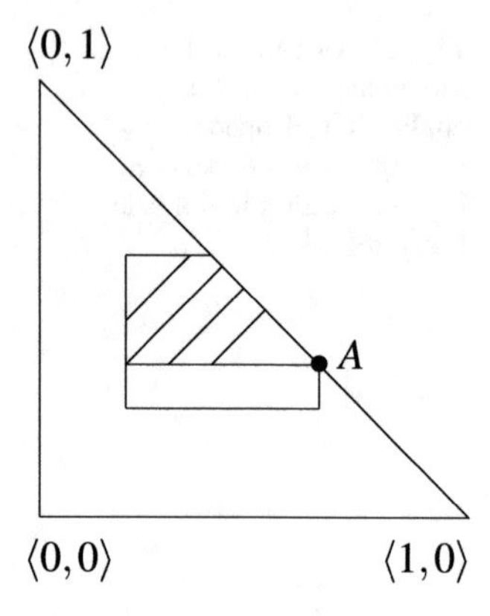

Fig. 25 Geometrical interpretations of the application of operator C_μ^*

⟨0,1⟩
A
⟨0,0⟩
⟨1,0⟩

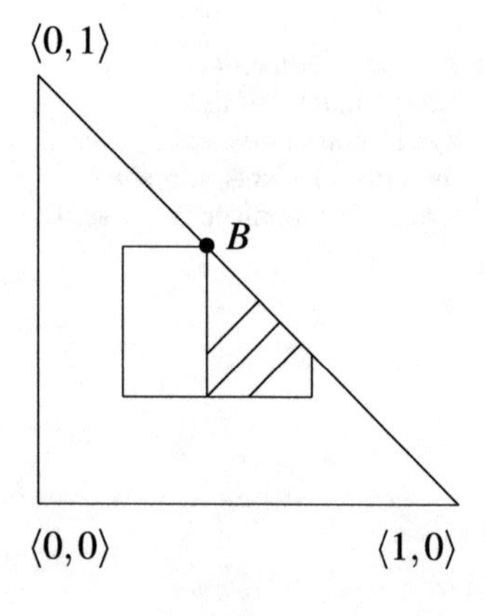

Fig. 26 Geometrical interpretations of the application of operator I_ν^*

⟨0,1⟩
B
⟨0,0⟩
⟨1,0⟩

In [7], the following equalities are proved for $\neg = \neg_1$ and for each IFS A:

$$\neg\mathcal{C}(\neg A) = \mathcal{I}(A),$$

$$\neg\mathcal{I}(\neg A) = \mathcal{C}(A),$$

$$\neg\mathcal{C}_\mu(\neg A) = \mathcal{I}_\nu(A),$$

$$\neg\mathcal{I}_\mu(\neg A) = \mathcal{C}_\nu(A),$$

$$\neg\mathcal{C}^*_\mu(\neg A) = \mathcal{I}^*_\nu(A),$$

$$\neg\mathcal{I}^*_\mu(\neg A) = \mathcal{C}^*_\nu(A).$$

The following **Open question** is interesting: Which other negations satisfy the above equalities and which of them satisfy these equalities in a modified form. For example, if $\neg = \neg_2$, the above equalities obtain the forms:

$$\neg\mathcal{C}(\neg A) = \neg\neg\mathcal{I}(A),$$

$$\neg\mathcal{I}(\neg A) = \neg\neg\mathcal{C}(A),$$

$$\neg\mathcal{C}_\mu(\neg A) = \neg\neg\mathcal{I}_\nu(A),$$

$$\neg\mathcal{I}_\mu(\neg A) = \neg\neg\mathcal{C}_\nu(A),$$

$$\neg\mathcal{C}^*_\mu(\neg A) = \neg\neg\mathcal{I}^*_\nu(A),$$

$$\neg\mathcal{I}^*_\mu(\neg A) = \neg\neg\mathcal{C}^*_\nu(A).$$

It is interesting whether we can define analogues of the above topological operators in the area of topology.

In [32], a lot of intuitionistic fuzzy norms and metrics are described. A part of them can be transformed to fuzzy or crisp sets.

Here, following [33], we discuss two non-standard norms, defined on the basis of one of the most important Georg Cantor's ideas in set theory and by this reason below we call them "*Cantor's intuitionistic fuzzy norms*". They are essentially different from the Euclidean and Hamming norms, existing in fuzzy set theory.

Let $x \in E$ be a fixed universe and let

$$\mu_A(x) = 0, a_1 a_2 \ldots,$$

$$\nu_A(x) = 0, b_1 b_2 \ldots.$$

Then, we bijectively construct the numbers

$$||x||_{\mu,\nu} = 0, a_1 b_1 a_2 b_2 \ldots,$$

and

$$||x||_{\nu,\mu} = 0, b_1 a_1 b_2 a_2 \ldots,$$

for which we see that

1. $||x||_{\mu,\nu}, ||x||_{\nu,\mu} \in [0,1]$
2. having both of them, we can directly reconstruct numbers $\mu_A(x)$ and $\nu_A(x)$.

Let us call numbers $||x||_{\mu,\nu}$ and $||x||_{\nu,\mu}$ Cantor norms of element $x \in E$. In some cases, these norms are denoted by $||x||_{2,\mu,\nu}$ and $||x||_{2,\nu,\mu}$ with aim to emphasize that they are related to the two-dimensional IFS-interpretation. In [7], the three-dimensional case is discussed, as well.

Following [7], a relation between the geometrical interpretation of the IFSs and element of complex analysis is discussed hereunder.

Two elements x and y of E are said to be in relation *negation* if

$$\mu_A(x) = \nu_A(y) \text{ and } \nu_A(x) = \mu_A(y).$$

Their geometrical interpretation in the intuitionistic fuzzy interpretation triangle is shown in Fig. 27.

It is well-known that numbers $a + \mathbf{i}b$ and $a - \mathbf{i}b$ are conjugate in the complex plane, where $\mathbf{i}$ is the imaginary unit. Let $a, b \in [0,1]$ and let the geometrical representation of both points is given in Fig. 28. Let sections OA, OB and OC in Fig. 28 have unit length.

We introduce formulae that transform the points of triangle ABC into triangle ABO. One of the possible forms of these formulae is:

$$f(a,b) = \begin{cases} \left(\frac{a}{2}, \frac{a}{2} + b\right), & \text{if } b \geq 0 \\ \left(\frac{a}{2} - b, \frac{a}{2}\right), & \text{if } b \leq 0 \end{cases},$$

where (a,b) are the coordinates of the complex number $a + \mathbf{i}b$ and $a \in [0,1], b \in [-1,1]$.

It can immediately be seen, that

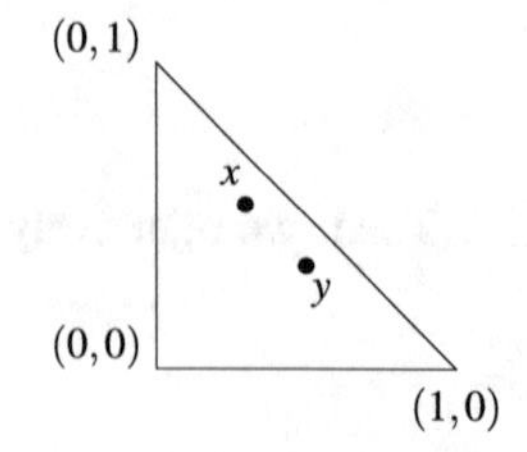

Fig. 27 Elements of x and y are in relation negation

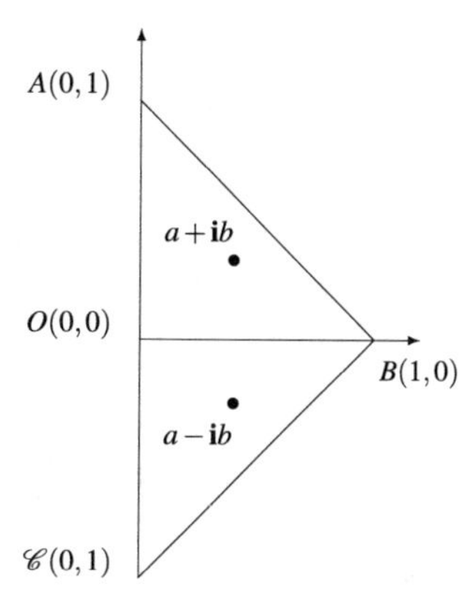

Fig. 28 a + ib and ai **i** b are conjugate in the complex plane

$$\begin{aligned} f(0,0) &= (0,0),\\ f(0,1) &= (0,1),\\ f(1,0) &= \left(\frac{1}{2},\frac{1}{2}\right),\\ f(0,-1) &= (1,0). \end{aligned}$$

It can be easily checked that function f is continuous and bijective.
Now, we directly see that

$$f^{-1}(a,b) = (2a, b-a).$$

Now, we see that for the arbitrary complex conjugate numbers $a + \mathbf{i}b$ and $a - \mathbf{i}b$ in triangle ABC (here $a, b, a+b \in [0,1]$):

$$\begin{aligned} f(a,b) &= \left(\frac{a}{2},\frac{a}{2}+b\right),\\ f(a,-b) &= \left(\frac{a}{2}+b,\frac{a}{2}\right). \end{aligned}$$

Therefore, if IFS-element x has degrees of membership and non-membership $\frac{a}{2}$ and $\frac{a}{2}+b$, respectively, then IFS-element y that has degrees of membership and non-membership $\frac{a}{2}+b$ and $\frac{a}{2}$, respectively, shares with x, a relation *negation*.

The above construction generates a lot of interesting (and currently open) problems. Some of them are the following:

- To develop complex analysis for the intuitionistic fuzzy interpretation triangle.
- To find transformations from the complex plane to the intuitionistic fuzzy interpretation triangle and vice versa in order to acquire more convenient

methods for studying different properties of given IFSs or of given objects in the complex plane.

- To find interpretations of the other (not published by the moment) intuitionistic fuzzy negations in the complex plane and study their behaviour.
- To construct complex analysis interpretations of the IFS-operators from modal, topological and level types.

3.3 Interval Analysis

When working with interval data, we can transform them to intuitionistic fuzzy form. Then, we can interpret them as points of the IFS-interpretation triangle. For example, let us have the set of intervals

$$[a_1, b_1],$$
$$[a_2, b_2],$$
$$\ldots$$
$$[a_n, b_n].$$

Let $A = \min_{1 \le i \le n} a_i < \max_{1 \le i \le n} b_i = B$. Of course, $A < B$, since otherwise for all i: $a_i = b_i$. Now, for the interval $[a_i, b_i]$ we can construct the numbers

$$\mu_i = \frac{a_i - A}{B - A},$$

$$\nu_i = \frac{B - b_i}{B - A}$$

which satisfy the condition $0 \le \mu_i + \nu_i \le 1$ and have the geometrical interpretation from Fig. 3. This idea was introduced for the first time in [34] and it has been used in a series of joint research of Kreinovich, Mukaidono, Nguyen, Wu, Koshelev, Rachamreddy, Yasemis and the author (see, [8, 35–37]).

3.4 Number Theory

In [7, 38] are given some examples for application of IFSs to representing arithmetical functions, e.g., the well-known Euler's totient function and Möbius's function. Here, we give one of these examples.

If for a fixed natural number $n \ge 2$ we define the set

$$F(n) = \{x \mid 1 \le x < n \ \& \ (x, n) = 1\},$$

then,

$$card(F(n)) = \varphi(n),$$

where (p, q) is the greatest common divisor of the natural numbers p and q and $card(X)$ is the cardinality of set X.

Let us define

$$\varphi_\mu(n) = card(\{x \mid 1 < x \le n \ \& \ x/n\}),$$

where x/n denotes that x divides n, and

$$\varphi_\nu(n) = \varphi(n).$$

Then, we can define

$$\begin{aligned}\varphi_\pi(n) &= n - \varphi_\mu(n) - \varphi_\nu(n)\\ &= n - card(F(n)) - card(\{x|1 \le x \le n \,\&\, x/n\})\\ &= card(\{x|\,1 < x < n \,\&\, 1 < (x, n) < n\}).\end{aligned}$$

Therefore, for the set $\mathcal{N}$ of all natural numbers we can construct the following IFS:

$$\mathcal{N}^* = \left\{\left\langle n, \frac{\varphi_\mu(n)}{n}, \frac{\varphi_\nu(n)}{n}\right\rangle \mid n \in \mathcal{N}\right\}.$$

It is an interesting question of future research which other arithmetical functions can obtain IFS-interpretations.

3.5 *Artificial Intelligence and Data Mining*

In [39], the author discussed in details the possible applications of the IFSs as tools for evaluation

- of parameters in pattern recognition procedures,
- of parameters in neural networks and evolutionary algorithms,
- of facts in expert systems, data bases, data warehouses, big data, OLAP-structures,
- of parameters in decision making procedures,
- of parameters in machine and e-learning,
- of data in clusterisation and classification procedures,

- of knowledge discovery processes and of processes for imputation (filling in) of missing data,
- of procedures for inductive reasoning.

3.6 InterCriteria Analysis

During the last year, a new approach for decision support, named InterCriteria Analysis, has been subject of research and development by the author and his research team [40]. Under this approach, arrays of data obtained by the measurement of multiple objects against multiple criteria are processed in order to obtain for each pair of criteria coefficient of correlation (termed here positive consonance, negative consonance of dissonance) in the form of an intuitionistic fuzzy pair of values in the [0;1]-interval. The goal is to detect dependences between criteria on the basis of previous measurement, in order to eliminate some of these criteria in future measurements.

The InterCriteria Analysis can be successfully applied to problems, where measuring according to some of the criteria is slower or more expensive, which results in delaying or raising the cost of the overall process of decision making. When solving such problems it is necessary to adopt an approach for justified and reliable elimination of these criteria, in order to achieve economy and efficiency. The approach has already demonstrated first evidences of its potential, when applied to economic and industrial data, and represents a viable direction of applied research using intuitionistic fuzzy sets.

4 Conclusion

Already 32 years, specialists from more than 50 countries have been working in the area of IFS theory, and each step forward opens new perspectives for theoretical and applied research, which is the best proof that this extension of the area of fuzzy sets has promising prospects for the future.

Acknowledgements This paper has been partially supported by the Bulgarian National Science Fund under the Grant Ref. No. DFNI-I-02-5 “InterCriteria Analysis: A New Approach to Decision Making”. The author is thankful to Evgeniy Marinov, Peter Vassilev and Vassia Atanassova for their valuable comments.

References

1. Atanassov, K.: Intuitionistic fuzzy sets, VII ITKR's Session, Sofia, June 1983 (Deposed in Central Sci.—Techn. Library of Bulg. Acad. of Sci. 1697/84) (in Bulgaria)
2. Atanassov, K., Stoeva S.: Intuitionistic fuzzy sets. In: Proceedings of polich symposium on interval & fuzzy mathematics, pp. 23–26, Poznan, (1983)
3. Zadeh, L.: Fuzzy sets. Inf. Cont. **8**, 338–353 (1965)
4. Kaufmann, A.: Introduction a la Theorie des Sour-ensembles Flous. Masson, Paris (1977)
5. Dubois, D., Prade, H.: Fuzzy Sets and Systems: Theory and Applications. Academic Press, New York (1980)
6. Buhaescu, T.: Some observations on intuitionistic fuzzy relations. Itinerant Seminar on Functional Equations, pp. 111–118, Approximation and Convexity, Cluj-Napoca (1989)
7. Atanassov, K.: On Intuitionistic Fuzzy Sets Theory. Springer, Berlin (2012)
8. Atanassov, K.: 25 years of intuitionistic fuzzy sets or the most interesting results and the most important mistakes of mine. Advances in Fuzzy Sets, Intuitionistic Fuzzy Sets, Generalized Nets and Related Topics, vol. I, pp. 1–35, Foundations, Academic Publishing House EXIT, Warszawa (2008)
9. Atanassov, K.: On some intuitionistic fuzzy negations. Notes Intuitionistic Fuzzy Sets **11**(6), 13–20 (2005)
10. Atanassov, K.: On the temporal intuitionistic fuzzy sets. In: Proceedings of the Ninth International Conference IPMU 2002, vol. III, pp. 1833–1837, Annecy, France, 1–5 July (2002)
11. Klir, G., Yuan, B.: Fuzzy Sets and Fuzzy Logic. Prentice Hall, New Jersey (1995)
12. Atanassov, K.: Intuitionistic fuzzy sets. Springer, Heidelberg (1999)
13. Antonov, I.: On a new geometrical interpretation of the intuitionistic fuzzy sets. Notes Intuitionistic Fuzzy Sets **1**(1), 29–31 (1995). http://ifigenia.org/wiki/issue:nifs/1/1/29-31
14. Danchev, S.: A new geometrical interpretation of some concepts in the intuitionistic fuzzy logics. Notes Intuitionistic Fuzzy Sets **1**(2), 116–118 (1995)
15. Yang, Y., Chiclana, F.: Intuitionistic fuzzy sets: spherical representation and distances. Int. J. Intell. Syst. **24**, 399–420 (2009)
16. Szmidt, E.: Applications of intuitionistic fuzzy sets in decision making. D.Sc. Dissertation, Technical University, Sofia (2000)
17. Szmidt, E., Kacprzyk, J.: Similarity of intuitionistic fuzzy sets and the Jaccard coefficient. In: Proceedings of Tenth International Conference IPMU'2004, vol. 2, 1405–1412, Perugia, 4–9 July 2004
18. Szmidt, E., Kacprzyk, J.: New measures of entropy for intuitionistic fuzzy sets. Notes Intuitionistic Fuzzy Sets **11**(2), 12–20 (2005). http://ifigenia.org/wiki/issue:nifs/11/2/12-20
19. Atanassova, V.: Representation of fuzzy and intuitionistic fuzzy data by Radar charts. Notes on Intuitionistic Fuzzy Sets **16**(1), 21–26 (2010). http://ifigenia.org/wiki/issue:nifs/16/1/21-26
20. Feys, R.: Modal Logics. Gauthier, Paris (1965)
21. Atanassov, K.: Temporal intuitionistic fuzzy sets. Comptes Rendus de l'Academie bulgare des Sciences, Tome 44(7), 5–7 (1991)
22. Atanassov, K.: On the temporal intuitionistic fuzzy sets. In: Proceedings of the ninth international conference IPMU 2002, vol. III, pp. 1833–1837, Annecy, France, 1–5 July 2002
23. Atanassov, K.: Temporal intuitionistic fuzzy sets (review and new results). In: Proceedings of the Thirty Second Spring of the Union of Bulgaria, pp. 79–88, Mathematicians, Sunny Beach, 5–8 Apr 2003
24. Atanassov, K.: A short remark on intuitionistic fuzzy operators $X_{a,b,c,d,e,f}$ and $x_{a,b,c,d,e,f}$. Notes Intuitionistic Fuzzy Sets **19**(1), 54–56 (2013) http://ifigenia.org/wiki/issue:nifs/19/1/54-56
25. Dencheva, K.: Extension of intuitionistic fuzzy modal operators ⊞ and ⊠. In: Proceedings of the Second International IEEE Symposium: Intelligent Systems, vol. 3, pp. 21–22, Varna 22–24 June 2004

26. Atanassov K., Çuvalc oğlu, G., Atanassova, V.: A new modal operator over intuitionistic fuzzy sets. Notes Intuitionistic Fuzzy Sets **20**(5), 1–8 (2014). http://ifigenia.org/wiki/issue:nifs/20/5/1-8
27. Çuvalcioğlu, G.: Some properties of $E_{\alpha,\beta}$ operator. Adv. Stud. Contemp. Math. **14**(2), 305–310 (2007)
28. Çuvalcioğlu, G.: On the diagram of one type modal operators on intuitionistic fuzzy sets: last expanding with $Z_{\alpha,\beta}{}^{\omega\theta}$. Iran. J. Fuzzy Syst **10**(1), 89–106 (2013)
29. Çuvalcioğlu, G.: The extension of modal operators' diagram with last operators. Notes Intuitionistic Fuzzy Sets **19**(3), 56–61 (2013). http://ifigenia.org/wiki/issue:nifs/19/3/56-61
30. Atanassov, K.: On four intuitionistic fuzzy topological operators. Mathware Soft. Comput. **8**, 65–70 (2001)
31. Atanassov, K.: On two topological operators over intuitionistic fuzzy sets. Issues Intuitionistic Fuzzy Sets Generalized Nets **8**, 1–7 (2010)
32. Atanassov, K., Vassilev, P., Tsvetkov, R.: Intuitionistic fuzzy sets, measures and integrals. In: Drinov, M. (ed.) Academic Publishing House, Sofia (2013)
33. Atanassov, K.: Cantor's norms for intuitionistic fuzzy sets. Issues Intuitionistic Fuzzy Sets Generalized Nets **8**, 36–39 (2010)
34. Asparoukhov, O., Atanassov K.: Intuitionistic fuzzy interpretation of confidential intervals of criteria for decision making. In: Lakov, D. (ed.) Proceedings of the First Workshop on Fuzzy Based Expert Systems, pp. 56-58, Sofia, 28–30 Sept
35. Koshelev, M., Kreinovich, V., Rachamreddy, B., Yasemis, H., Atanassov, K.: Fundamental justification of intuitionistic fuzzy logic and interval-valued fuzzy methods. Notes Intuitionistic Fuzzy Sets **4**(2), 42–46 (1998). http://ifigenia.org/wiki/issue:nifs/4/2/42-46
36. Kreinovich, V., Mukaidono, M., Atanassov, K.: From fuzzy values to intuitionistic fuzzy values, to intuitionistic fuzzy intervals etc.: can we get an arbitrary ordering? Notes Intuitionistic Fuzzy Sets **5**(3), 11–18 (1999). http://ifigenia.org/wiki/issue:nifs/5/3/11-18
37. Kreinovich, V., Nguyen, H., Wu, B, Atanassov, K.: Fuzzy justification of heuristic methods in inverse problems and in numerical computations, with applications to detection of business cycles from fuzzy and intuitionistic fuzzy data. Notes Intuitionistic Fuzzy Sets **4**(2), 47–56 (1998). http://ifigenia.org/wiki/issue:nifs/4/2/47-56
38. Atanassov, K.: Remark on an application of the intuitionistic fuzzy sets in number theory. Adv. Stud. Contemp. Math. 5(1), 49–55 (2002)
39. Atanassov, K.: Intuitionistic fuzzy logics as tools for evaluation of data mining processes. Knowl-Based Syst. **80**, 122–130 (2015)
40. InterCriteria Research Portal.: http://www.intercriteria.net/publications. Accessed 1 Aug 2015

Some Comments on Ordinary Reasoning with Fuzzy Sets

Enric Trillas and Adolfo R. de Soto

Abstract The main goal of Computing with Words is essentially a calculation allowing to automate a part of the reasoning done thanks to the natural language. Fuzzy Logic is the main tool to perform this calculation because it is be able to represent the most common kind of predicates in natural language, graded predicates, in terms of functions, and to calculate with them. However there is still not an adequate framework to perform this task, commonly referred to as commonsense reasoning. This chapter proposes a general framework to model a part of this type of reasoning. The fundamental fact of this framework is its ability to adequately represent noncontradiction, the minimum condition for considering a reasoning as valid. Initially, the characteristics of the commonsense reasoning are analyzed, and a model for the crisp case is shown. After that the more general case in which graded predicates are taken under consideration is studied.

Keywords Fuzzy sets · Computing with words · Conjectures · Commonsense reasoning

1 Introduction: In the Way Towards Computing with Words

Fuzzy sets were introduced by Zadeh in 1965 [12] and first related with natural language, and latter on with commonsense reasoning. Onwards from 1996 the same Zadeh, once the theoretical armamentarium of fuzzy logic allowed it after more than thirty years of development, and basing his thinking on the concept of a linguistic variable previously introduced by himself [13], turned back to his original idea by introducing the new field of 'Computing with Words' (CwW) [11]. This new field

E. Trillas (✉)
Oviedo (Asturias), Spain
e-mail: etrillasetrillas@gmail.com

A.R. de Soto
University of León, León, Spain

C. Kahraman et al. (eds.), *Fuzzy Logic in Its 50th Year*,
Studies in Fuzziness and Soft Computing 341,
DOI 10.1007/978-3-319-31093-0_4

basically deals with the representation of commonsense reasoning expressed in Natural Language by fuzzy sets and fuzzy relations, and means a step ahead with respect to the original fuzzy logic capabilities.

Commonsense, or ordinary, reasoning is scarcely deductive and hence monotonic; it essentially consists in conjecturing and refuting. It is by means of reasoning that people tries not only to deduce what is implicitly hidden in the premises, but to enlarge their informational content for either reaching explanations (hypotheses), or just to lucubrate for going further than what is known (speculations). In addition, in ordinary situations the laypeople's "I deduce that" does not exactly mean the same that when a mathematician states "I deduce that" with the help of a formal proof. In the second case the proof consists in some argumentation made in a formal framework, the reasoning can be reduced to a finite chain of elementary steps, each one following the former by applying a known rule of inference, without hidden jumps between these steps, and it can be reproduced by someone managing mathematics and knowing the corresponding subject [5]. Plainly speaking, the correction of the reasoning can be checked by an expert just using "paper and pencil"; that possibility shows a great advantage for what concerns the safety of the reasoning. In the first case, it is difficult to do always it with the help of a formal framework and by reducing the reasoning to a chain of elementary steps and, consequently, being sure that there are no jumps; even, it is often the case that some of the intermediate steps cannot be endowed with the tautological character of those in a mathematical proof. Additionally, not always the premises of the reasoning are neither well known, nor its truth is well known; this modality of ordinary deduction lacks the safe character of formal deduction.

Mathematical modeling is what allows to introduce formal deductive reasoning in science, and since it just consists in "models", the Eugene Wigner's "unreasonable effectiveness of Mathematics in the Natural Sciences" can be appreciated. Deductive reasoning should be classified in formal and informal, being the first a strengthening of the second that, hence, lacks a total safety for the validity of its conclusions. If, usually, mathematical models cannot capture all the nuances of reality and thinking, they allow to figure out, black over white, their basic concepts; to clarify complex problems the "exact thinking" advocated by Menger [3] is, at least, a good and fruitful strategy.

For what concerns hypotheses and speculations, usually they neither can be directly developed from the premises, that is, no forward deduction can always be used for reaching them (for instance, to find a hypothesis) although in some cases backwards deduction can allow it. Both kinds of deduction are monotonic, even if in the informal case the conclusions are not safe enough to be taken as new premises. In the case of abductive reasoning (the search for hypotheses) neither the conclusions, or hypotheses, can generally be taken as new premises, nor the reasoning is monotonic but anti-monotonic. In the case of speculative reasoning (the search for speculations), neither the conclusions, or speculations, can be taken as new premises, nor it is monotonic but without a fixed law of growing, or no-monotonic. All reasoning starts from the conjunction, or résumé, of their premises. In addition, among speculations there are some that can be backwards deductively reached from the negation of the résumé, and others that cannot and that can be called "creative" speculations. It is the reasoning conducting to creative speculations, as well as some non deductive ones conducting to hypotheses, those that can be truly called "inductive" reasoning. In any case, and among conjectures, hypotheses and speculations are those that are non-decidable.

Since ordinary reasoning is basically conducted through words in natural language, it is important to open the possibility of its study by modeling the involved terms by fuzzy sets and not only in classical terms, since much of those words are imprecise or show non-random uncertainty. This is the goal of this chapter and, as it will be shown, what will be presented is reducible to the classical but insufficient case of Boolean algebras that, useful for representing reasoning with precise terms, enjoys too many laws for modeling the totality of reasoning; for instance, the commutative law of Boolean conjunction is not generally valid when time intervenes as it often the case in ordinary reasoning. To obtain new types of conclusions, the study of the case in which the conclusions are graded is actually important and, specially, towards approaching some modalities of informal deduction.

This chapter tries to serve as a theoretical support for CwW, but, and specially, to motivate young researchers to go further in its extension towards the different types of commonsense reasoning.

2 Descriptive Aspects of Ordinary Reasoning

Some descriptive aspects that can partially characterize commonsense reasoning (CR), are the following [8]:

1. Neither CR, nor formal deductive reasoning, is possible without starting from either some previous information, evidence, suppositions, or axioms.
2. Often the previous information from which CR starts is not only partial, but partially liable, and expressed in natural language containing ambiguous, precise and imprecise linguistic terms or also numbers, functions, pictures, etc.
3. Non-deductive CR lacks monotony, that is, when the number of significant items or premises collecting the previous information increases, then either the number of conclusion items decreases or there is no law for its variation. That is, more premises mean less conclusions, or there is no way of knowing what happens with the variation in the number of conclusions.
4. In CR, people essentially lucubrate to extract as conclusions from the previously given "knowledge", "evidence", or "information", either consequences, refutations or explanations of what it is supposed to be known, to follow up with the safest possible conclusions to get new ideas or to just deploy what is still hidden in the premises.
5. It is usually accepted in CR that the previous knowledge should be non-contradictory, that is, without existing self-contradictory premises, and also without premises contradictory with other premises. Hence, CR is done under

the minimal limitation of trying to keep some kind of consistency, or non-contradiction, with the information conveyed by the premises.

6. Typically, in CR there are "jumps" from the premises to some initial, or to the final, conclusions, and contrarily to formal deduction where the deployment of conclusions is done in a step-by-step manner and under precisely known rules. Very often, such jumps cannot be reduced to a chain of elementary steps.
7. In general, CR is done under some not always well-known "rules of thumb" with which people often lucubrate for extracting either consequences, "new" ideas, or to reject some "old" ones. Such rules of thumb do not enjoy the tautological character of the rules of inference typical of formal deduction.
8. More often than not, people uses metaphors or similarities for getting conclusions, at the extreme that metaphorical reasoning is considered as a part of CR.
9. It is typical of CR that the premises, or the conclusions, are taken as graded statements. That is, there are premises and conclusions that are not universally valid but only partially liable.
10. Contrarily to what happens in formal deductive reasoning, the conclusions of CR cannot be taken as new premises. For instance, it is often the case of counting with two contradictory hypotheses or speculations.
11. Usually, CR cannot guarantee the "safety" of its conclusions, and, in any case, such safety cannot always follow from that of the premises.

From these characteristics of CR, it follows that the Boolean structure of classical logic or even, for instance, the weaker one of Quantum logic, are not able to capture all the nuances presented by the imprecision and uncertainty shown by the involved statements. In what follows the only that will be not considered is the presence, sometimes pervasive, of ambiguous linguistic terms. For what concerns uncertainty, it should be recalled that it mainly belongs to either the random type or to the non-random or singular type; the first can be dealt with the mathematical concept of probability, and for the second there are other mathematical concepts that are more or less useful, like it is the theory of possibility introduced by Zadeh.

3 Imprecision, Uncertainty, Ambiguity, and Analogy

1. The use of most words in dictionaries is not describable by, formal, "if and only if" definitions, that is, its use cannot be characterized by means of such a definition. Very few entries are like "even number: integer multiple of two". Even the entry "odd" not only refers to numbers, in which case it is characterized by "if and only if, divided by two the rest is one", but it also can refer to people and then the only possibility for telling on its use lies in informally describing it, like it also happens with other uses of "even". Natural languages are full of imprecise linguistic terms, that is, terms that can be understood as those that do not collectivize in the form of a classical set in the universe of

discourse, but in the form of a cloudy entity like, for instance, the collective of "young Londoners", since the word "young" cannot classify the set of London inhabitants in just two classical sets. The "linguistic" collectives generated by imprecise words have no a single representation in the universe of discourse, like it happens with precise words. The collective of the even numbers is uniquely represented by the classical subset $\{2, 4, \ldots, 2n, \ldots\}$ of $\mathbb{N}$, but there is not any classical set able to represent the collective of young Londoners; there are only families of fuzzy sets, each one depending on the information available on the term young in London, and that can be viewed as "states" of the collective. There is just a single state when the predicate is made precise, for instance, when equating "young" with "less than 30 years old" that, anyway, is just a possible state of the collective like it is with "less than 31 years old", etc. Currently, the only systematic way to deal with the imprecision that, through words, permeates language is made by fuzzy sets.

2. Almost all statements in natural language show some kind of uncertainty, for instance, "John is between 30 and 40 years old", "Rita's face is beautiful", "My neighbor is rich", "In throwing this dice five points will be obtained", "In next election candidate X will not be elected", "The garden is wide enough", etc. The uncertainty of these statements is manifested by questioning if they are "probable". For instance, "Is it probable that my neighbor is rich?", "Is it probable to get five points?", etc. After accepting such a questioning, it goes the question of "Up to which extent is it probable?", whose answer depends on the type of uncertainty to which the statement can be ascribed. For instance, the one referring to throwing a dice can be seen as part of an "experiment", repeatable in the same, and controlled, conditions and from which, for instance, the successive results can be annotated, but that referring to candidate X does not admit this view since the only that can be known is what happened either for X, or for other candidates, but in past elections. The first type of uncertainty is called "random uncertainty", and the second can be called non-random, singular, or linguistic. Random uncertainty is measurable by means of probabilities, in the Kolmogorov sense if the "events" of the experiment can be supposed to belong to a Boolean algebra, or in that of Quantum Probability if the events are in a weaker structure like it is, for instance, an Orthomodular lattice [1].

For linguistic uncertainty there are, at least, two ways of measuring its extent, namely the measures of possibility-and-necessity [14], and those of fuzzy probability [16] that, both introduced by Lotfi A. Zadeh, the last are applicable when the statement can be represented by a fuzzy set in $\mathbb{R}^n$ [9, 16].

3. Usually the meaning of words is not unique but at each context one of them can be clearly taken, and the problem appears in those cases where the corresponding meaning is not clearly recognizable. Very often either joking, or confusion, are based on this linguistic ambiguity like, for instance, when someone says "I am looking for a bank", and it is understood as a search for a bank to seat in, instead of searching for a bank's office.

If imprecision and uncertainty can be taken into account thanks to fuzzy logic and probability theory, the formal study of ambiguity is still an open problem.

4. People use analogy, likeness, or similarity, either to explain, or to capture, how things are, or what can follow from something. That is, for identifying new things, for conjecturing how they are, or for lucubrating some conclusions from them. The richness of metaphors in all natural language is due to the usefulness of comparing things by similarity; the common use in English of the word "like" as, for instance, in "A dress like that of a cap", or "A house like that of a maharaja", help to have a first, although inexact, comprehension of either a dress, or a house saving the speaker of a, perhaps, unnecessary and too detailed precision in the description. Descriptions by similarity just help to acquire a first idea of what, being new, is tried to be described.

For what concerns reasoning, analogy is also commonly employed in commonsense reasoning. If something was successfully conjectured in a given situation, and the reasoning deals with a similar situation, it is often supposed that something similar can be conjectured in the new situation. In the field of Artificial Intelligence, techniques like case-based reasoning are used for doing reasoning by analogy. Since the similarity of situations is often difficult to be measured, an important problem in analogical reasoning is the numerical control of the extent up to which things are similar. For instance, when it was initially considered the atom's structure as similar to the solar system, it conducted to misleading conclusions not corresponding with some real atomic phenomena. Nevertheless, such atomic model is still useful at High Schools for some didactical purposes.

4 Frame of Representation

There is no science without, at least, counting with a process of schematization followed by abstraction. That is, isolating the essential elements of a given subject, and the basic relationships between them, allowing for a sufficient identification of what intervenes in the problem, as well as for the measuring of the main quantities that, originated by them, are observable. It is through such measuring how a theory obtained by abstraction can be tested against the reality at which it refers to [10].

For such goal it is relevant and useful to fix, at each case, a *frame of formal representation*; for instance, the vector space $\mathbb{R}^3$ for the 3-dimensional geometry, sigma-algebras for the theory of probability, Hilbert spaces for quantum physics, the real line for the infinitesimal calculus, De Morgan algebras for the theory of possibility, etc. That is, once a problem is isolated and identified, it is with the help of a formal frame for representing what intervenes in it, allowing to consider the problem as one of scientific type, that helps to contrast the theory with the corresponding reality. It is the establishment of a formal frame what allows the mathematical modeling of real problems, with these models is which mathematics gifts science with the possibility of introducing deduction, the safest type of reasoning, among its reasoning. It is just in this way that science can differentiate from, for instance, metaphysics.

Hence and for trying to scientifically study commonsense reasoning, it is important to initially fix a formal framework in which, avoiding ambiguity, imprecision, uncertainty and analogy can be represented. That is, doing it by means of fuzzy sets structured in a form sufficiently weak for not only help to appreciate very general treats of reasoning, but also to know which properties are suitable for the verification of some specific laws. For such purpose, the structure of a **Basic Algebra** (BA) [5, 8], a partially ordered set endowed with two binary operations, and a unary one, satisfying a minimal number of properties, seems to be sufficient. Such structure, is defined in $\mathcal{F}(X) = [0,1]^X$ with

- The pointwise ordering, $\mu \leq \sigma \Leftrightarrow \mu(x) \leq \sigma(x)$ for all x in X, with which $\mathcal{F}(X)$ is a poset,
- A (binary) conjunction $\cdot : (X) \times \mathcal{F}(X) \to \mathcal{F}(X)$, and a (binary) disjunction $+ : \mathcal{F}(X) \times \mathcal{F}(X) \to \mathcal{F}(X)$,
- A (unary) negation $' : \mathcal{F}(X) \to \mathcal{F}(X)$,

where $\mu_0(x) = 0$ and $\mu_1(x) = 1$ for all x in X are the minimum and the maximum, respectively, but in which neither associativity, nor commutativity, nor distributivity, nor double negation, nor functional expressibility, etc., are previously supposed, just should verify the properties:

1. (a) $\mu_0 \cdot \mu = \mu \cdot \mu_0 = \mu_0 \;\&\; \mu_0 + \mu = \mu + \mu_0 = \mu$
   (b) $\mu_1 \cdot \mu = \mu \cdot \mu_1 = \mu \;\&\; \mu_1 + \mu = \mu + \mu_1 = \mu_1$, for all μ in $\mathcal{F}(X)$.
2. (a) $\mu \leq \sigma \Rightarrow \sigma' \leq \mu'$
   (b) $\mu_0' = \mu_1 \;\&\; \mu_1' = \mu_0$
3. If $\mu \leq \sigma$, then
   (a) $\mu \cdot \gamma \leq \sigma \cdot \gamma \;\&\; \gamma \cdot \mu \leq \gamma \cdot \sigma$
   (b) $\mu + \gamma \leq \sigma + \gamma \;\&\; \gamma + \mu \leq \gamma + \sigma$, for all γ in $\mathcal{F}(X)$.
4. The restrictions of $\cdot$, $+$, and $'$, to $\{0,1\}^X$, are $\cdot = \min$, $+ = \max$, and $' = 1 - id$.

From these very weak laws of $\Omega = (\mathcal{F}(X);\, \leq, \mu_0,\, \mu_1;\, \cdot;\, +;')$, it easily can be proven the following few laws,

I. Among conjunctions, the largest is $\cdot = \min$, and the smallest is that given by

$$\mu[\cdot]\sigma := \begin{cases} \sigma & \text{if } \mu = \mu_1, \\ \mu & \text{if } \sigma = \mu_1, \\ \mu_0 & \text{otherwise.} \end{cases}$$

Among disjunctions, the smallest is $+ = \max$, and the largest is that given by

$$\mu[+]\sigma := \begin{cases} \sigma & \text{if } \mu = \mu_0, \\ \mu & \text{if } \sigma = \mu_0, \\ \mu_1 & \text{otherwise.} \end{cases}$$

Only with the pair (min, max) is Ω a lattice and, provided the negation is involutive, or strong, ($\mu'' = \mu$, for all μ in $\mathcal{F}(X)$), Ω is a De Morgan algebra verifying the Kleene Law $\mu \cdot \mu' \leq \sigma + \sigma'$, for all μ and σ.
There is no Ω that, with the full set $\mathcal{F}(X)$, can be neither an ortholattice nor, *a fortiori*, a Boolean algebra.

II. The definitions $\mu \underline{+} \sigma = (\mu' \cdot \sigma')'$, and $\mu \underline{\cdot} \sigma = (\mu' + \sigma')'$, give, respectively, a disjunction, and a conjunction, not always verifying laws of duality, like $\mu \underline{+} \sigma = (\mu' \underline{\cdot} \sigma')'$. The corresponding BA is called the "dual" of Ω.

III. Ω_0, the BA obtained by restricting the operations of Ω to the subset $\{0, 1\}^X$, is a Boolean algebra isomorphic to that of the crisp subsets of X. When Ω is the De Morgan-Kleene algebra given by min and max, Ω_0 consists in its Boolean elements.

IV. It is always:

(a) $\mu \leq \sigma \;\&\; \alpha \leq \beta \Rightarrow \mu \cdot \alpha \leq \sigma \cdot \beta \;\&\; \mu + \alpha \leq \sigma + \beta$
(b) $\mu' \cdot \sigma' \leq (\mu \cdot \sigma)' \;\&\; (\mu + \sigma)' \leq \mu' + \sigma'$
(c) If $+ = \max$, and regardless of $\cdot$ and $'$,

$$(\mu + \sigma)' \leq \mu' \cdot \sigma' \;\&\; \mu \cdot (\sigma + \gamma) = \mu \cdot \sigma + \mu \cdot \gamma.$$

(d) If $\cdot = \min$, and regardless of $+$ and $'$,

$$(\mu \cdot \sigma)' \leq \mu' + \sigma' \;\&\; \mu + (\sigma \cdot \gamma) = (\mu + \sigma) \cdot (\mu + \gamma).$$

V. Characterization of the BAs in which hold the laws of Non-contradiction (NC) and Excluded-middle (EM), when expressed in the typical form they take with crisp sets, namely, $\mu \cdot \mu' = \mu_0$ and $\mu + \mu' = \mu_1$, respectively, is not known. Nevertheless, it can be proven that they hold in the forms [7] $\mu \cdot \mu' \leq (\mu \cdot \mu')'$ for NC—namely, $\mu \cdot \mu'$ is self-contradictory, and $(\mu + \mu')' \leq ((\mu + \mu')')'$ for EM—namely, $(\mu + \mu')'$ is self-contradictory. Hence, it is not true that fuzzy sets violate these laws.

VI. When the operations $\cdot$, $+$, and $'$, are functionally expressible, the possible validity of additional laws to those proper of the BA, can be analyzed through functional equations. For instance, and for a conjunction that is functionally expressible by means of a numerical function F, and a negation expressible by a numerical function N, the validity of the NC's law $\mu \cdot \mu' = \mu_0$, strictly depends on the verification by F of the functional equation $F(a, N(a)) = 0$, for all a in [0, 1]. It should be pointed out that, the continuity of the functions expressing the three operations, without which solving functional equations is extremely difficult, is very important for not adding more discontinuities than those presented by the fuzzy sets being operate with. Since continuity just appears as a way for translating into fuzzy sets the flexibility usually required for representing imprecise predicates, or linguistic labels, it is not always suitable that, for instance, $\mu \cdot \sigma$ can show more discontinuity points than those eventually shown by μ or by σ.

A very particular case of functionally expressible BAs is obtained by supposing that the three operations are of the forms

$$\mu \cdot \sigma = T \circ (\mu \times \sigma),\ \mu + \sigma = S \circ (\mu \times \sigma), \text{ and } \mu' = N \circ \mu,$$

with T a continuous t-norm, S a continuous t-conorm, and N a strong negation function. Then, the corresponding Ω is called a Standard algebra (SA) of fuzzy sets that carries, obviously, with the associative and commutative laws not always present in language. In the setting of Standard algebras, the laws NC and EM can be characterized in its classical forms; for instance, NC holds with T equal to the Lukasiewicz t-norm $W(a, b) = \max(0, a + b - 1)$, and with the negation $N_0(a) = 1 - a$, and EM holds with this same N_0 and S equal to the dual t-conorm $W^*(a, b) = \min(1, a + b)$. Of course, with the triplet (W, W^*, N_0) both laws hold.

Even if SAs do not verify all the laws holding in Boolean algebras, they verify a sufficient number that make them too strong for constructing a mathematical model of reasoning inside them. In what follows such a model is presented inside of just a BA Ω, and for both the crisp and the graded reasoning.

5 Modeling Reasoning as Conjecturing and Refuting

This section tries to present a mathematical model for precise or crisp reasoning, with non-ambiguous and either imprecise, or precise, statements. By "crisp reasoning" it is understood the reasoning with fuzzy and/or crisp sets as premises, and also fuzzy and/or crisp sets as conclusions, but without considering any degree on their respective validity [5, 9].

1. Basically, reasoning consists in conjecturing and refuting as, for instance, when after having some information on the way in which John is living, it is either concluded that he is rich, or that he is not rich. Analogously, when knowing that someone will win the lottery's award, a person buys a ticket of it just by saying himself, "Why not me", even knowing that the probability of winning is very low, or when a woman after having acquired some acquaintance with a man convinces herself that he is not a good candidate to be married with, or when a young girl decides to study Literature at the university after thinking that her

ability in writing assures that she will be a famous writer. Most of the people's ordinary reasoning just tries to reach conclusions characterized by the following:

(a) What is known, or believed, consists on statements $p_1, \ldots, p_n$, supposed no one is self-contradictory (it is not "If p_k, then not p_k"), and no one is contradictory with another one (it is not, "If p_k, then not p_j"). Statements p_k are the premises of the reasoning, and their conjunction, $p = p_1$ and p_2 and ... and p_n, is its résumé.
(b) A statement q is a conclusion of the reasoning if: Either it can be stated "If p, then not q", or it can be stated the negation of this conditional statement, "It is not (If p, then not q)". In the second case the conclusion is a conjecture of the premises, and in the first is a refutation of them.

To formalize these two ideas, in the way of having a first mathematical model of (non-graded, or crisp) ordinary reasoning, let's suppose that all the premises, their résumé and conclusions, belong to $\mathcal{F}(X)$, that is do correspond to non-ambiguous and either imprecise, or precise, statements. Let's denote $P = \{p_1, \ldots, p_n\}$ the set of premises. Then,

1. $p = p_1 \cdot (p_2 \cdot (p_3 \cdot \cdots (p_{n-1} \cdot p_n)) \cdots)$
2. q is a conjecture of $P \Leftrightarrow p \nleq q'$
3. q is a refutation of $P \Leftrightarrow p \leq q'$,

with,

$$\text{Conj } P = \{q \in \mathcal{F}(X);\ p \nleq q'\}, \text{ and Ref } P = \{q \in \mathcal{F}(X);\ p \leq q'\},$$

the sets of conjectures and refutations of P. These two sets constitute a partition of $\mathcal{F}(X)$ since, obviously,

$$\mathcal{F}(X) = \text{Conj } P \cup \text{Ref } P, \text{ and Conj } P \cap \text{Ref } P = \emptyset.$$

It should be pointed out that not having presumed the conjunction $\cdot$ is associative, p depends on the numbering, or ordering, of the premises. The parentheses only can be skipped once $\cdot$ is associative, and then the ordering of the premises is not important. Because $\leq$ is a partial ordering, it is clear that conjectures can be partitioned in the two classes $\{q \in \text{Conj } P;\ q' < p\}$, and $\{q \in \text{Conj};\ q$ is not $\leq$-comparable with $p\}$, of which the second is called the class of speculations and is denoted by Sp(P), and the first is, at its turn, classified in the two disjoint subclasses,

$$\{q \in \mathcal{F}(X);\ p \nleq q' \;\&\; p \leq q\}, \text{ and } \{q \in \mathcal{F}(X);\ p \nleq q' \;\&\; \mu_0 < q < p\},$$

of which the first are the consequences of P, the second the hypotheses of P, and are respectively denoted by C(P) and Hyp(P). Hence, from the partition, Conj $P = \text{C}(P) \cup \text{Hyp}(P) \cup \text{Sp}(P)$, and finally it is obtained the partition,

$$\mathcal{F}(X) = \text{Ref}\, P \cup \text{Conj}\, P = \text{Ref}\, P \cup \text{C}(P) \cup \text{Hyp}(P) \cup \text{Sp}(P),$$

telling that once P and p are given, there are no other elements in $\mathcal{F}(X)$ than refutations, consequences, hypotheses and speculations. Obviously, among them, the set of the undecidable ones is $\text{Hyp}(P) \cup \text{Sp}(P)$.

At its turn, $\text{Sp}(P)$ can be classified in the two classes

$$\text{Sp}_1(P) = \{q \in \mathcal{F}(X);\ p \text{ nc } q \ \&\ q' < p\},$$

and,

$$\text{Sp}_2(P) = \{q \in \mathcal{F}(X);\ p \,\text{nc}\, q \,\&\, p \,\text{nc}\, q'\},$$

shortening "not ≤-comparable" by "nc". Hence, speculations either are of type-one, or of type-two.

Remarks

(a) The obvious equalities $\text{Conj}\, P = \text{Conj}\, \{p\}$, and $\text{Ref}\, P = \text{Ref}\, \{p\}$, highlight the relevance of the résumé p, as well as the model's dependence on it. Hence, and provided the conjunction $\cdot$ is non-associative, special care should be placed to "define" the résumé, i.e., in the ordering of the premises.

(b) Since ≤ is transitive, any consequence q can be reached through forward paths starting at the résumé: $p \le q_1,\ q_1 \le q_2,\ \ldots,\ q_n \le q$. That is, by means deductive steps that, only in the case of mathematical formal deduction, are chains of elementary steps in which never can be jumps, in the sense that each q_k follows from q_{k-1} by applying "Modus Ponens".

(c) For what concerns hypotheses h, it is possible to reach them by backwards deductive paths starting at the résumé: $p \ge h_1,\ h_1 \ge h_2,\ \ldots,\ h_n \ge h$, provided at least one of the ≥ is >. Whenever these paths are chains that is, such that each h_k follows by h_{k-1} by applying "Modus Tollens", it can be said that h is deductively reached. Hence, in principle, only some hypotheses can be deductively obtained.

(d) Type-one speculations q such that $q' \le p$, can be deductively reached by backwards paths starting at the résumé and ending at q'. Nevertheless, no type-two speculation can be reached through a path and, hence, by a deductive chain. Hence, very few type-one speculations can be obtained by deduction, and type-two speculations can be never deductively obtained. Those hypotheses, type-one speculations that cannot be reached by deductive chains, as well as type-two speculations, only can be obtained through induction, that is, in a heuristic form by means of jumps. Actually, only type-two speculations can be viewed as those that mainly contribute to the reasoning with "new ideas" and, for this reason, can be called "creative speculations".

(e) Since it is always $\mu_1 \in \text{C}(P)$, it is also $\mu_1 \in \text{Conj}\, P$, but neither μ_1 is in $\text{Hyp}(P)$, nor in $\text{Sp}(P)$. For what concerns μ_0, it is never in $\text{Conj}\, P$, but, since its negation is μ_1, it is $\mu_0 \in \text{Ref}\, P$.

2. What happens with the growing of the number conclusions when that of premises grows? It is not difficult to prove what follows. Let's suppose that P and Q are sets of premises, and that $P \subseteq Q$, with the corresponding résumés verifying $p \leq q$. Then,

   (a) $\mathrm{Conj}\, Q \subseteq \mathrm{Conj}\, P$, that is, the operator Conj is anti-monotonic.
   (b) $\mathrm{Ref}\, P \subseteq \mathrm{Ref}\, Q$, that is, Ref is monotonic.
   (c) $\mathrm{C}(P) \subseteq \mathrm{C}(Q)$, C is monotonic.
   (d) $\mathrm{Hyp}(Q) \subseteq \mathrm{Hyp}(P)$, Hyp is anti-monotonic.
   (e) It is neither $\mathrm{Sp}(P) \subseteq \mathrm{Sp}(Q)$, nor $\mathrm{Sp}(Q) \subseteq \mathrm{Sp}(P)$, that is, Sp is just non-monotonic, there is not a general law for the growing of speculations. Anyway, it is $\mathrm{Sp}_1(Q) \subseteq \mathrm{Sp}_1(P)$, Sp_1 is anti-monotonic but it is Sp properly non-monotonic.

The only of these sets that can be taken as a new set of premises is $\mathrm{C}(P)$, since it neither can contain self-contradictory consequences, nor contradictory pairs of them. Hence, the only of these operators that can be always re-applied is C: It can be considered $\mathrm{C}(\mathrm{C}(P)) = \mathrm{C}^2(P)$ as a set of premises, and since $P \subseteq \mathrm{C}(P)$, it follows $\mathrm{C}(P) \subseteq \mathrm{C}^2(P)$, and more concretely $\mathrm{C}(P) = \mathrm{C}^2(P)$. The new set of premises do not enlarge those consequences obtained from P. C is a closure operator.

From $P \subseteq \mathrm{C}(P)$, follows $P \subseteq \mathrm{Conj}\, P$, but P is neither contained in $\mathrm{Hyp}(P)$, nor in $\mathrm{Sp}(P)$, nor in $\mathrm{Ref}\, P$. Only consequences are able to "extent" the premises.

Remark Actually, the name "creative" not only can be identified with type-two speculations. Also those hypotheses and those type-one speculations that cannot be reached through deductive paths, can be involved in "creative reasoning".

For what concerns to refutation, it should be pointed out that those refutations q such that $q' \leq q$, can be also reached by paths starting at the résumé of P, and, eventually, by deductive chains. Hence, also the refutations that are not deductively reachable can deserve to be called "creative refutations".

3. Let's show a way in which speculations can be very useful. Consider $q \in \mathrm{Sp}(P)$.

   - From $p \leq p + q$, it follows $p + q \in \mathrm{C}(P)$.
   - Provided $p \cdot q \neq \mu_0$, since $p \cdot q \leq p$, it suffices that $p \cdot q \neq p$ to obtain $p \cdot q \in \mathrm{Hyp}(P)$. Notice that, in the case in which $\cdot = \min$, it were $p \cdot q = p$, in this case equivalent to $p \leq q$, it will suffice $p \nleq q$, for reaching a hypothesis $p \cdot q$.

Consequently, the disjunction between the résumé and a speculation is always a consequence, and a non-empty conjunction of the résumé with a speculation, provided such conjunction is different from the résumé, is a hypothesis. Speculating not only serves for creating new ideas, but also for explaining what is known, and for deploying hidden consequences from it.

4. A capability the presented model shows is that of proving two well known criteria for falsifying that a statement h is a hypothesis of P. That is, the falsification of hypothesis.

   - Provided it is presumed that $h \in \text{Hyp}(P)$, from $h < p$, it follows $C(\{p\}) = C(P) \subseteq C(\{h\})$. Hence, provided it were found a consequence of P that cannot be a consequence of h, it will be $h \notin \text{Hyp}(P)$: h is deductively falsified as a hypothesis of P.
   - Since $\text{C}(\{h\}) \subseteq \text{Conj}\{h\}$, for falsifying h it suffices to find a consequence of P that cannot be conjectured from h.

The first is the falsification used in mathematics, and the second is used in the experimental sciences.

6 Analogy in Ordinary Reasoning

As it was said before, analogy, likeness, or similarity appears very often in ordinary reasoning. If a set of premises P can be seen as analogous to another Q, it can be supposed that their respective résumés p and q are also similar, and a first worry to be considered is if analogy can be considered as a different kind of reasoning than conjecturing or refuting. In other words, the question is if the images by analogy can give, or cannot, a new type of conclusions; if analogy, being very important, is more than a tool for reasoning.

Provided there is a mapping A: $\mathcal{F}(X) \to \mathcal{F}(X)$, transforming each fuzzy set q in another fuzzy set $A(q)$, and preserving some features of a given problem, it can be said that "q and $A(q)$ are analogous" [5]. Then, the set $\text{Conj}\, A(P)$, with $A(P) = \{A(q);\ q \in \text{Conj}\, P\}$, can be seen as analogous to the set Conj P, provided its résumé is $A(p) = A(p_1 \cdot (p_2 \cdot (\ldots(p_{n-1} \cdot p_n)\ldots)$, that is, $\text{Conj}\, A(P) = \text{Conj}\, A(p)$. At this respect, provided $\cdot$ is associative and A is a morphism for the conjunction $\cdot$, from $A(p) = A(p_1) \cdot A(p_2) \cdot \cdots \cdot A(p_n)$, it can consistently be taken $A(P) = \{A(p_1), \ldots, A(p_n)\}$ with résumé $A(p)$.

Defining the function A', by $A'(q) = (A(q))'$, for all q in $\mathcal{F}(X)$, there only exists the two possibilities of being, either $Id_{\mathcal{F}(X)} \leq A'$, or $Id_{\mathcal{F}(X)} \nleq A'$. Thus, and for all $q \in \mathcal{F}(X)$, it is either $q \leq A(q)'$ and only refutations can be obtained, or it is $q \nleq A(q)'$, and only conjectures can be obtained.

In the second case, and since as its turn function A, only can verify:

1. $A \leq id_{\mathcal{F}(X)}$, in which case it is $A(q) \leq q$, for all q in $\mathcal{F}(X)$, and only consequences can be obtained.
2. $Id_{\mathcal{F}(X)} < A$, and $q < A(q)$, and only hypotheses can be obtained.
3. $Id_{\mathcal{F}(X)}$ and A are not comparable, i.e., function $Id_{\mathcal{F}(X)}$ crosses with function A, and there are cases in which it is $q < A(q)$, and others in which it is $A(q) \leq q$. In this case, only speculations can be obtained.

It does not seem that analogy can give a different type of reasoning, but that it is a precious helping methodology for it. For instance, if q is a creative speculation with which it was found the hypothesis $p \cdot q$, and it is $A < Id_{\mathcal{F}(X)}$, from $A(p \cdot q) < p \cdot q \leq p$, follows that the fuzzy set $A(p \cdot q)$, analogous to the old hypothesis $p \cdot q$, is a new hypothesis for P.

Metaphors are but statements or, sometimes, images or drawings, reflecting something analogous to what is being considered, and are not at all avoidable for ordinary reasoning. Greek philosophy is full of metaphors, and Plato's dialogs constitute a paramount example on the use of linguistic metaphorical reasoning. In the learning of elementary mathematics, for instance, the drawing of a particular triangle can serve as a metaphor for inspiring how to solve a question on triangles. Metaphors can be very useful, are used from very old, and their danger just lies in the "distance" between the metaphor and the current problem, in what was believed when, a metaphor was created in the past, and what is currently known. The good use of metaphors lies, at its turn and at the end, in the experience and, especially, the specific knowledge of who is using them, and there are many examples in the history of science of metaphors that conducted to new and fruitful knowledge as it is, for instance, the "ouroboros" metaphor that allowed August Kekulé to find the benzene's chemical structure.

Last remark, It should be pointed out that almost all that has been presented can be done, *mutatis mutandis*, in a BA different of Ω. Hence, the model can be exported to different settings than that of fuzzy sets.

7 Graded Ordinary Reasoning with Fuzzy Sets

The modeling of reasoning through the classification and study of statements in conjectures and refutations, along its subsequent subdivisions in hypothesis, consequences and speculations is a model that accounts for many of the characteristics of human reasoning. However, the model presented in Sect. 5 does not take into account that, in common sense reasoning, new sentences are obtained from "possibilistic" rather than "safe" knowledge. Reasoning has much more of "may" than "definitively is". So our knowledge progresses, when using reasoning mechanisms, incorporating judgments that are possible given our state of knowledge. But there are many possible sentences and some of them usually contradict each other, at least with some intensity. We can say that from our current knowledge a tree of diverse possibilities is open and some branches are clearly contradictory with others. However, if we want an appropriate reasoning mechanism, choosing any of these branches should require a minimum of coherence between its elements.

Thus on the one hand, the model should incorporate the ability of deducing with a graduated measuring of the strength in the reasoning and, on the other hand, it should be established a principle of "minimum consistency" so the steps of reasoning should not include judgments that would lead to absurd. To achieve this, the framework of representation given in Sect. 4 should be expanded.

Remark Different models are possible to perform this task. In all of them, it seems natural to introduce a graded relationship as a way of spreading conclusions. However the definition of the relevant sets in this new context is difficult to assess. In what follows, the sets of conjectures and refutations together with their posterior divisions in consequences, hypotheses and speculations are crisp sets defined from levels of chaining reasoning but it must be understood that is certainly not the only possibility, and can not even be the most interesting. Therefore the following should be considered as provisional, pending of a deeper development and being checked with various examples.

In the above formulation, a BA uses the pointwise ordering between fuzzy sets, which of course is a crisp ordering between fuzzy sets. From now, a graded relation between statements through a more general ordering is defined. So a fuzzy relation $\preccurlyeq : \mathcal{F}(X) \times \mathcal{F}(X) \to [0,\ 1]$ between fuzzy sets will be adopted.

The relation $\mu \preccurlyeq \eta$ represents a conditional fuzzy relation between propositions μ, η from the habitual interpretation in mathematical logic

$$(\mu \preccurlyeq \eta) = r \Leftrightarrow \text{if } \mu, \text{ then } \eta, \text{ up to the degree } r \in [0,\ 1].$$

Natural requirements for relations $\preccurlyeq$ are the reflexive property and the T-transitive property:

$$T(\mu \preccurlyeq \eta,\ \eta \preccurlyeq \sigma) \leq \mu \preccurlyeq \sigma.$$

Specially last one gives the quite basic requirement of knowing some minimum level to which a reasoning with several propositions can be chained. A fuzzy relation with both properties is called a T-preorder.

From the relation $\preccurlyeq$, classical relations between fuzzy sets can be defined fixing a threshold r:

$$\mu \preccurlyeq_r \sigma \text{ iff } (\mu \preccurlyeq \sigma) \geq r.$$

In particular, the classical relation $\preccurlyeq_1$, called the kernel of $\preccurlyeq$, is a classical preorder.

In this new framework, when such a fuzzy relation is used in a Basic Algebra, we will call this structure a Basic Fuzzy Algebra (BFA). The properties of a Basic Fuzzy Algebra are the equivalent ones to the properties given in Sect. 4 for BA. In particular the conjunction and disjunction will be monotonic with respect to the relation $\preccurlyeq$ by the left and the right. Important enough is also the behavior of the negation with this new ordering.

As it is said in point V of Sect. 4, in a Basic Fuzzy Algebra there is a representation of contradiction. In the classical sense of the non-contradiction principle in boolean logic, i.e. that the degree of truth of $p \wedge p^c$ is always 0, is not valid in general in standard fuzzy logic algebras [6]. However in any human reasoning the non-contradiction principle appears as an important principle; the relation of being contradictory plays a fundamental role. Our reasoning goes well if, at least, there is not a contradiction in it. In a BFA, a fuzzy set μ is said contradictory to degree r with another fuzzy set ρ if $(\mu \preccurlyeq \rho') > r > 0$.[1] That means the negation of ρ is derived from μ in some degree. When $(\mu \preccurlyeq \rho') = 0$, μ and ρ are said non-contradictory fuzzy sets. A fuzzy set μ will be named self-contradictory if it is the case that $(\mu \preccurlyeq \mu') > 0$. The fuzzy set μ_0 is self-contradictory because $\mu_0 \preccurlyeq_1 \mu_1 \equiv \mu_0'$. In classical logic, the empty set is the unique self-contradictory set, but in Fuzzy Logic with a different negation function can exist many self-contradictory fuzzy sets [7]. In fact, in any BFA an expression of the principle of Non-contradiction is always valid because $\mu \cdot \mu'$ is self-contradictory, i.e.

$$\mu \cdot \mu' \preccurlyeq_1 (\mu \cdot \mu')' \tag{1}$$

because from $\mu \cdot \mu' \preccurlyeq_1 \mu$ it is valid that $\mu' \preccurlyeq_1 (\mu \cdot \mu')'$ and with $\mu \cdot \mu' \preccurlyeq_1 \mu'$ results (1).

1. In the context of Basic Fuzzy Algebras, the set of admissible premises are considered as non-contradictory and hence as a valid set to represent a state of knowledge. In a BFA Ω, let's consider those sets $P = \{\rho_1, \ldots, \rho_n\} \subseteq \mathcal{F}(X)$, such that:
   1. for all $i, j \in 1, \ldots, n : (\rho_i \preccurlyeq \rho_j) = 0$;
   2. its résumé, $\rho_P \equiv \rho_1 \cdot (\rho_2 \cdot (\ldots(\rho_{n-1} \cdot \rho_n)\ldots)$, verifies $(\rho \preccurlyeq \rho') = 0$, and $(\rho' \preccurlyeq \rho) = 0$.[2]

   Last definition tries to fix a kind of minimal precaution on the available information from which some conclusions should be extracted based on forbidding self-contradictory information.[3] It can be proven that in a set of admissible premises, no pair of contradictory premises can exist. It should be recalled that in lack of associativity, premises must be numerated to properly define p

2. In the crisp case, if ρ represents the résumé, a conjecture is a proposition μ whose negation μ' is not deduced from the résumé, i.e. $\rho \not\leq \mu'$. This means that a conjecture is any proposition which is not contradictory with the knowledge generated by the résumé, or which is compatible with that knowledge. Or put another way, a conjecture is a candidate to in some way increase consistently the knowledge. Refutations, on the contrary, are propositions which are in contradiction to the current knowledge. In a graded context, it is possible to try to catch these ideas by means of following definitions.

$$\mathrm{Conj}^r P = \{\mu \in \mathcal{F}(X);\ (\rho \preccurlyeq \mu') < r\} \tag{2}$$

---

[1]If the negation is strong, if μ is contradictory with ρ is equivalent to ρ is contradictory with μ because $0 < r < (\mu \preccurlyeq \rho') \leq (\rho'' \preccurlyeq \mu') = (\rho \preccurlyeq \mu')$.

[2]If P is clear from the context we use ρ instead of ρ_P for the résumé of P.

[3]Note that in a framework of a graded ordering relation, first condition is a strong one but necessary to avoid self-contradiction.

$$\text{Ref}^r P = \{\mu \in \mathcal{F}(X);\ (\rho \preccurlyeq \mu') \geq r\}, \tag{3}$$

where r represents a considered acceptable degree of refutation. Observe that when $r = 1$, μ is a conjecture if $(\rho \preccurlyeq_1 \mu')$. As in the crisp case, both sets constitute a partition of $\mathcal{F}(X)$.

The degree r is, in some sense, a level of admissible contradiction. With these definitions, a refutation introduces some level of contradiction in the information given by the résumé, because it is easy to prove that if $\mu \in \text{Ref}^r(P)$, then $r \leq (\mu \cdot \rho \preccurlyeq (\mu \cdot \rho)')$. So from $0 < r \leq (\rho \preccurlyeq \mu')$ we obtain a level r of contradiction between μ and the résumé, because $0 < (\mu \cdot \rho \preccurlyeq (\mu \cdot \rho)')$. It shows that a refutation added by the left to the set of premises causes a level of contradiction in our knowledge.

Operators Ref^r are not extensive for any $r \in (0, 1]$, and they are monotonic, i.e. if the set of admissible premises increases, the set of refutations also increases. However the conjecture operator is anti-monotonic with respect to the set of admissible premises. Also the behavior of both operators with respect to the grade r is inverse, the sets of refutations are anti-monotonic with respect to the grades r, that is if $s < r$, then $\text{Ref}^r \subseteq \text{Ref}^s$ and the sets of conjectures are monotonic with respect to the grades.

3. Again, in the framework of BFA, it is possible to consider the set of consequences of a set of admissible premises.

$$\text{C}^r(P) = \{\mu \in \mathcal{F}(X);\ (\rho \preccurlyeq \mu) \geq r\}. \tag{4}$$

The difference with the crisp case lies in that the set of consequences is not always a subset of conjectures. It is necessary to establish a couple of extra conditions to that goal: It can be proven that, for any non-nilpotent t-norm T and any strong negation, any set of admissible premises P and any $r \in (0, 1]$, is $\text{C}^r(P) \subseteq \text{Conj}^r(P)$.

The sets of consequences are anti-monotonic with respect to the grade r and they satisfy two properties of the Tarski' consequence operators: they are extensible, $P \subseteq \text{C}^r P$, and they are monotonic, if $P \subseteq Q$, then $\text{C}^r P \subseteq \text{C}^r Q$.

The T-transitivity, together with the two conditions $(\rho \preccurlyeq \rho') = 0$ and $(\rho' \preccurlyeq \rho) = 0$ constitute, in the case of non-nilpotent t-norms, a strong condition. Using the first one, T-transitivity and the basic property of the operator of negation, the relation

$$T(\rho \preccurlyeq \mu', \rho \preccurlyeq \mu) \leq T(\rho \preccurlyeq \mu', \mu' \preccurlyeq \rho') \leq \rho \preccurlyeq \rho' = 0, \tag{5}$$

is always true for any fuzzy set μ, so either $(\rho \preccurlyeq \mu) = 0$, i.e. μ is not a consequence of ρ at any level, or $(\rho \preccurlyeq \mu') = 0$, i.e. ρ is not contradictory with μ'. Of course, also both cases are simultaneously possible.

Analogously, using the condition $(\rho' \preccurlyeq \rho) = 0$, for any μ, either ρ is not a consequence of μ, or it is not a consequence of μ'. Observe that this is not necessary true for nilpotent t-norms. But in some cases a threshold could be establish between the values of $(\mu \preccurlyeq \rho), (\mu' \preccurlyeq \rho), (\rho \preccurlyeq \mu)$, and $(\rho \preccurlyeq \rho')$.

Note that in case of a non-nilpotent t-norm T, a refutation never is a consequence up to any degree because

$$T(\rho \preccurlyeq \mu', \rho \preccurlyeq \mu) \leq T(\rho \preccurlyeq \mu', \mu' \preccurlyeq \rho') \leq (\rho \preccurlyeq \rho'),$$

and it is not possible to have $(\rho \preccurlyeq \mu') > 0$ and $(\rho \preccurlyeq \mu) > 0$, simultaneously. However with nilpotent t-norms it can be the case that a refutation to some degree can be a consequence to a non-zero degree. In particular we have for a non-nilpotent t-norm T, that, if $\mu \in \mathrm{C}^r(P)$, then $\mu' \notin \mathrm{C}^r(P)$.

4. Take into consideration the difference-set:

$$\mathrm{Conj}^r(P) - \mathrm{C}^r(P) = \{\mu \in \mathcal{F}(X) : (\rho \preccurlyeq \mu) < r \ \& \ (\rho \preccurlyeq \mu') < r\} \tag{6}$$

Remember that the set $\mathrm{C}^r(P)$ is not necessarily a subset of $\mathrm{Conj}^r(P)$, since in fact, $\mathrm{Conj}^r(P)$ can be empty. The set given in (6), if exists, represents the conjectures that are not consequences.

Consider next two sets:

$$\mathrm{Hyp}^r(P) = \{\mu \in \mathcal{F}(X) : \rho \preccurlyeq \mu < r \ \& \ \rho \preccurlyeq \mu' < r \ \& \ \mu \preccurlyeq \rho \geq r\},$$
$$\mathrm{Sp}^r(P) = \{\mu \in \mathcal{F}(X) : \rho \preccurlyeq \mu < r \ \& \ \rho \preccurlyeq \mu' < r \ \& \ \mu \preccurlyeq \rho < r\}.$$

With these definitions, if $\mathrm{Conj}^r(P) \neq \emptyset$, then

$$\mathrm{Conj}^r(P) = \mathrm{C}^r(P) \cup \mathrm{Hyp}^r(P) \cup \mathrm{Sp}^r(P).$$

Newly, the conjectures, if they exist, are classifies in consequences, hypothesis and speculations.

With respect to the grade r, if $r \leq r'$, $\mathrm{Hyp}^r(P)$ is incomparable with $\mathrm{Hyp}^{r'}(P)$, hypotheses can grow or not. Conjectures increase with r consequences decrease, and speculations increase as it is easy to prove. For the set of hypothesis nothing can be said.

A natural requirement for a hypothesis is that they explain the consequences, in particular, any consequence of a premise must be a consequence of a hypothesis. In our case this depends of the t-norm, because if $\sigma \in \mathrm{Hyp}^r(P)$ and $\mu \in \mathrm{C}^r(P)$, then

$$T(r,r) \leq T(\sigma \preccurlyeq \rho, \rho \preccurlyeq \mu) \leq \sigma \preccurlyeq \mu.$$

So μ is a consequence of σ to level $T(r,r)$.

It also and newly shows that, for a fixed r, operators Hyp^r are anti-monotonic, while operators $\mathrm{Sp}^r(P)$ are neither monotonic nor anti-monotonic.

8 Conclusions

It is a strong believing of the authors that CwW [11, 15] can offer more potentiality than that of being just applied to practical problems, and even if this is of a paramount importance. As far as CwW deals with reasoning expressed in natural language, it can count with all the armamentarium either existing, or in course of development in fuzzy logic, for expanding its possibilities in, at least, two directions. The first concerns the scientific study of language, and the second the exploration and modeling of human commonsense reasoning. It should be remarked that "thinking" is nothing else, but and notwithstanding nothing less, than a natural phenomenon made inside the brain, and of which the layman is familiar with some of its external manifestations like, for instance, the most apparent of language and reasoning. If the study of thinking and speaking do correspond to natural science and, in particular, to neurobiology, that of representing all the nuances of language and that of analyzing and modeling reasoning, is a proper subject for not only Linguistics but also for the Sciences of Computation; it is not to be forgotten that these two problems are considered in the 'Gordian Knot' of Artificial Intelligence. Thinking is not only done through language; in it, memory of images plays an important role and, at the end, language is the natural tool the brain developed for representing what is both internal and external to it.

This chapter deals, indeed, with reasoning and it should be noticed the interest of, for instance, the proven conclusion that if p is the résumé of a set of premises, and s a speculation, and provided $p \cdot s \neq p$, then $h = p \cdot s$ is a hypothesis, and that, analogously, and in any case, $p + s$ is a consequence of P. Hence, the presented model allows to show a possible reason for why people speculate on some problems; speculation helps for both explaining and reaching consequences of P.

Section 5, devoted to crisp reasoning with either imprecise or precise statements, does not show anything not holding in the classical case of a Boolean algebra. Nevertheless, all in this section is proven under very weak conditions, and if speculations only did appear by the first time in weak algebraic structures, like that of a BA [9], the "creative" ones were not be seen until the classification of conjectures was made explicit. It should be noticed that the setting of BAs is so general that all what is presented holds without most of the laws of classical logic, and is applicable, in particular, to the so-called case of Quantum Logic, the non-distributive logic of Quantum Physics [1]. Notice also, that the lack of the associative and commutative laws force to any process of either conjecturing or refuting, to previously assigning an ordering to the premises, something that should be done by considering a suitable, and perhaps external character, concerning how they were obtained. All that means that nothing actually "fuzzy" can appear in this crisp reasoning apart of dealing with fuzzy sets and that, at the end, just consists in seeing ordinary reasoning *à la* Popper [4], that is, by considering that the reasoning's conclusions are either conjectures or refutations. The deep relationship of the model with Popper's thinking is reinforced by the formalization of hypothesis' falsification, and also, even if done under the perhaps debatable view of analogy presented in Sect. 6, by proving that analogical reasoning only can conduct to conjecture or to refute.

To reach the "fuzzy" singularities reasoning shows, it seems a reasonable way that of introducing the concept of graded conjectures, and graded refutations. To such an important goal, and of which only graded consequences were previously considered [2], it is shown in this chapter what is in Sect. 7 that, nevertheless, is just nothing else than a first approach still requiring not only a deeper study, but, specially, a systematic testing against real cases in ordinary reasoning. This last section only tries to open a window for looking at a possible way for studying the "fuzzy" reasoning with fuzzy sets.

The study of language and reasoning, once viewed, as they are actually, some natural phenomena, cannot be pursued by only abstract methods inspired in those of logic and mathematics. It could remember, indeed, the study of the universe based on the geometry of spheres and polyhedrons jointly with eye nude's observations and the simple Aristotelian syllogism. That study requires of wise observation of reality, and controlled experimentation against it, as well as of mathematical models that can supply with numerical 'parameters' the measurability of their basic variables for giving credibility to what could be provisory concluded [10].

It is the evolution of Computing with Words and Perceptions towards a new experimental science of language and reasoning by specially attending to imprecision, uncertainty and ambiguity, something like a "physics" of language and reasoning, what can be the big jump of Fuzzy Logic in the XXI Century.

Acknowledgments This author acknowledges the support of the Spanish Ministry of Economy and Competitiveness and the European Regional Development Fund (ERDF/FEDER) under grant TIN2014-56633-C3-1-R.

References

1. Bodiou, G.: *Théorie diallectique des probabilités*. Gauthier-Villars (1964)
2. Castro, J.L., Trillas, E., Cubillo, S.: On consequence in approximate reasoning. J. Appl. Non-Class. Logics **4**(1), 91–103 (1994)
3. Menger, K.: Morality, Decision and Social Organization: Toward a Logic of Ethics. Springer Science & Business Media (1974)
4. Popper, K.R.: Conjectures and Refutations. Routledge and Kegan Paul, London (1965)
5. Trillas, E.: Glimpsing at Guessing. Fuzzy Sets and Systems **281**, 32–43 (2015)
6. Trillas, E., Alsina, C.: Elkan's theoretical argument, reconsidered. Int. J. Approximate Reasoning **26**(2), 145–152 (2001)
7. Trillas, E.: Non contradiction, excluded middle, and fuzzy sets. In: Di Gesù, V., et al. (eds.) Fuzzy Logic and Applications, pp. 1–11. Springer (2009)
8. Trillas, E.: A model for "crisp reasoning" with fuzzy sets. Int. J. Intell. Syst. **27**(10), 859–872 (2012)
9. Trillas, E.: Some uncertain reflections on uncertainty. Arch. Philos. Hist. Soft Comput. **1**, 1–16 (2013)
10. Trillas, E.: How science domesticates concepts? Arch. Philos. Hist. Soft Comput. **1**, 1–17 (2014)

11. Zadeh, L.A.: Fuzzy logic = computing with words. IEEE Trans. Fuzzy Syst. **4**(2), 103–111 (1996)
12. Zadeh, L.A.: Fuzzy sets. Inf. Control **8**, 338–353 (1965)
13. Zadeh, L.A.: The concept of a linguistic variable and its application to approximate reasoning. Part I. Inf. Sci. **8**, 199–249 (1975)
14. Zadeh, L.A.: A theory of approximate reasoning. Mach. Intell. **9**, 149–194 (1979)
15. Zadeh, L.A.: Computing with Words. Principal Concepts and Ideas, volume 277 of Studies in Fuzziness and Soft Computing. Springer (2012)
16. Zadeh, L.A.: Probability measures of fuzzy events. J. Math. Anal. Appl. **23**(2), 421–427 (1968)

A Review of Hesitant Fuzzy Sets: Quantitative and Qualitative Extensions

Rosa M. Rodríguez, Luis Martínez, Francisco Herrera and Vicenç Torra

Abstract Since the concept of fuzzy set was introduced, different extensions and generalizations have been proposed to manage the uncertainty in different problems. This chapter is focused in a recent extension so-called hesitant fuzzy set. Many researchers have paid attention on it and have proposed different extensions both in quantitative and qualitative contexts. Several concepts, basic operations and its extensions are revised in this chapter.

Keywords Hesitant fuzzy set · Operations · Extensions

1 Introduction

Fuzzy sets were introduced by Zadeh [69]. Since then theory on fuzzy sets and fuzzy logic has been developed in parallel with a large number of successful applications.

Fuzzy sets permit to represent that elements have partial membership to a set, and they can be modeled to represent graduality between non-membership and complete membership to a set. A fuzzy set is represented mathematically by means of membership functions which generalize characteristic functions. While the latter

R.M. Rodríguez (✉) · F. Herrera
University of Granada, Granada, Spain
e-mail: rosam.rodriguez@decsai.ugr.es

F. Herrera
e-mail: herrera@decsai.ugr.es

L. Martínez
University of Jaén, Jaén, Spain
e-mail: martin@ujaen.es

V. Torra
University of Skövde, Skövde, Sweden
e-mail: vtorra@his.se

C. Kahraman et al. (eds.), *Fuzzy Logic in Its 50th Year*,
Studies in Fuzziness and Soft Computing 341,
DOI 10.1007/978-3-319-31093-0_5

are functions that given an element of the reference set return a value that is either 0 or 1, membership functions return a value in the interval [0,1].

Therefore, the definition of a fuzzy set $\tilde{A}$ requires the definition of the membership function of $\tilde{A}$. This implies that we need to assign a number in the interval [0,1] to all elements of the reference set.

At present there exist several generalizations of fuzzy sets. Some of them have been introduced in order to ease the definition of fuzzy sets by means of relaxing the requirement that the membership function needs a value for each element in the reference set. In particular we can mention type 2 fuzzy sets [27] where the membership of an element is a fuzzy set instead of a single number. In this way, we can model the uncertainty on the number we need to assign. Interval valued fuzzy sets (IFS) [45] and Atanassov's intuitionistic fuzzy sets (A-IFS) [1] are two other examples. In this case, the value assigned to an element is an interval. Differences from the point of view of interpretation and discussion about the terminology between these two extensions of fuzzy sets can be found in [11]. Then, the concept of type 2 fuzzy sets can be further generalized into type n fuzzy set. Informally speaking, a type n fuzzy sets corresponds to a type (n-1) fuzzy set in which the membership values of the type (n-1) fuzzy sets are a fuzzy set.

Goguen introduced [13] L-fuzzy sets which also generalize fuzzy sets. The idea is that while in a fuzzy set the membership assigns values in the range [0,1] which is a total order, we can consider the assignment of membership values in partial orders (posets).

Recently, hesitant fuzzy sets (HFSs) were introduced in [40]. In this type of fuzzy sets, the membership value of an element is a subset of [0,1] and typically a finite set of values in [0,1]. As stated in [43], the motivation for introducing this type of fuzzy sets "is that when defining the membership of an element, the difficulty of establishing the membership degree is not because we have a margin of error (as in A-IFS), or some possibility distribution (as in type 2 fuzzy sets) on the possible values, but because we have a set of possible values".

In this chapter some of the results found in the literature on HFSs are revised. Its structure is as follows. Section 2 reviews the concept of HFS, some basic operations and hesitant fuzzy relations. Sections 3 and 4 revise extensions of HFS in quantitative and qualitative contexts. Section 5 introduces some discussions and trends of the hesitant context, and finally some conclusions are pointed out in Sect. 6.

2 Hesitant Fuzzy Sets

The concept of HFS was recently introduced as an extension of fuzzy sets with the goal of modeling the uncertainty provoked by the hesitation when it is necessary to assign the degree of membership of an element to a fuzzy set. This section revises some basic concepts and operations about HFSs.

2.1 Concepts of Hesitant Fuzzy Sets

As briefly it is stated in the introduction, a HFS is a generalization of a fuzzy set in which the membership function returns a subset of values in [0,1]. This is formalized in the following definition.

Definition 1 [40] Let X be a reference set. Then, a HFS on X is a function h that returns a subset of [0,1] to elements $x \in X$.

$$h : X \rightarrow \wp([0,1]) \tag{1}$$

Xia and Xu [61] call $h(x)$ a hesitant fuzzy element (HFE). Note that a hesitant fuzzy element is a set of values in [0,1], and a HFS is a set of HFEs, one for each element in the reference set. That is, if $h(x)$ is the HFE associated to x then $\cup_{x \in X} h(x)$ is a HFS.

A typical hesitant fuzzy set [2] is when $h(x)$ is a finite nonempty subset of [0,1] for all $x \in X$, i.e., HFEs are finite nonempty sets.

The literature presents several papers in which operators on HFS are defined (or can be defined) through operators on HFEs. The extension principle introduced in [44] is one of them.

Definition 2 Let $\{H_1, \ldots, H_n\}$ be n HFSs on a reference set X, let ϕ a function on n HFEs (i.e., ϕ combines n sets into a new set). Then,

$$\phi'(H_1, \ldots, H_n)(x) = \phi(H_1(x), \ldots, H_n(x))$$

defines an operation ϕ' on HFSs.

The extension principle is defined as follows.

Definition 3 Let $\{H_1, \ldots, H_n\}$ be n HFSs on a reference set X, and let Θ be a function $\Theta : [0,1]^n \rightarrow [0,1]$, we export Θ to HFSs defining the HFS Θ_E as follows

$$\Theta_E(x) = \cup_{\gamma \in H_1(x) \times \cdots \times H_n(x)} \{\Theta(\gamma)\}.$$

Note that for a given function Θ we can define the function,

$$\phi(S_1, \ldots, S_n) = \cup_{\gamma \in S_1 \times \cdots \times S_n} \{\Theta(\gamma)\},$$

which permits to express the extension principle in terms of Definition 2.

Both Definition 2 and the extension principle cause that the properties of the operator ϕ and Θ are inherited by ϕ' and Θ_E. As reported by Rodríguez et al. [34], it is trivial to prove the commutativity and associativity of Θ_E from the ones of Θ.

The definition of operations for HFS from operations for HFEs (as in Definition 2) is not the only way to do so. There are operations on HFS that cannot be represented in this way. The following operator is an example that illustrates this fact.

Definition 4 [34] Let $\{H_1, \ldots, H_n\}$ be n HFSs on a reference set X, then

$$\phi(H_1, \ldots, H_n)(x) = \frac{(\max_i \max(H_i) + \min_i \min(H_i))}{2} \wedge \cup_i H_i(x)$$

where for any α in [0,1], $\alpha \wedge h$ corresponds to the set $\{s | s \in h, s \leq \alpha\}$.

This operation ϕ cannot be represented in terms of another function ϕ' on HFEs.

2.2 *Basic Operations of Hesitant Fuzzy Sets*

In [40] were introduced some basic operations to manage HFEs. These definitions follow the approach of the Definition 2, that is, a function for HFSs defined in terms of a function for HFEs.

Definition 5 [40] Given a HFE, h, its lower and upper bounds are:

$$h^- = inf\{\gamma | \gamma \in h\} \quad (2)$$

$$h^+ = sup\{\gamma | \gamma \in h\} \quad (3)$$

Definition 6 [40] Let h be a HFE, the complement of h is defined as follows:

$$h^c = \bigcup_{\gamma \in h} \{1 - \gamma\} \quad (4)$$

Definition 7 [40] Let h_1 and h_2 be two HFEs, the union of two HFEs $h_1 \cup h_2$, is defined as:

$$h_1 \cup h_2 = \bigcup_{\gamma_1 \in h_1, \gamma_2 \in h_2} \{\max\{\gamma_1, \gamma_2\}\} \quad (5)$$

Definition 8 [40] Let h_1 and h_2 be two HFEs, the intersection of two HFEs $h_1 \cap h_2$ is defined as:

$$h_1 \cap h_2 = \bigcup_{\gamma_1 \in h_1, \gamma_2 \in h_2} \{\min\{\gamma_1, \gamma_2\}\} \quad (6)$$

The relation between HFS and A-IFS was discussed in [40]. Let us start recalling the definition of interval valued fuzzy sets (IFS) and A-IFS. As both are mathematically equivalent, we only give one definition.

Definition 9 Let X be a reference set. Then, an IFS on X is represented by means of two functions $\mu : X \rightarrow [0,1]$ and $v : X \rightarrow [0,1]$ such that $0 \leq \mu(x) + v(x) \leq 1$ for all $x \in X$.

It is easy to prove the following.

Proposition 1 [43] *All IFSs are HFS.*

Definition 10 [43] Given a HFE h, we define the envelope of h as the IFS represented by μ and v defined by $\mu(x) = h^{-}(x)$ and $v(x) = 1 - h^{+}(x)$, respectively.

It can be proven that this is the smallest IFS that includes the HFE h.

Proposition 2 *All HFS are L-fuzzy sets.*

This follows from the fact that subsets of [0,1] define a partial order. Note that IFS and type n fuzzy sets can also be seen as L-fuzzy sets.

Sometimes, it is necessary to compare two HFEs to establish an order between them. Different proposals have been introduced in the literature to compare HFEs. Xia and Xu defined a score function to compare HFEs [61], however Farhadinia pointed out that this score function could not distinguish between two HFEs in some cases. Thus, a new score function was presented by Farhadinia [12]. Despite the new score function can compare HFEs when the score function proposed by Xia and Xu cannot, Rodríguez et al. shown a counterexample [34] in which the new function cannot either discriminate some HFEs. Recently, Xia and Xu have presented a variance function [23] to improve the comparison law proposed in [61].

Definition 11 [61] Let h be a HFE, the score function $s(h)$ is defined as follows:

$$s(h) = \frac{1}{l(h)} \sum_{\gamma \in h} \gamma, \tag{7}$$

being $l(h)$ the number of elements in h.

Definition 12 [23] Let h be a HFE, the variance function $v(h)$, is defined as follows:

$$v(h) = \frac{1}{l(h)} \sqrt{\sum_{\gamma_i, \gamma_j \in h} (\gamma_i - \gamma_j)^2}. \tag{8}$$

From the Definitions 11 and 12 of $s(\cdot)$ and $v(\cdot)$ respectively, the following comparison law was defined.

Definition 13 Let h_1 and h_2 be two HFEs,

If $s(h_1) < s(h_2)$, then $h_1 < h_2$,
If $s(h_1) = s(h_2)$, then

If $v(h_1) < v(h_2)$, then $h_1 > h_2$,
If $v(h_1) = v(h_2)$, then $h_1 = h_2$.

More operations and their properties have been presented in [29].

2.3 *Hesitant Fuzzy Relations*

Fuzzy relations have been used in several contexts. They generalize crisp relations permitting fuzzy membership. For example, binary relations on a reference set X are generalized to fuzzy relations by means of a membership degree of each pair $(x_1, x_2) \in X \times X$. When X is finite, say $X = \{x_1, \ldots, x_n\}$, a fuzzy relation R is represented by a matrix $R = \{r_{ij}\}_{ij}$ where r_{ij} is the membership degree of (x_i, x_j) into the relationship R. A fuzzy relation can be understood as a weighted graph with nodes X and weights R.

Literature discusses additive preference relations (APR), which are fuzzy relations where $\mu(x_i, x_j) + \mu(x_j, x_i) = 1$ for all $x_i, x_j \in X$.

In addition, the literature discusses multiplicative preference relations (MPR). They diverge from fuzzy relations because they are functions from $X \times X$ into [1/9,9]. They require $\mu(x_i, x_j) \cdot \mu(x_j, x_i) = 1$. The range [1/9,9] is based on Saaty's scale for the Analytic Hierarchy Process [35, 36]. In multicriteria decision making problems, multiplicative preference relations are used to represent users' preferences on the criteria from which weights on the criteria are extracted using prioritization methods.

Additive and multiplicative relations are isomorphic. Note that given a MPR $R = \{r_{ij}\}_{ij}$, with values in the range [1/9,9], then when R' is defined in terms of $r'_{ij} = 0.5(1 + \log_9 r_{ij})$ as $R' = \{r'_{ij}\}_{ij}$, we have that R' is an APR, and given an APR we can define the corresponding MPR using the inverse of the function $f(x) = 0.5(1 + \log_9 x)$ (see [8] for details).

Fuzzy preference relations and multiplicative preference relations have been extended in order to include hesitancy on the value assigned in the matrix.

A hesitant fuzzy relation is defined by means of a function μ which assigns a finite subset of [0,1] to each pair (x_i, x_j). Constraints are added on the possible values of $\mu(x_i, x_j)$. Formally, if $\mu(x_i, x_j)$ is a finite set, we can denote the hesitant fuzzy relation by a *matrix*-like structure

$$C(i,j) = \{a_{ij}^1, \ldots, a_{ij}^{k_{ij}}\},$$

for all $i, j = 1, \ldots, |X|$ and where k_{ij} is the number of elements in the pair (x_i, x_j). A hesitant fuzzy relation is defined requiring $k_{ij} = k_{ji}$, $C(i,i) = \{1/2\}$ and that if elements of $C(i,j)$ and $C(j,i)$ are ordered the first set in increasing order, and the second one in decreasing order the pairs in the kth positions should sum one. That is, assume that elements are ordered

$$C(i,j) = \left(a_{ij}^1 \leq \cdots \leq a_{ij}^{k_{ij}}\right)$$
$$C(j,i) = \left(a_{ji}^1 \geq \cdots \geq a_{ji}^{k_{ji}}\right)$$

then, $a_{ij}^k + a_{ji}^k = 1$ (see [76] for details).

A hesitant multiplicative preference relation (HMPR) was defined in [62, 71] requiring that values a_{ij}^k are in the [1/9,9] interval, that $C(i,j)$ and $C(j,i)$ have the same number of elements (i.e., $k_{ij} = k_{ji}$), and that the elements in $C(i,j)$ and $C(j,i)$ can be matched, so that the multiplication of any matched pair is one.

In [41, 42] the conditions on the number of elements and pairing elements are not considered. This definition is isomorphic to the definition of numerical preference relations introduced in [78].

A consistent hesitant multiplicative matrix is then defined as follows. Here, a $n \times n$ structure is the function $C(i,j)$ defined above where $C(i,j)$ is a finite set of values. No constraints are given on the possible values in $C(i,j)$.

Definition 14 Let M be a $n \times n$ structure (or hesitant matrix). We say that M is a consistent hesitant multiplicative preference relation (cHMPR) if it satisfies

C1. $C(i,i) = \{1\}$ for all i,
C2. For all i, j if $a_{ij} \in C(i,j)$ then there is $1/a_{ij} \in C(j,i)$,
C3. For all i, j if $a_{ij} \in C(i,j)$ then there exists k such that $a_{ij} = a_{ik}a_{kj}$ and $a_{ik} \in C(i,k)$ and $a_{kj} \in C(k,j)$.

Prioritization methods have been obtained to derive weights from hesitant matrices. See e.g. [41, 42] inspired in the geometric mean approach introduced in [9].

In [42] an algorithm was introduced to build a cHMPR for any given $n \times n$ structure (or hesitant matrix). The algorithm can also be applied to any standard real-valued matrix which is not consistent (i.e., which is not a multiplicative preference relation) obtaining a hesitant matrix. Algorithm 1 corresponds to this process.

Algorithm 1: Reconcile

```
Data: C : n × n structure
Result: n × n structure
C0 = C(i,j) ∪_{a_ij ∈ C(j,i)} 1/a_ji ;
for all i,j such that i > j do
    mult(i,j) = there exists k such that a_ij = a_ik a_kj? ;
    if false(mult(i,j)) then
        for some k (the selection of k is arbitrary) ;
            C0(k,j) = C0(k,j) ∪ a_ij/a_ik ;
            C0(j,k) = C0(j,k) ∪ a_ik/a_ij ;
    return (C' = C0)
```

3 Extensions of Hesitant Fuzzy Sets in Quantitative Settings

We have stated in Sect. 2.1 that a typical hesitant fuzzy set [2] is when $h(x)$ is a finite nonempty subset of [0,1] for all $x \in X$. Several extensions and generalizations of HFS which diverge from this typical type of HFS have been proposed to deal with the hesitation in quantitative settings. These extensions are introduced in this section.

3.1 Dual Hesitant Fuzzy Sets

The concept of Dual Hesitant Fuzzy Set (DHFS) [79] is an extension of HFS based on A-IFS that deals with the hesitation both for the membership and non-membership degrees. Therefore, a DHFS is defined in terms of two functions that return two sets of membership and non-membership values respectively for each element in the domain:

Definition 15 [79] Let X be a set, a DHFS D on X is defined as:

$$D = \{\langle x, h(x), g(x)\rangle | x \in X\} \tag{9}$$

being $h(x)$ and $g(x)$ two sets of values in the interval [0,1], that denote the possible membership and non-membership degrees of the element $x \in X$ to the set D respectively, with the following conditions,

$$0 \leq \gamma, \eta \leq 1, 0 \leq \gamma^+ + \eta^+ \leq 1$$

where $\gamma \in h(x)$, $\eta \in g(x)$, $\gamma^+ = \max_{\gamma \in h(x)}\{\gamma\}$, and $\eta^+ = \max_{\eta \in g(x)}\{\eta\} \forall x \in X$.

The pair $d(x) = (h(x), g(x))$ is called Dual Hesitant Fuzzy Element (DHFE) and by simplicity it is noted $d = (h, g)$.

Example 1 Let $X = \{x_1, x_2\}$ be a reference set, a DHFS D, is defined as follows:

$$D = \{\langle x_1, \{0.4, 0.5\}, \{0.3\}\rangle, \langle x_2, \{0.2, 0.4\}, \{0.3, 0.5\}\rangle\}$$

Some basic operations, such as the complement of a DHFE, the union and intersection of two DHFEs were introduced in [79]. A score function and accuracy function were also defined with the goal of proposing a comparison law to compare DHFEs. Recently, different aggregation operators to aggregate DHFEs have been defined. In [50] it has been introduced Dual Hesitant Fuzzy Weighted Average (DHFWA), Dual Hesitant Fuzzy Weighted Geometric (DHFWG), Dual Hesitant

Fuzzy Ordered Weighted Average (DHFOWA), Dual Hesitant Fuzzy Ordered Weighted Geometric (DHFOWG), Dual Hesitant Fuzzy Hybrid Average (DHFHA) and Dual Hesitant Fuzzy Hybrid Geometric (DHFHG). These operators have been used to propose some generalized dual hesitant fuzzy aggregation operators [46, 67, 68]. The Hamacher operations have been extended to propose some aggregation operators for DHFEs [17, 70]. The Choquet integral has been also used to develop several aggregation operators for DHFEs [16].

Several approaches to compute the correlation coefficient of DHFEs have been defined [7, 53, 65] and some properties have been studied.

A similarity measure that considers the membership and non-membership degrees of DHFEs has been introduced in [39].

3.2 *Interval-Valued Hesitant Fuzzy Sets*

Sometimes, in real-world decision making problems, it is difficult for experts to express their assessments by using crisp values, because of the lack of information about the problem. In these situations, an interval value belonging to [0,1] could be used. Keeping in mind the concept of HFS, Chen et al. introduced the definition of Interval-Valued Hesitant Fuzzy Set (IVHFS) [5] where the membership degrees are given by several possible interval values.

An IVHFS is defined as follows.

Definition 16 [5] Let X be a reference set, and I([0,1]) be a set of all closed subintervals of [0,1]. An IVHFS on X is,

$$\tilde{A} = \{\langle x_i, \tilde{h}_A(x_i)\rangle | x_i \in X, \quad i = 1, \ldots, n\} \tag{10}$$

where $\tilde{h}_A(x_i) : X \to \wp(I([0,1]))$ denotes all possible interval-valued membership degrees of the element $x_i \in X$ to the set $\tilde{A}$.

$\tilde{h}_A(x_i)$ is called an Interval-Valued Hesitant Fuzzy Element (IVHFE), where each $\tilde{\gamma} \in \tilde{h}_A(x_i)$ is an interval and $\tilde{\gamma} = [\tilde{\gamma}^L, \tilde{\gamma}^U]$, being $\tilde{\gamma}^L$ and $\tilde{\gamma}^U$ the lower and upper limits of $\tilde{\gamma}$, respectively.

Example 2 Let $X = \{x_1, x_2\}$ be a reference set, a IVHFS $\tilde{A}$, could be as follows,

$$\tilde{A} = \{\langle x_1, \{[0.2, 0.3], [0.4, 0.5]\}\rangle, \langle x_2, \{[0.1, 0.4], [0.5, 0.6], [0.8, 0.9]\}\rangle\}$$

When the upper and lower limits of all the interval values are equal, the IVHFS is a HFS.

Some basic operations, such as the union, intersection and complement were introduced in [5]. A score function to compare two IVHFEs was also defined [5].

Several aggregation operators for IVHFEs such as, the Interval-Valued Hesitant Fuzzy Weighted Averaging (IVHFWA), Interval-Valued Hesitant Fuzzy Weighted Geometric (IVHFWG), Interval-Valued Hesitant Fuzzy Ordered Weighted Averaging (IVHFOWA), Interval-Valued Hesitant Fuzzy Ordered Weighted Geometric (IVHFOWG) and their generalizations were defined in [5, 59]. Different Einstein aggregation operators for IVHFEs have been presented in [58, 80]. In [20] was studied the Hamacher t-norms to extend and generalize the Hamacher operations for IVHFEs. Two induced generalized hybrid operators based on Shapley for IVHFEs have been defined in [28]. A set of continuous aggregation operators for IVHFEs are introduced in [30]. Some operations for IVHFEs based on Archimedean t-norms and t-conorms are presented in [4] as well as their properties.

Different correlations coefficient for IVHFEs have been introduced in [60].

In order to calculate the distance between two IVHFEs the Hamming, Euclidean and Hausdorff distances are extended to propose a variety of distance measures for IVHFEs [5, 56].

3.3 *Generalized Hesitant Fuzzy Sets*

Another extension of HFS is the Generalized Hesitant Fuzzy Set (GHFS) [31] which consists of representing the membership as the union of some A-IFS [1].

Definition 17 [31] Given a set of n membership functions:

$$M = \{\alpha_i = (\mu_i, \upsilon_i) | 0 \leq \mu_i, \upsilon_i \leq 1, 0 \leq \mu_i + \upsilon_i \leq 1, \quad i = \{1, \ldots, n\}\}, \tag{11}$$

the GHFS associated to M, $\mathfrak{h}_M$, is defined as follows:

$$\mathfrak{h}_M(x) = \cup_{(\mu_i(x), \upsilon_i(x)) \in M} (\mu_i(x), \upsilon_i(x)). \tag{12}$$

Remark 1 Notice that a GHFS extends slightly the concept of DHFS [79] as we can see in the following example.

Example 3 Let $X = \{x_1\}$ be a reference set, then

$$\mathfrak{h}_M(x_1) = \{(0.5, 0.3), (0.6, 0.3), (0.4, 0.5)\}$$

is a GHFS.

In this example, $\gamma^+ = 0.6$ and $\eta^+ = 0.5$, therefore $0.6 + 0.5 > 1$, it does not achieve the restriction to be a DHFS.

The complement, union and intersection of GHFSs, as well as, the envelope of a GHFS were presented in [31]. Some properties and relationships with HFSs were also discussed [31]. A comparison law was introduced to compare two GHFSs according to the score and consistency functions defined for this type of information. It was also proposed an extension principle which extends the operations for A-IFSs to GHFSs.

3.4 Hesitant Triangular Fuzzy Sets

Some authors [57, 66, 74] point out that in many decision making problems due to the increasing complexity of the socioeconomic environment and the uncertain information, it is difficult for experts to express the membership degrees of an element to a given set only by means of crisp values. Therefore, the concept of Hesitant Triangular Fuzzy Set (HTFS) was introduced as an extension of HFS where the membership degrees of an element to a fuzzy set are expressed by several triangular fuzzy numbers [18]. This concept has been proposed by different authors [57, 66, 74] with different names. Here it will be used HTFS.

Definition 18 [57, 66, 74] Let X be a fixed set, a HTFS $\tilde{E}$ on X is defined in terms of a function $\tilde{f}_{\tilde{E}}(x)$ that returns several triangular fuzzy values,

$$\tilde{E} = \{\langle x, \tilde{f}_{\tilde{E}}(x)\rangle | x \in X\} \tag{13}$$

where $\tilde{f}_{\tilde{E}}(x)$ is a set of several triangular fuzzy numbers which express the possible membership degrees of an element $x \in X$ to a set $\tilde{E}$. $\tilde{f}_{\tilde{E}}(x)$ is called Hesitant Triangular Fuzzy Element (HTFE) and noted $(\tilde{f})_{\tilde{E}}(x_i) = \{(\tilde{\xi}^L, \tilde{\xi}^M, \tilde{\xi}^U) | \tilde{\xi} \in \tilde{f}_{\tilde{E}}(x_i)\}$.

Example 4 Let $X = \{x_1, x_2\}$ be a reference set, a HTFS $\tilde{E}$, is defined by

$$\tilde{E} = \{\langle x_1, \{(0.1, 0.3, 0.5), (0.4, 0.6, 0.8)\}\rangle, \langle x_2, \{(0.1, 0.2, 0.3)\}\rangle\}.$$

Note that if $\tilde{\xi}^L = \tilde{\xi}^M = \tilde{\xi}^U$, then the HTFS is a HFS.

Some basic operations such as, the addition and multiplication of HTFEs were defined in [66]. A score function and an accuracy function were defined to propose a comparison law for HTFEs [57, 66].

Different aggregation operators for HTFEs such as, Hesitant Triangular Fuzzy Weighted Averaging (HTFWA), Hesitant Triangular Fuzzy Ordered Weighted Averaging (HTFOWA), Hesitant Triangular Fuzzy Weighted Geometric (HTFWG), Hesitant Triangular Fuzzy Ordered Weighted Geometric (HTFOWG), Hesitant Triangular Fuzzy Hybrid Average (HTFHA), Hesitant Triangular Fuzzy Hybrid Geometric (HTFHG) have been defined [57, 66]. A set of aggregation operators based on Bonferroni Mean have been introduced in [47]. The Einstein operation has been extended to propose a family of aggregation operators for HTFEs [38, 74]. Two different aggregation operators based on Choquet integral have been also proposed for HTFEs [21, 75].

This type of information has been applied to solve evaluation problems [21, 66].

4 Extensions of Hesitant Fuzzy Sets in Qualitative Settings

The previous section revises extensions of HFS defined in quantitative contexts, but the use of numbers to represent uncertain information is not always appropriate, and usually it is difficult to provide numerical values when the knowledge is vague and imprecise. Usually, experts involved in this type of problems use linguistic information to express their assessments regarding the uncertain knowledge that they have about the problem [26]. Therefore, different extensions about HFS have been proposed to model the experts' hesitancy in qualitative contexts. This section revises such extensions.

4.1 Hesitant Fuzzy Linguistic Term Sets

Most the linguistic approaches model the information by means of just one linguistic term, but sometimes experts might hesitate among several values to express their assessments because of the lack of information and knowledge about the problem. In order to cope with these hesitant situations Rodríguez et al. proposed the concept of Hesitant Fuzzy Linguistic Term Set (HFLTS) [32].

A HFLTS is defined as follows

Definition 19 [32] Let $S = \{s_0, \ldots, s_g\}$ be a linguistic term set, a HFLTS H_s, is defined as an ordered finite subset of consecutive linguistic terms of S:

$$H_S = \{s_i, s_{i+1}, \ldots, s_j\} \ \ such\ that\ \ s_k \in S,\ \ k \in \{i, \ldots, j\} \tag{14}$$

Example 5 Let $S = \{s_0 : nothing, s_1 : very\ low, s_2 : low, s_3 : medium, s_4 : high, s_5 : very\ high, s_6 : perfect\}$ be a linguistic term set and ϑ be a linguistic variable, then $H_S(\vartheta)$ defined by

$$H_S(\vartheta) = \{very\ low, low, medium\}$$

is a HFLTS.

Remark 2 The use of consecutive linguistic terms in HFLTS is because of a cognitive point of view in which in a discrete domain with a short number of terms (usually not more than 9) makes not sense to hesitate among arbitrary and total different linguistic terms, $\{low, high, very\ high\}$, and not hesitate in their middle terms. The use of comparative linguistic expressions [33] is a clear example of human beings' hesitancy. The natural representation of such comparative linguistic expressions in decision making is HFLTS.

Some basic operations for HFLTS, such as the complement, union and intersection and diverse properties were defined in [32]. It was also introduced the envelope of a HFLTS that was used to propose a comparison law for HFLTSs. Two symbolic aggregation operators, *min_upper* and *max_lower* were developed to aggregate HFLTSs [32].

The concept of HFLTS was introduced as something that can be used directly by experts to elicit several linguistic terms, but usually human beings do not provide their assessments in such a way. Therefore, Rodríguez et al. proposed the use of context-free grammars to generate linguistic expressions close to the natural language used by human beings that are easily represented by HFLTS. A context-free grammar G_H, that generates comparative linguistic expressions similar to the expressions used by experts in decision making problems was proposed in [33].

Definition 20 [33] Let G_H be a context-free grammar and $S = \{s_0, \ldots, s_g\}$ a linguistic term set. The elements of $G_H = (V_N, V_T, I, P)$ are defined as follows:

$$
\begin{aligned}
&V_N = \{\langle \textit{primary term} \rangle, \langle \textit{composite term} \rangle, \langle \textit{unary relation} \rangle, \langle \textit{binary relation} \rangle, \\
&\langle \textit{conjunction} \rangle\} \\
&V_T = \{\textit{lower than}, \textit{greater than}, \textit{at least}, \textit{at most}, \textit{between}, \textit{and}, s_0, s_1, \ldots, s_g\} \\
&I \in V_N \\
&P = \{I ::= \langle \textit{primary term} \rangle | \langle \textit{composite term} \rangle \\
&\langle \textit{composite term} \rangle ::= \langle \textit{unary relation} \rangle \langle \textit{primary term} \rangle | \langle \textit{binary relation} \rangle \\
&\langle \textit{primary term} \rangle \langle \textit{conjunction} \rangle \langle \textit{primary term} \rangle \\
&\langle \textit{primary term} \rangle ::= s_0 | s_1 | \ldots | s_g \\
&\langle \textit{unary relation} \rangle ::= \textit{lower than} | \textit{greater than} | \textit{at least} | \textit{at most} \\
&\langle \textit{binary relation} \rangle ::= \textit{between} \\
&\langle \textit{conjunction} \rangle ::= \textit{and}\}
\end{aligned}
$$

A transformation function E_{G_H} to obtain HFLTS from the comparative linguistic expressions was defined [32].

Even though the concept of HFLTS is quite novel, it has received a lot of attention by other researchers and different proposals based on this concept have been already presented in the literature. Wei et al. [55] have proposed a comparison method for HFLTS and two aggregation operators, the Hesitant Fuzzy Linguistic Weighted Averaging (HFLWA) and the Hesitant Fuzzy Linguistic Ordered Weighted Averaging (HFLOWA). A different comparison method and more aggregation operators were introduced in [19]. Some authors have studied consistency measures for Hesitant Fuzzy Linguistic Preference Relation (HFLPR) [73, 77]. A variety of distance and similarity measures for HFLTS has been also defined [14, 15, 22] and applied to multicriteria decision making problems.

Liu and Rodríguez pointed out [25] that the semantics of the comparative linguistic expressions based on a context-free grammar and HFLTSs should be represented by fuzzy membership functions instead of linguistic intervals [32]. Therefore, a new fuzzy representation for comparative linguistic expressions based on a fuzzy envelope has been introduced [25]. By using a consensus measure for

HFLTS, an optimization-based consensus model that minimizes the number of adjusted single terms in the consensus process has been recently proposed in [10].

Furthermore, different decision making approaches dealing with HFLTS have been proposed, such as TOPSIS [3], outranking [51], TODIM [54] and so on [6, 33]. And some real applications have been already presented, Sahu et al. [37] used HFLTS to classify documents and Yavuz et al. [64] presented a hierarchical multi-criteria decision making approach using HFLTS to manage complex problems, such as alternative-fuel vehicle selection.

4.2 Extended Hesitant Fuzzy Linguistic Term Sets

Recently, the concept of HFLTS has been generalized to deal with non-consecutive linguistic terms. This generalization has been presented in [48, 72] with different names, Extended Hesitant Fuzzy Linguistic Term Set (EHFLTS) and Hesitant Fuzzy Linguistic Set (HFLS) respectively. Although the names are different its definition is the same. Here we will use EHFLTS to refer this extension.

An EHFLTS is built by the union of several HFLTS. It is formally defined as follows.

Definition 21 [48] Let S be a linguistic term set, an ordered subset of linguistic terms of S, that is,

$$EH_S = \{s_i | s_i \in S\}, \tag{15}$$

is an EHFLTS.

Example 6 Let $S = \{s_{-3} : very\ poor, s_{-2} : poor, s_{-1} : slightly\ poor, s_0 : fair, s_1 : slightly\ good, s_2 : good, s_3 : very\ good\}$ be a linguistic term set and ϑ be a linguistic variable, then $EH_S(\vartheta)$ defined by

$$EH_S(\vartheta) = \{fair, good, very\ good\}$$

is an EHFLTS.

Some basic operations, such as the union and intersection of EHFLTSs, the complement of an EHFLTS and its envelope have been defined in [48]. Two families of aggregation operators to aggregate a set of EHFLTSs where weighting vectors take the form of real numbers and linguistic terms are also proposed [48]. Wang and Xu have studied the additive and weak consistency of a Extended Hesitant Fuzzy Linguistic Preference Relation (EHFLPR) [49].

Remark 3 It is worthy to note that the concept of EHFLTS is not used directly by experts, but experts involved in a decision making problem provide their preferences by using HFLTS and instead of carrying out an aggregation process, the group's preferences are formed by the union of such HFLTSs obtaining as result a EHFLTS.

4.3 Other Linguistic Extensions

Recently, different linguistic extensions of HFS, DHFS and IVHFS have been introduced. Although they are not well known, their concepts and an example to understand them easily are presented.

In [24] was presented the concept of Hesitant Fuzzy Linguistic Set (HFLS) as follows.

Definiton 22 Let X be a reference set, a HFLS on X is a function that returns a subset of values in [0,1]. It is expressed by a mathematical symbol as follows:

$$A = (\langle x, s_{\theta(x)}, h_A(x)\rangle | x \in X) \tag{16}$$

where $h_A(x)$ is a set of some values in [0,1] denoting the possible membership degrees of the element $x \in X$ to the linguistic term $s_{\theta(x)}$.

Example 7 Let $X = \{x_1, x_2\}$ be a reference set and $S = \{s_0 : nothing, s_1 : very\ low, s_2 : low, s_3 : medium, s_4 : high, s_5 : very\ high, s_6 : perfect\}$ be a linguistic term set, then A defined by

$$A = \{\langle x_1, s_1, \{0.3, 0.4, 0.5\}\rangle, \langle x_2, s_3, \{0.3, 0.5\}\rangle\}$$

is a HFLS.

The concept of Interval-Valued Hesitant Fuzzy Linguistic Set (IVHFLS) has been proposed as an extension of IVHFS based on linguistic term sets.

Definition 23 [52] Let X be a reference set, a IVHFLS on X is an object:

$$B = (\langle x, s_{\theta(x)}, \Gamma_B(x)\rangle | x \in X) \tag{17}$$

where $\Gamma_B(x)$ is a set of finite numbers of closed intervals belonging to (0,1] and it denotes the possible interval-valued membership degrees that x belongs to $s_{\theta(x)}$.

Example 8 Let $X = \{x_1, x_2\}$ be a reference set and $S = \{s_0 : nothing, s_1 : very\ low, s_2 : low, s_3 : medium, s_4 : high, s_5 : very\ high, s_6 : perfect\}$ be a linguistic term set, a IVHFLS might be

$$B = \{\langle x_1, s_5, \{[0.4, 0.5], [0.6, 0.7], [0.7, 0.8]\}\rangle, \langle x_2, s_6, \{[0.1, 0.3], [0.5, 0.6]\}\rangle\}.$$

Yang and Ju have extended also the concept DHFS by using linguistic terms.

Definition 24 [63] Let X be a reference set, a DHFLS on X is described as:

$$C = (\langle x, s_{\theta(x)}, h(x), g(x)\rangle | x \in X) \tag{18}$$

where $s_{\theta(x)} \in S$, $h(x)$ and $g(x)$ are two sets of some values in [0,1] denoting the possible membership degrees and non-membership degrees of the element $x \in X$ to the linguistic term $s_{\theta(x)}$ with the following conditions:

$$0 \leq \gamma, \eta \leq 1, 0 \leq \gamma^+ + \eta^+ \leq 1,$$

where $\gamma \in h(x)$, $\eta \in g(x)$, $\gamma^+ = \max_{\gamma \in h(x)}\{\gamma\}$, and $\eta^+ = \max_{\eta \in g(x)}\{\eta\} \forall x \in X$.

Example 9 Let $X = \{x_1, x_2\}$ be a reference set and $S = \{s_0 : \textit{nothing}, s_1 : \textit{very low}, s_2 : \textit{low}, s_3 : \textit{medium}, s_4 : \textit{high}, s_5 : \textit{very high}, s_6 : \textit{perfect}\}$ be a linguistic term set, a DHFLS might be

$$C = \{\langle x_1, s_3, \{0.4, 0.5\}, \{0.3, 0.4\}\rangle, \langle x_2, s_4, \{0.3, 0.5\}, \{0.2, 0.3\}\rangle\}.$$

5 Trends and Discussions

Due to the usefulness of modelling hesitancy uncertainty in real-world problems, the research of HFS and its extensions have been intensive and extensive researched in the recent years despite the young of this concept. Because of this interest many proposals related with its use and extensions have been developed in the literature. However, some critical points and comments about the new concepts and tools based on HFS must be pointed out in order to clarify better the trends and future direction on this topic:

- Any HFS extension should be clearly justified from a theoretical or practical point of view and solve real-problems with uncertainty. So far, the usefulness of some extensions of HFS is debatable because it is not clear either why are necessary? or in what type of real-world problems can be used?
- It is remarkable that the same concepts have been published by different authors in different papers. Despite the quick growth of research on this topic, it is necessary to make a deep revision of the related literature in order to avoid repeating several times the same concepts, operators and so forth.
- As it has been argued in Sect. 2, some operators for HFS defined in a straightforward way from operators on fuzzy sets inherit their properties. The study of the properties have to take into account this fact. In addition, it is possible to define new operators that are not just extensions of operators on fuzzy sets. This line has not yet been much explored.

6 Conclusions

The introduction of HFS by Torra (see Sect. 2) has attracted the attention of many researchers that have found a new way to model the uncertainty related to the hesitation of human beings in real-world problems when they are not sure about their knowledge.

This chapter reviews the ground concepts and ideas of HFS, its basic operations and the most important extensions that have been presented in the literature about it both quantitative and qualitative way. After such a revision, it points out some important aspects that must be taken into account when new operators and extensions about HFS are developed in order to avoid some disfunctions that have been found in some of the current proposals. Eventually some hints about future directions in HFS research are sketched.

Acknowledgements This work is partially funded by the Spanish research projects TIN2011-27076-C03-03, TIN2015-66524-P, the Spanish Ministry of Economy and Finance Postdoctoral Training (FPDI-2013-18193) and ERDF.

References

1. Atanassov, K.T.: Intuitionistic fuzzy sets. Fuzzy Sets Syst. **20**, 87–96 (1986)
2. Bedregal, B., Reiser, R., Bustince, H., López-Molina, C., Torra, V.: Aggregating functions for typical hesitant fuzzy elements and the action of automorphisms. Inf. Sci. **256**(1), 82–97 (2014)
3. Beg, I., Rashid, T.: TOPSIS for hesitant fuzzy linguistic term sets. Int. J. Intell. Syst. **28**, 1162–1171 (2013)
4. Chen, N., Xu, Z.S.: Properties of interval-valued hesitant fuzzy sets. J. Intell. Fuzzy Syst. **27** (1), 143–158 (2014)
5. Chen, N., Xu, Z.S., Xia, M.M.: Interval-valued hesitant preference relations and their applications to group decision making. Knowl. Based Syst. **37**(1), 528–540 (2013)
6. Chen, S.M., Hong, J.A.: Multicriteria linguistic decision making based on hesitant fuzzy linguistic term sets and the aggregation of fuzzy sets. Inf. Sci. **286**, 63–74 (2014)
7. Chen, Y., Penga, X., Guanb, G., Jiangb, H.: Approaches to multiple attribute decision making based on the correlation coefficient with dual hesitant fuzzy information. J. Intell. Fuzzy Syst. **26**(5), 2547–2556 (2014)
8. Chiclana, F., Herrera, F., Herrera-Viedma, E.: Integrating multiplicative preference relations in a multipurpose decision-making model based on fuzzy preference relations. Fuzzy Sets Syst. **122**(2), 277–291 (2001)
9. Crawford, G., Williams, C.: A note on the analysis of subjective judgment matrices. J. Math. Psychol. **29**(4), 387–405 (1985)
10. Dong, Y., Chen, X., Herrera, F.: Minimizing adjusted simple terms in the consensus reaching process with hesitant linguistic assessments in group decision making. Inf. Sci. **297**, 95–117 (2015)
11. Dubois, D., Gottwald, S., Hajek, P., Kacprzyk, J., Prade, H.: Terminological difficulties in fuzzy set theory—the case of "intuitionistic fuzzy sets". Fuzzy Sets Syst. **156**(3), 485–491 (2005)

12. Farhadinia, B.: Information measures for hesitant fuzzy sets and interval-valued hesitant fuzzy sets. Inf. Sci. **240**, 129–144 (2013)
13. Goguen, J.A.: L-fuzzy sets. J. Math. Anal. Appl. **18**(1), 145–174 (1967)
14. Hesamian, G., Shams, M.: Measuring similarity and ordering based on hesitant fuzzy linguistic term sets. J. Intell. Fuzzy Syst. **28**(2), 983–990 (2015)
15. Huang, H.C., Yang, X.: Pairwise comparison and distance measure of hesitant fuzzy linguistic term sets. Math. Probl. Eng. **1–8**, 2014 (2014)
16. Ju, Y., Yang, S., Liu, X.: Some new dual hesitant fuzzy aggregation operators based on Choquet integral and their applications to multiple attribute decision making. J. Intell. Fuzzy Syst. **27**(6), 2857–2868 (2014)
17. Ju, Y., Zhang, W., Yang, S.: Some dual hesitant fuzzy hamacher aggregation operators and their applications to multiple attribute decision making. J. Intell. Fuzzy Syst. **27**(5), 2481–2495 (2014)
18. Klir, G.J., Yuan, B.: Fuzzy Sets and Fuzzy Logic: Theory and Applications. Prentice-Hall PTR (1995)
19. Lee, L.W., Chen, S.M.: Fuzzy decision making based on likelihood-based comparison relations of hesitant fuzzy linguistic term sets and hesitant fuzzy linguistic operators. Inf. Sci. **294**, 513–529 (2015)
20. Li, L.G., Peng, D.H.: Interval-valued hesitant fuzzy hamacher synergetic weighted aggregation operators and their application to shale gas areas selection. Math. Probl. Eng. **1–25**, 2014 (2014)
21. Li, Y.B., Zhang, J.P.: Approach to multiple attribute decision making with hesitant triangular fuzzy information and their application to customer credit risk assessment. J. Intell. Fuzzy Syst. **26**(6), 2853–2860 (2014)
22. Liao, H., Xu, Z.S., Zeng, X.J.: Distance and similarity measures for hesitant fuzzy linguistic term sets and their application in multi-criteria decision making. Inf. Sci. **271**, 125–142 (2014)
23. Liao, H.C., Xu, Z.S.: A VIKOR-based method for hesitant fuzzy multi-criteria decision making. Fuzzy Optim. Decis. Making **12**, 373–392 (2013)
24. Lin, R., Zhao, X., Wei, G.: Models for selecting an ERP system with hesitant fuzzy linguistic information. J. Intell. Fuzzy Syst. **26**(5), 2155–2165 (2014)
25. Liu, H., Rodríguez, R.M.: A fuzzy envelope for hesitant fuzzy linguistic term set and its application to multicriteria decision making. Inf. Sci. **258**, 266–276 (2014)
26. Martínez, L., Liu, J., Yang, J.B., Herrera, F.: A multigranular hierarchical linguistic model for design evaluation based on safety and cost analysis. Int. J. Intell. Syst. **20**(12), 1161–1194 (2005)
27. Mendel, J.M., John, R.I.: Type-2 fuzzy sets made simple. IEEE Trans. Fuzzy Syst. **10**(2), 117–127 (2002)
28. Meng, F., Chen, X.: An approach to interval-valued hesitant fuzzy multi-attribute decision making with incomplete weight information based on hybrid Shapley operators. Informatica **25**(4), 617–642 (2014)
29. Pei, Z., Yi, L.: A note on operations of hesitant fuzzy sets. Int. J. Comput. Intell. Syst. **8**(2), 226–239 (2015)
30. Peng, D.H., Wang, T.D., Gao, C.Y., Wang, H.: Continuous hesitant fuzzy aggregation operators and their application to decision making under interval-valued hesitant fuzzy setting. Sci. World J. **1–20**, 2014 (2014)
31. Qian, G., Wang, H., Feng, X.: Generalized hesitant fuzzy sets and their application in decision support system. Knowl. Based Syst. **37**(1), 357–365 (2013)
32. Rodríguez, R.M., Martínez, L., Herrera, F.: Hesitant fuzzy linguistic term sets for decision making. IEEE Trans. Fuzzy Syst. **20**(1), 109–119 (2012)
33. Rodríguez, R.M., Martínez, L., Herrera, F.: A group decision making model dealing with comparative linguistic expressions based on hesitant fuzzy linguistic term sets. Inf. Sci. **241** (1), 28–42 (2013)
34. Rodríguez, R.M., Martínez, L., Torra, V., Xu, Z.S., Herrera, F.: Hesitant fuzzy sets: state of the art and future directions. Int. J. Intell. Syst. **29**(6), 495–524 (2014)

35. Saaty, L.T.: The Analytic Hierarchy Process. McGraw-Hill (1980)
36. Saaty, R.W.: The analytic hierarchy process-what it is and how it is used. Math. Model. **9**(3–5), 161–176 (1987)
37. Sahu, S.K., Sahu, N., Thakur, R.S., Thakur, G.S.: Hesitant fuzzy linguistic term set based document classification. In: Proceedings of the International Conference on Communication Systems and Network Technologies, pp. 586–590, Gwalior, India (2013)
38. Shi, J., Meng, C., Liu, Y.: Approach to multiple attribute decision making based on the intelligence computing with hesitant triangular fuzzy information and their application. J. Intell. Fuzzy Syst. **27**(2), 701–707 (2014)
39. Singh, P.: A new method for solving dual hesitant fuzzy assignment problems with restrictions based on similarity measure. Appl. Soft Comput. **24**, 559–571 (2014)
40. Torra, V.: Hesitant fuzzy sets. Int. J. Intell. Syst. **25**(6), 529–539 (2010)
41. Torra, V.: Artificial intelligence research and development, chapter on the derivation of weights using the geometric mean approach for set-valued matrices. In: Frontiers in Artificial Intelligence and Applications, pp. 193–201 (2014)
42. Torra, V.: Derivation of priorities and weights for set-valued matrices using the geometric mean approach. Appl. Artif. Intell. **29**(5), 500–513 (2015)
43. Torra, V., Narukawa, Y.: Modeling decisions: information fusion and aggregation operators. Springer, Heidelberg (2007)
44. Torra, V., Narukawa, Y.: On hesitant fuzzy sets and decision. In: Proceedings of the 18th IEEE International Conference on Fuzzy Systems, pp. 1378–1382 (2009)
45. Türksen, I.B.: Interval valued fuzzy sets based on normal forms. Fuzzy Sets Syst. **20**, 191–210 (1986)
46. Wang, C., Li, Q., Zhou, X.: Multiple attribute decision making based on generalized aggregation operators under dual hesitant fuzzy environment. J. Appl. Math. **1–12**, 2014 (2014)
47. Wang, C., Li, Q., Zhou, X., Yang, T.: Hesitant triangular fuzzy information aggregation operators based on bonferroni means and their application to multiple attribute decision making. Sci. World J. **1–15**, 2014 (2014)
48. Wang, H.: Extended hesitant fuzzy linguistic term sets and their aggregation in group decision making. Int. J. Comput. Intell. Syst. **8**(1), 14–33 (2015)
49. Wang, H., Xu, Z.S.: Some consistency measures of extended hesitant fuzzy linguistic preference relations. Inf. Sci. **297**, 316–331 (2015)
50. Wang, H., Zhao, X., Wei, G.: Dual hesitant fuzzy aggregation operators in multiple attribute decision making. J. Intell. Fuzzy Syst. **26**(5), 2281–2290 (2014)
51. Wang, J.Q., Wang, J., Chen, Q.H., Zhang, H.Y., Chen, X.H.: An outranking approach for multi-criteria decision-making with hesitant fuzzy linguistic term sets. Inf. Sci. **280**, 338–351 (2014)
52. Wang, J.Q., Wu, J.T., Wang, J., Zhang, H.Y., Chen, X.H.: Interval-valued hesitant fuzzy linguistic sets and their applications in multi-criteria decision-making problems. Inf. Sci. **288**, 55–72 (2014)
53. Wang, L., Ni, M., Zhu, L.: Correlation measures of dual hesitant fuzzy sets. J. Appl. Math. **1–12**, 2013 (2013)
54. Wei, C., Ren, Z., Rodríguez, R.M.: A hesitant fuzzy linguistic TODIM method based on a score function. Int. J. Comput. Intell. Syst. **8**(4), 701–712 (2015)
55. Wei, C., Zhao, N., Tang, X.: Operators and comparisons of hesitant fuzzy linguistic term sets. IEEE Trans. Fuzzy Syst. **22**(3), 575–585 (2014)
56. Wei, G., Lin, R., Wang, H.: Distance and similarity measures for hesitant interval-valued fuzzy sets. J. Intell. Fuzzy Syst. **27**(1), 19–36 (2014)
57. Wei, G., Wang, H., Zhao, X., Lin, R.: Hesitant triangular fuzzy information aggregation in multiple attribute decision making. J. Intell. Fuzzy Syst. **26**(3), 1201–1209 (2014)
58. Wei, G., Zhao, X.: Induced hesitant interval-valued fuzzy Einstein aggregation operators and their application to multiple attribute decision making. J. Intell. Fuzzy Syst. **24**(4), 789–803 (2013)

59. Wei, G., Zhao, X., Lin, R.: Some hesitant interval-valued fuzzy aggregation operators and their applications to multiple attribute decision making. Knowl. Based Syst. **46**, 43–53 (2013)
60. Wei, G., Zhao, X., Lin, R.: Models for hesitant interval-valued fuzzy multiple attribute decision making based on the correlation coefficient with incomplete weight information. J. Intell. Fuzzy Syst. **26**(4), 1631–1644 (2014)
61. Xia, M.M., Xu, Z.S.: Hesitant fuzzy information aggregation in decision making. Int. J. Approx. Reason. **52**, 395–407 (2011)
62. Xia, M.M., Xu, Z.S.: Managing hesitant information in GDM problems underfuzzy and multiplicative preference relations. Int. J. Uncertain. Fuzziness Knowl. Based Syst. **21**(06), 865–897 (2013)
63. Yang, S., Ju, Y.: Dual hesitant fuzzy linguistic aggregation operators and their applications to multi-attribute decision making. J. Intell. Fuzzy Syst. **27**(4), 1935–1947 (2014)
64. Yavuz, M., Oztaysi, B., Cevik Onar, S., Kahraman, C.: Multi-criteria evaluation of alternative-fuel vehicles via a hierarchical hesitant fuzzy linguistic model. Expert Syst. Appl. **42**(5), 2835–2848 (2015)
65. Ye, J.: Correlation coefficient of dual hesitant fuzzy sets and its application to multiple attribute decision making. Appl. Math. Model. **38**(2), 659–666 (2014)
66. Yu, D.: Triangular hesitant fuzzy set and its application to teaching quality evaluation. J. Inf. Comput. Sci. **10**(7), 1925–1934 (2013)
67. Yu, D.: Some generalized dual hesitant fuzzy geometric aggregation operators and applications. Int. J. Uncertain. Fuzziness Knowl. Based Syst. **22**(3), 367–384 (2014)
68. Yu, D., Li, D.F.: Dual hesitant fuzzy multi-criteria decision making and its application to teaching quality assessment. J. Intell. Fuzzy Syst. **27**(4), 1679–1688 (2014)
69. Zadeh, L.: Fuzzy sets. Inf. Control **8**, 338–353 (1965)
70. Zhang, Y.: Research on the computer network security evaluation based on the DHFHCG operator with dual hesitant fuzzy information. J. Intell. Fuzzy Syst. **28**(1), 199–204 (2015)
71. Zhang, Z., Wu, C.: A decision support model for group decision making with hesitant multiplicative preference relations. Inf. Sci. **282**, 136–166 (2014)
72. Zhang, Z., Wu, C.: Hesitant fuzzy linguistic aggregation operators and their applications to multiple attribute group decision making. J. Intell. Fuzzy Syst. **26**(5), 2185–2202 (2014)
73. Zhang, Z., Wu, C.: On the use of multiplicative consistency in hesitant fuzzy linguistic preference relations. Knowl. Based Syst. **72**, 13–27 (2014)
74. Zhao, X., Lin, R., Wei, G.: Hesitant triangular fuzzy information aggregation based on Einstein operations and their application to multiple attribute decision making. Expert Syst. Appl. **41**(4, Part 1), 1086–1094
75. Zhong, G., Xu, L.: Models for multiple attribute decision making method in hesitant triangular fuzzy setting. J. Intell. Fuzzy Syst. **26**(5), 2167–2174 (2014)
76. Zhu, B., Xu, Z.S.: Regression methods for hesitant fuzzy preference relations. Technol. Econ. Dev. Econ. 19, S214–S227 (2013)
77. Zhu, B., Xu, Z.S.: Consistency measures for hesitant fuzzy linguistic preference relations. IEEE Trans. Fuzzy Syst. **22**, 35–45 (2014)
78. Zhu, B., Xu, Z.S.: Stochastic preference analysis in numerical preference relations. Eur. J. Oper. Res. **237**(2), 628–633 (2014)
79. Zhu, B., Xu, Z.S., Xia, M.M.: Dual hesitant fuzzy sets. J. Appl. Math. **1–13**, 2012 (2012)
80. Zhu, J.Q., Fu, F., Yin, K.X., Luo, J.Q., Wei, D.: Approaches to multiple attribute decision making with hesitant interval-valued fuzzy information under correlative environment. J. Intell. Fuzzy Syst. **27**(2), 1057–1065 (2014)

Type-1 to Type-n Fuzzy Logic and Systems

M.H. Fazel Zarandi, R. Gamasaee and O. Castillo

Abstract In this chapter, the motivation for using fuzzy systems, the mathematical concepts of type-1 to type-n fuzzy sets, logic, and systems as well as their applications in solving real world problems are presented.

Keywords Type-n fuzzy sets · Type-2 fuzzy sets · T-norm · S-norm · Takagi-Sugeno-Kang (TSK) fuzzy system

1 Introduction

Fuzzy logic and sets has been proposed by Zadeh [1]. According to Zadeh, fuzzy logic is a precise logic of imprecision and approximate reasoning. It is used to reason and make logical decisions in the presence of uncertainty, imprecision, and imperfect information. In addition, it is capable of modeling problems in case of no measurements [2]. Since real world problems are mostly involved with uncertainty and imperfect information, fuzzy logic and fuzzy sets are required to model and solve those problems.

Therefore, Zadeh [1] introduced the first type of fuzzy logic to be used in the abovementioned problems. In type-1 fuzzy logic, the traditional view to a set which believes that membership of an element to a set is either zero or one is relaxed. In other words, membership of an element to a set is a matter of degree. This mem-

M.H. Fazel Zarandi (✉) · R. Gamasaee
Department of Industrial Engineering, Amirkabir University of Technology, Tehran, Iran
e-mail: zarandi@aut.ac.ir

M.H. Fazel Zarandi
Knowledge Intelligent Systems Laboratory, University of Toronto, Toronto, Canada

O. Castillo
Division of Graduate Studies, Tijuana Institute of Technology, Calzada Tecnologico S/N, Fracc., Tomas Aquino Tijuana, B.C., Mexico

C. Kahraman et al. (eds.), *Fuzzy Logic in Its 50th Year*,
Studies in Fuzziness and Soft Computing 341,
DOI 10.1007/978-3-319-31093-0_6

bership degree is a certain single value for an element. However, there are some situations in which uncertainty of the data is too high. When the degree of information vagueness is too high, membership functions (MFs) are not certain, and they encounter uncertainty and volatility. Therefore, type-1 fuzzy membership functions are not applicable in this case since they are a certain value.

In order to model and solve problems with higher degree of vagueness and uncertain MFs, Zadeh [3] introduced type-2 fuzzy sets as an extension to type-1 fuzzy sets. In type-2 fuzzy sets, each MF is represented by another MF known as secondary MF. Type-2 fuzzy sets have been proposed by Zadeh [3] to tackle the drawbacks of type-1 fuzzy sets when uncertainty is too high. Three ways in which such uncertainty can occur are: (1) the words that are used in antecedents and consequents of rules can mean different things to different people, (2) consequents obtained by polling a group of experts will often be different for the same rule because the experts will not necessarily be in agreement, and (3) noisy training data (Liang and Mendel [4]). However, due to the various degrees of uncertainty, different kinds of type-2 fuzzy sets are required.

There are two kinds of type-2 fuzzy sets: (I) interval type-2 fuzzy sets (IT2 FSs); (II) general type-2 fuzzy sets (GT2 FSs). In IT2 FSs, MFs are interval instead of a single value. They are special type of GT2 FSs whose secondary membership values for the entire members of the primary domain are one. GT2 FSs are extensions of IT2 FSs, and they are bivariate interval valued [0, 1] supporting more degrees of freedom [5]. However, IT2 FSs are less computationally expensive than GT2 FSs.

In this chapter, fundamental concepts of type-1 fuzzy sets and mathematical operations on type-1 fuzzy sets are briefly reviewed in Sects. 2 and 3 respectively. Then fuzzy logic is described in Sect. 4. Thereafter, since approximate reasoning is required to apply fuzzy set theory in real world problems, it is presented in Sect. 5. In order to model real world problems, the use of fuzzy systems is inevitable, so they are illustrated in Sect. 6. Type-2 fuzzy sets and type-2 fuzzy systems are described in Sects. 7 and 8 respectively. Then, the applications of type-2 fuzzy systems are reviewed in Sect. 9. Finally, type-n fuzzy systems are presented in Sect. 10.

2 Type-1 Fuzzy Sets

Type-1 fuzzy logic assigns a membership degree in the interval of [0, 1] to the objects indicating the degree to which each object belongs to the fuzzy set. Consider two fuzzy sets X and A, where, X is a universe of discourse whose element is shown by x. A membership function of element x is denoted by $\mu_A(x)$ as it is stated in (1), the following equations are mostly adapted from Celikyilmaz and Türksen [6].

$$\mu_A(x): X \to [0, 1] \tag{1}$$

Fuzzy sets are represented in discrete and continuous forms. If fuzzy set A is defined on a discrete universe of discourse X, it is represented by (2).

$$A = \sum_{x \in X} \mu_A(x)/_x \tag{2}$$

where, $\sum_{x \in X}$ is an aggregation operator indicating all membership degrees pertain to a fuzzy set A and universe of discourse X. If fuzzy set A is defined on a continuous universe of discourse X, it is defined by (3).

$$A = \int_{x \in X} \mu_A(x)/_x \tag{3}$$

where, $\int_{x \in X}$ is not an integration operation but an aggregation operation. Other important operations in fuzzy sets are α-cut (α_A) and strong α-cut (α_A^+) as shown in the following equations.

$$\alpha_A = \{x | \mu_A(x) \geq \alpha\} \alpha \in [0, 1] \tag{4}$$

$$\alpha_A^+ = \{x | \mu_A(x) > \alpha\} \alpha \in [0, 1] \tag{5}$$

α-cut decomposition of fuzzy set A on U is shown by (6).

$$\mu_A(U) = sup \min\{\alpha, \mu_{\alpha_A}(u)\}, \forall u \in U \quad \alpha \in [0, 1]$$

where, $\mu_{\alpha_A}(u) \in \{0, 1\}$ is a MF of a crisp set.

Another useful definition is the "support" of a fuzzy set. Let X be a universe of discourse and A be a fuzzy set. "Support" of A shows all the elements of X with non-zero membership grades in A, as it is indicated in (6).

$$Supp(A) = \{x \in X | \mu_A(x) > 0\} \tag{6}$$

3 Mathematical Operations on Type-1 Fuzzy Sets

In this section, mathematical operations on fuzzy sets are reviewed.

Let A and B be two fuzzy sets; then, the "equality of fuzzy sets" is defined by the following equation.

$$A = B \Leftrightarrow \mu_A(x) = \mu_B(x) \tag{7}$$

Equation (8) shows "inclusion of fuzzy sets" A and B.

$$A \subseteq B \Leftrightarrow \mu_A(x) \leq \mu_B(x), \forall x \in S \tag{8}$$

Equation (9) indicates the "union of fuzzy sets" A and B.

$$A \cup B : \mu_{A \cup B}(x) = \max[\mu_A(x), \mu_B(x)] \tag{9}$$

Equation (10) shows the "intersection of fuzzy sets" A and B.

$$A \cap B : \mu_{A \cap B}(x) = \min[\mu_A(x), \mu_B(x)] \tag{10}$$

Equation (11) is the "Complement" of the fuzzy set A (Zadeh [1]).

$$\mu_{AC}(x) = 1 - \mu_A(x) \tag{11}$$

3.1 *Fuzzy Intersection (T-norm)*

Different types of logical operators other than "min" have been defined in literature known as "t-norm" operators. The following axioms have been defined by Klir and Yuan [7] for fuzzy t-norm in which $a, b, d \in [0, 1]$.

(I) $i(a, 1) = a$ (boundary condition)
(II) $b \leq d$ implies $i(a, b) \leq i(a, d)$ (monotonicity).
(III) $i(a, b) = i(b, a)$ (commutativity)
(IV) $i(a, i(b, d)) = i(i(a, b), d)$ (associativity)
Sometimes considering other restrictions on defining fuzzy t-norms leads to better results. Therefore, the following additional axioms should be satisfied for defining a fuzzy t-norm (Klir and Yuan [7]).
(V) i is a continuous function (continuity).
(VI) $i(a, b) < a$ (subidempotency)
(VII) $a_1 < a_2$ and $b_1 < b_2$ implies $i(a_1, b_1) < i(a_2, b_2)$ (strict monotonicity)

Table 1 shows different classes of t-norms applied in literature, and Fig. 1 depicts some of them.

The boundary condition in axiom (I) means if an argument of a t-norm is 1, membership degree of the t-norm is equal to the other argument. Commutativity means the fuzzy t-norm is symmetric. Monotonicity of t-norm indicates that any decrease in membership degrees in set A or B will not increase the membership value in t-norm. The axiom (IV) shows that order of numbers in t-norm is not important.

Table 1 Different types of fuzzy t-norms (Klir and Yuan [7])

Authors	Reference	Year	t-norm
Dombi	[8]	1982	$\left\{\left[\left(\frac{1}{a}-1\right)^{\lambda}+\left(\frac{1}{b}-1\right)^{\lambda}\right]^{1/\lambda}\right\}^{-1}$
Frank	[9]	1979	$\log[1+\frac{(s^{a}-1)(s^{b}-1)}{s-1}]$
Hamacher	–	1987	$\frac{ab}{r+(1-r)(a+b-ab)}$
Schweizer and Sklar	[10]	1963	$\{\max(0,a^{p}+b^{p}-1)\}^{\frac{1}{p}}$
Schweizer and Sklar	[11]		$1-[(1-a)^{p}+(1-b)^{p}-(1-a)^{p}(1-b)^{p}]^{\frac{1}{p}}$
Schweizer and Sklar	[12]		$\exp(-(\lvert\ln a\rvert^{p}+\lvert\ln b\rvert^{p})^{\frac{1}{p}})$
Schweizer and Sklar	[13]		$\frac{ab}{[a^{p}+b^{p}-a^{p}b^{p}]^{\frac{1}{p}}}$
Yager	[14]	1980	$1-\min\{1,[(1-a)^{\omega}+(1-b)^{\omega}]^{\frac{1}{\omega}}\}$
Dubois and Prade	[15]	1980	$\frac{ab}{\max(a,b,a)}$
Weber	[16]	1983	$\max\left(0,\frac{a+b+\lambda ab-1}{1+\lambda}\right)$
Yu	[17]	1985	$\max[0,(1+\lambda)(a+b-1)-\lambda ab]$

Some of the common t-norms applied in literature are as follows:

- Standard t-norms: $i(a,b)=\min(a,b)$
- Algebraic product: $i(a,b)=ab$
- Bounded difference: $i(a,b)=\max(0,a+b-1)$
- Drastic t-norm: $i(a,b)=\begin{cases} a & when\ b=1 \\ b & when\ a=1 \\ 0 & otherwise \end{cases}$

3.2 *Fuzzy Union (T-conorm)*

Different types of logical operators other than “max” have been defined in literature known as “t-conorm” operators. The following axioms have been defined by Klir and Yuan [7] for fuzzy t-conorm in which $a,b,d\in[0,1]$.

(I) $u(a,0)=a$ (boundary condition).
(II) $b\leq d$ implies $u(a,b)\leq u(a,d)$ (monotonicity).
(III) $u(a,b)=u(b,a)$ (commutativity).
(IV) $u(a,u(b,d))=u(u(a,b),d)$ (associativity).
There are some additional axioms that must be satisfied to form a t-conorm as follows. Table 2 shows different types of t-conorms, and Fig. 2 depicts some of them.
(V) u is a continuous function (continuity).
(VI) $u(a,a)>a$ (superidempotency).
(VII) $a_1<a_2$ and $b_1<b_2$ implies $u(a_1,b_1)<u(a_2,b_2)$ (strict monotonicity)

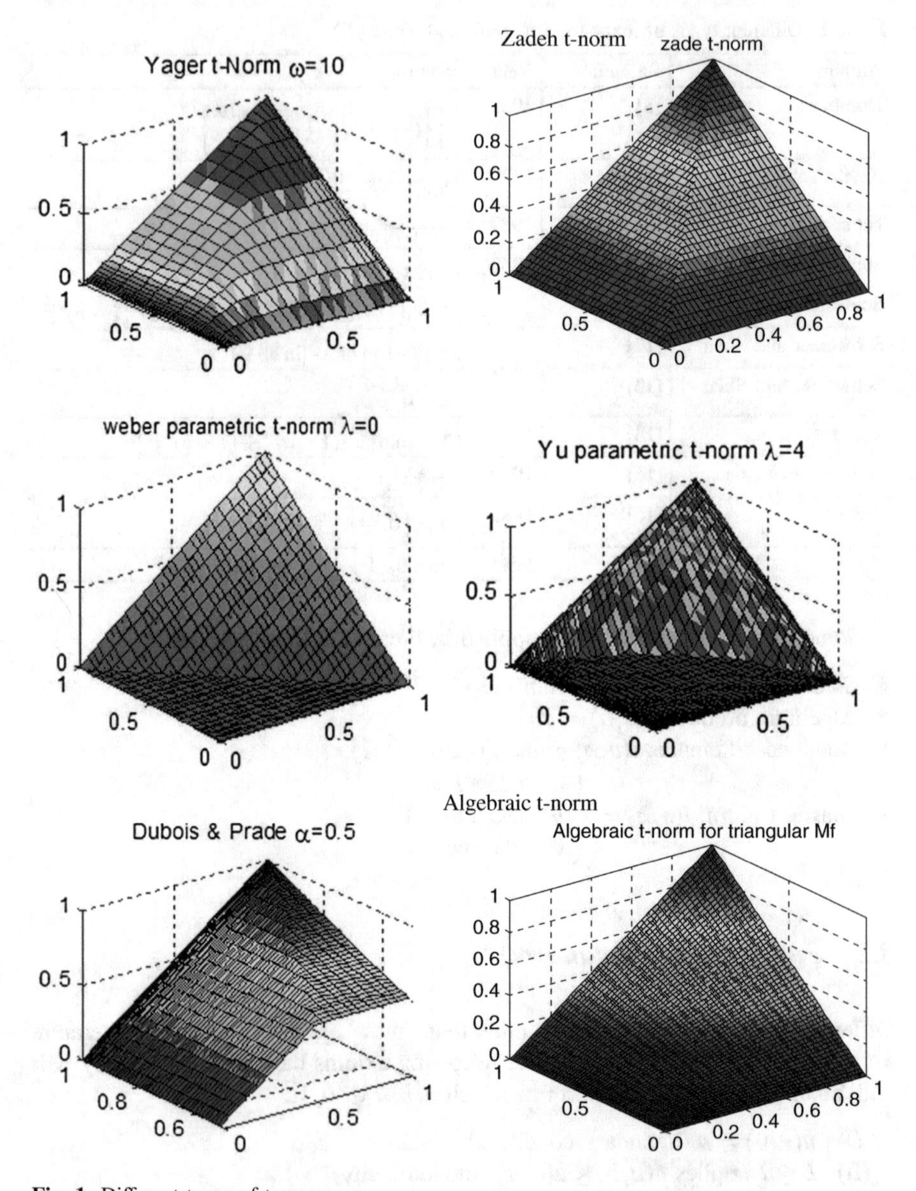

Fig. 1 Different types of t-norms

Some of the common t-conorms applied in literature are as follows:

- Standard t-conorms: $u(a,b) = \max(a,b)$
- Algebraic sum: $u(a,b) = a + b - ab$

Table 2 Different types of fuzzy t-conorms (Klir and Yuan [7])

Authors	Reference	Year	
Dombi	[8]	1982	$\left\{1+\left[\left(\frac{1}{a}-1\right)^{\lambda}+\left(\frac{1}{b}-1\right)^{\lambda}\right]^{\frac{-1}{\lambda}}\right\}^{-1}$
Frank	[9]	1979	$1-\log_s\left[1+\frac{\left(s^{1-a}-1\right)\left(s^{1-b}-1\right)}{s-1}\right]$
Hamacher	–	1978	$\frac{a+b+(r-2)ab}{r+(r-1)ab}$
Schweizer and Sklar	[10]	1963	$1-\{\max(0,(1-a)^p+(1-b)^p-1)\}^{\frac{1}{p}}$
Schweizer and Sklar	[11]		$[a^p+b^p-a^pb^p]^{\frac{1}{p}}$
Schweizer and Sklar	[12]		$1-\exp\left(-\left(\lvert\ln(1-a)\rvert^p+\lvert\ln(1-b)\rvert^p\right)^{\frac{1}{p}}\right)$
Schweizer and Sklar	[13]		$1-\frac{(1-a)(1-b)}{[(1-a)^p+(1-b)^p-(1-a)^p(1-b)^p]^{\frac{1}{p}}}$
Yager	[14]	1980	$\min\left[1,\left(a^{\lambda}+b^{\lambda}\right)^{\frac{1}{\lambda}}\right]$
Dubois and Prade	[15]	1980	$1-\frac{(1-a)(1-b)}{\max((1-a),(1-b),a)}$
Weber	[16]	1983	$\min\left(1,a+b-\frac{\lambda}{1-\lambda}ab\right)$
Yu	[17]	1985	$\min(1,a+b+\lambda ab)$

- Bounded sum: $u(a,b)=\min(1,a+b)$
- Drastic t-conorm: $u(a,b)=\begin{cases} a & \textit{when } b=0 \\ b & \textit{when } a=0 \\ 1 & \textit{otherwise} \end{cases}$

4 Fuzzy Logic

Fuzzy logic is different from classical logic because the truth or falsity of a proposition is stated with a degree of truth represented by a number in the interval [0, 1]. Fuzzy logic was introduced by Zadeh [1] along with fuzzy set theory. According to Turksen [18], fuzzy logic is an application of fuzzy set theory. "Degree of membership" used in fuzzy set theory is applied as "degree of truth" in fuzzy logic. Let A be a fuzzy set. The truth of fuzzy proposition "x is a member of A" is stated in two different ways Turksen [18]: (I) using classical logic, the degree of "truth" of proposition 'x is A' is a single value: $\tau(A(x))=1$; (II) using fuzzy logic, the degree of "truth" is not a single value: $\tau(A(x))=\tau[0,1]$. Fuzziness of propositions, include different combinations of linguistic terms. Fuzzy implications are used generally which includes fuzzy propositions as antecedent and consequent as presented in (12).

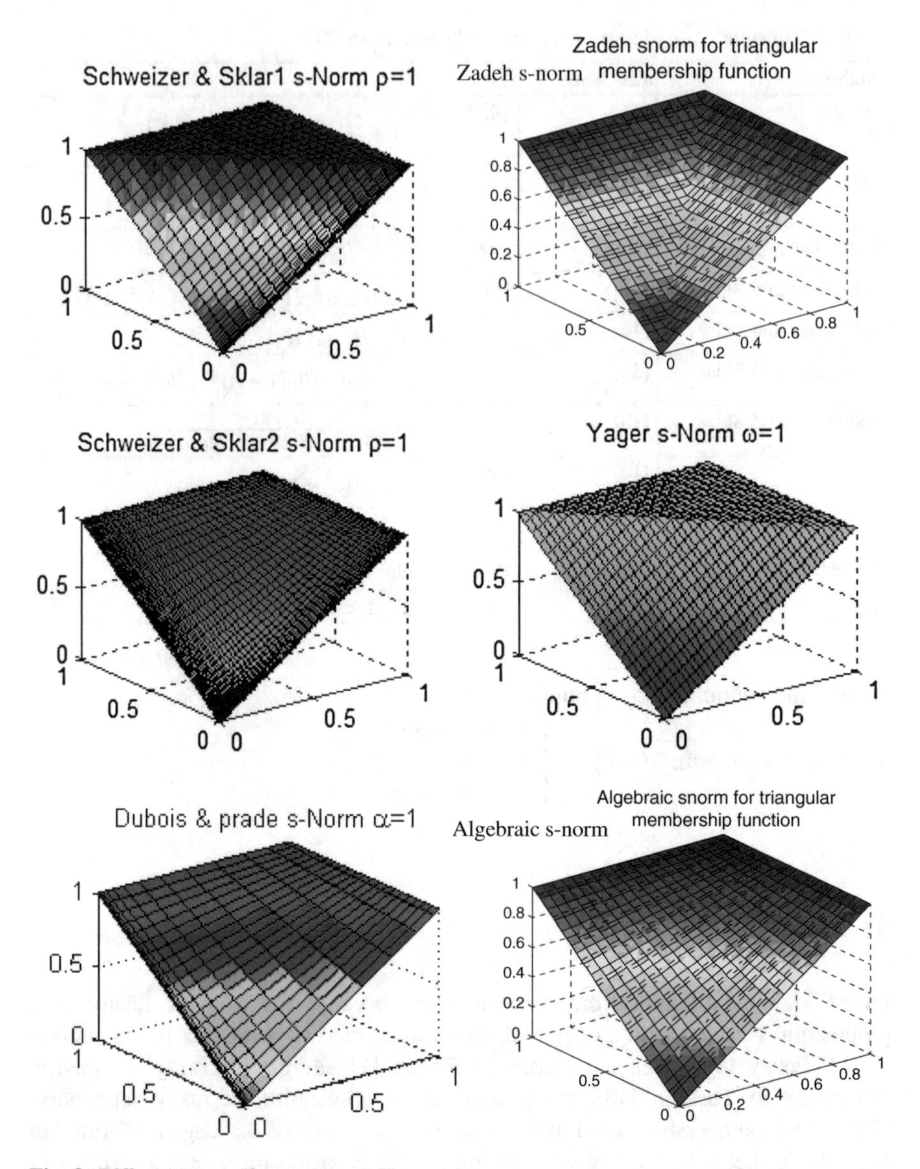

Fig. 2 Different types of t-conorms (s-Norms)

$$P\text{: IF } X \text{ is } A \text{ is true, THEN } Y \text{ is } B \text{ is true} \tag{12}$$

where X and Y are variables that take values x and y from sets X and Y, respectively, A and B are linguistic variables [6]. Implication between $A(x)$ and $B(x)$ are represented as follows:

$$A(x) \Rightarrow B(x) \tag{13}$$

Using an appropriate function, the degree to which the fuzzy proposition is true will be calculated. Lukasiewicz implication is generally used in literature and it is presented in (14).

$$I(P_{xy}) = I[A(x), B(y)] = \min[1, 1 - A(x) + B(x)] \tag{14}$$

Equation (14) indicates that calculating the degree of truth values $A(x)$ and $B(x)$ leads to the degree of truth value of (12). I is used to connect fuzzy sets to fuzzy propositions. The fuzzy truth value is indicated by a value of $\tau \in [0, 1]$ which is shown in (15).

$$P\text{:IF } X \text{ is } A \text{ is true to the degree } \tau_1, \text{ THEN } Y \text{ is } B \text{ is true to the degree } \tau_2 \tag{15}$$

Two states "X is A" is τ_1 and "Y is B" is τ_2 are indicated by $(A(x), \tau_1)$ and $(B(y), \tau_2)$ and the implication of them is presented by (16).

$$(A(x), \tau_1) \Rightarrow (B(y), \tau_2) \text{ is true} \tag{16}$$

Using Lukasiewicz implication the following equation is obtaines.

$$I(P_{xy}) = I[(A(x), \tau_1), (B(y), \tau_2)] = [\min[1, 1 - A(x) + B(x)], \min[1, 1 - \tau_1 + \tau_2]] \tag{17}$$

5 Approximate Reasoning

In order to apply fuzzy set theory in real world problems, approximate reasoning is required. One of the most common techniques of reasoning are inference rules in classical logic which use Modus Ponens for inference process. Generally speaking, Modus Ponens have been extended to be used in fuzzy logic. Zadeh [19] introduced Generalized Modus Ponens (GMP) for fuzzy aet and logic theory [6]. Using Compositional Rule of Inference (CRI) for infering fuzzy consequents with GMP, he developed fuzzy reseaning from classical reasoning. GMP is presented by the following equations [6].

$$\begin{array}{l} \text{Premise 1: A} \rightarrow \text{B} \\ \underline{\text{Premise 2: A}'} \\ \text{Deduction: B}^* \end{array} \tag{18}$$

where A and A$'$ are two linguistic variableson the universe of discourse of variable x with membership functions $\mu_A(x): x \in X \rightarrow [0, 1]$. Moreover, B and B* are linguistic terms on the universe of discourse of consequent variable y with membership functions $\mu_B(y)\colon y \in Y \rightarrow [0,1]$. Using membership degrees, (18) is calculated by the following equation proposed by Zadeh [1].

$$\mu_{B^*}(y) : sup_{x\in X} AND(\mu_A(x), \mu_{A\rightarrow B}(x,y)) \tag{19}$$

Using Zadeh rule base, "AND" operator is used for representing "MIN" and "MAX" operator is utilized as "sup" [6]. Table 3 shows different types of implications.

6 Fuzzy Systems

There are two main types of fuzzy systems: (I) Mamdani fuzzy systems; (II) Takagi-Sugeno-Kang (TSK) fuzzy systems.

6.1 Mamdani Fuzzy System

$$\text{IF } X \text{ isr } B_i \text{ THEN } Y \text{ isr } D_i \; i = 1, \ldots, m \tag{20}$$

$$R_i = B_i \cap D_i \text{ and } R = \bigcup_{i=1}^{m} R_i \tag{21}$$

$$\mu_{R_i}(x,y) = \mu_{B_i}(x) \bigwedge \mu_{D_i}(y) \tag{22}$$

$$\mu_R(x,y) = \bigvee_{i=1}^{m} \mu_{R_i}(x,y) \tag{23}$$

$$\mu_F(y) = \bigvee_x \left(\mu_A(x) \bigwedge \mu_R(x,y)\right) = \bigvee_{i=1}^{m} \left[\bigvee_x \left(\mu_A(x) \bigwedge \mu_{B_i}(x)\right)\right] \bigwedge \mu_{D_i}(y) \tag{24}$$

where X and Y are two fuzzy variables; B_i and D_i are two linguistic labels; i is the number of rules.The intersection and union operators are represented by $\cap$ and $\cup$ respectively in (21). Membership functions for antecedent variable (X) and consequent variable (Y) are represented by $\mu_{B_i}(x)$ and $\mu_{D_i}(y)$ respectively in (22); $\wedge$ and $\vee$ are t-norm and t-conorm operators explained in Sects. 3.1 and 3.2 respectively. $\mu_{R_i}(x,y)$ shows the aggregated MFs of antecedents and consequents in each rule. $\mu_R(x,y)$ indicates the aggregated MFs of all rules, and $\mu_F(y)$ represents the final aggregated MF. Figure 3 depicts Mamdani fuzzy inference system.

Table 3 Different types of implications (Klir and Yuan [7])

Name	References	Function *I*(*a, b*)	Year
Zadeh	[20]	$\max[1-a, \min(a,b)]$	1973
Gaines Rescher	[21]	$\begin{cases} 1 & a \le b \\ 0 & a > b \end{cases}$	1969
Godel	–	$\begin{cases} 1 & a \le b \\ b & a > b \end{cases}$	1976
Goguen	[22]	$\begin{cases} 1 & a \le b \\ b/a & a > b \end{cases}$	1969
Kleene-Dienes	–	$\max(1-a, b)$	1938–1949
Lukasiewicz	[23]	$\min(1, 1-a+b)$	1920
Smets and Magrez	[24]	$\min[1, \frac{1-a+(1+\lambda)b}{1+\lambda a}]$	1987
Smets and Magrez	[24]	$\min[1, (1-a^{\omega}+b^{\omega})^{\frac{1}{\omega}}]$	1987
Reichenbach	[25, 26]	$1-a+ab$	1935–1949
Willmott	[27]	$\min[\max(1-a,b), \max(a,1-a), \max(b,1-b)]$	1980
Wu	[28]	$\begin{cases} 1 & a \le b \\ \min(1-a,b) & a > b \end{cases}$	1986
Yager	[15]	$\begin{cases} 1 & a=b=0 \\ b^a & others \end{cases}$	1980
Klir and Yuan	[29]	$1-a+a^2b$	1994
Klir and Yuan	[29]	$\begin{cases} b & a=1 \\ 1-a & a \ne 1, b \ne 1 \\ 1 & a \ne 1, b=1 \end{cases}$	1994

After finding the final aggregated MF as an output of the fuzzy rule based system, deffuzification is used for obtaining a crisp and single-valued result of the system. There are several deffuzification methods in literature, and two of them are listed herein. They are the Center-of-Area (CoA) or Center-of-Gravity method, and the Middle-of-Maxima (MoM) defuzzification methods. Figures 4 and 5 show CoA and MoM defuzzification techniques. Equations (25) and (26) indicate CoA method in discrete and continuous cases [30].

$$y^* = \frac{\sum_{i=1}^{l} y_i \mu_Y(y_i)}{\sum_{i=1}^{l} \mu_Y(y_i)} = \frac{\sum_{i=1}^{l} y_i \max_k \mu_{CLY}(k)(y_i)}{\sum_{i=1}^{l} \max_k \mu_{CLY}(k)(y_i)} \tag{25}$$

$$u^* = \frac{\int_y u \mu_Y(y) dy}{\int_y \mu_Y(y) dy} = \frac{\int_y y \max_k \mu_{CLY}(k)(y) dy}{\int_y \max_k \mu_{CLY}(k)(y) dy} \tag{26}$$

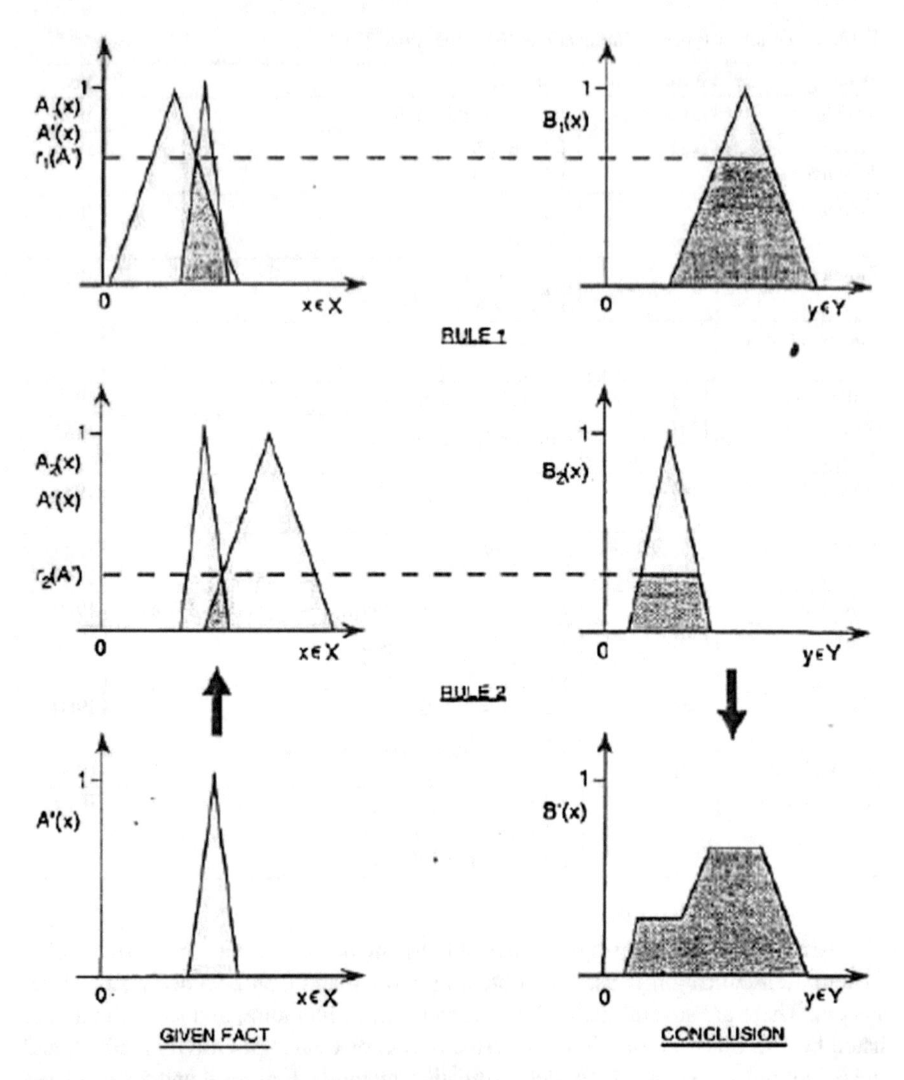

Fig. 3 Mamdani fuzzy inference system (Klir and Yuan [7])

where $\mu_{CLY}(k)$ is defined the same as $\mu_{R_i}(x, y)$ in (22), and "$\int$" is the classical integral.

Equation (27) shows MoM defuzzification technique. It determines the first and the last from of all values where Y has maximal membership degree and then takes the average of these two values [30].

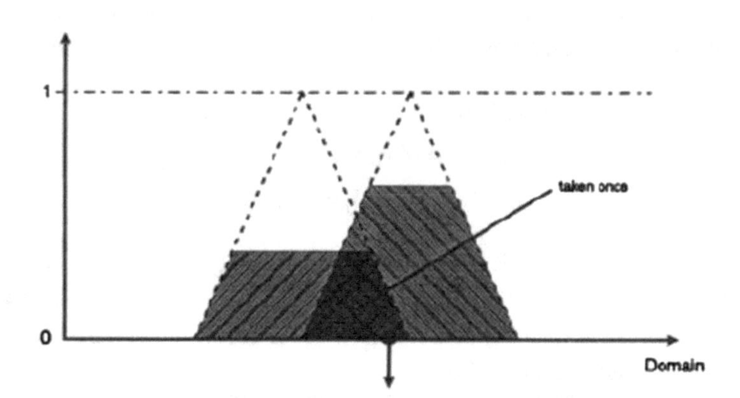

Fig. 4 CoA deffuzification method (Driankov and Saffiotti [30])

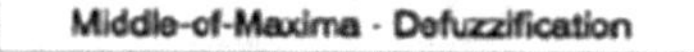

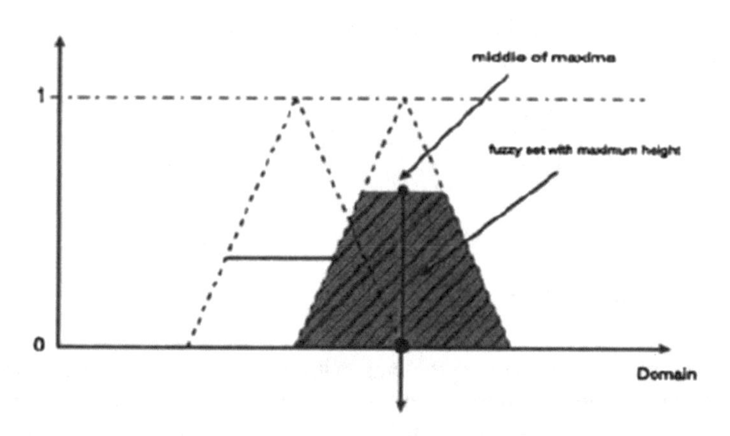

Fig. 5 MoM deffuzification method (Driankov and Saffiotti [30])

$$y^* = \frac{inf_{y \in Y}\{y \in Y|\mu_y(y) = hgt(Y)\} + sup_{y \in Y}\{y \in Y|\mu_y(y) = hgt(Y)\}}{2} \quad (27)$$

where $hgt(Y) = sup_{y \in X}\mu_B(y)$.

6.2 Takagi-Sugeno-Kang (TSK) Fuzzy System

In TSK fuzzy systems, consequent of each rule includes a function instead of the linguistic variable(s). Equation (28) shows the TS fuzzy model including R rules.

$$\begin{aligned}&\text{Rule } i\text{:if } z_1 \text{ is } A_1^{i,k_1}, z_2 \text{ is } A_2^{i,k_2}, \ldots, \text{ and } z_m \text{ is } A_m^{i,k_m}\\&\text{then } y^i = a_1^i x_1 + a_2^i x_2 + \cdots + a_q^i x_q\\&i = 1, 2, \ldots, R.\ k_j = 1, 2, \ldots, r_j.\end{aligned} \tag{28}$$

where R is the number of rules in the TS fuzzy model. $z_j\ (j = 1, 2, \ldots, m)$ is the jth variable. $x_l\ (l = 1, 2, \ldots, q)$ is the lth model input. y^i is the output of the ith rule. For the ith rule, A_j^{i,k_j} is the k_j^{th} fuzzy subset of z_j. a_l^i is the coefficient of the consequent. r_j is the fuzzy partition number of z_j.

The output of the TSK model is calculated by taking the weighted-average of output of each rule as follows:

$$y = \sum_{i=1}^{R} \mu^i y^i \Big/ \sum_{i=1}^{R} \mu^i \tag{29}$$

where y^i is determined by the equation in the consequent of the ith rule. Moreover, (29) can be restated by (30).

$$y = \left(\sum_{i=1}^{R} \mu^i a_1^i x_1 + \cdots + \sum_{i=1}^{R} \mu^i a_q^i x_q \right) \Big/ \sum_{i=1}^{R} \mu^i \tag{30}$$

7 Type-2 Fuzzy Sets

In order to model and solve problems with higher degree of vagueness and uncertain MFs, Zadeh [3] introduced type-2 fuzzy sets as an extension to type-1 fuzzy sets. In type-2 fuzzy sets, each MF is represented by another MF known as secondary membership function. Type-2 fuzzy sets have been proposed by Zadeh [3] to tackle the drawbacks of type-1 fuzzy sets when uncertainty is too high. Three ways in which such uncertainty can occur are: (1) the words that are used in antecedents and consequents of rules can mean different things to different people, (2) consequents obtained by polling a group of experts will often be different for the same rule because the experts will not necessarily be in agreement, and (3) noisy training data (Liang and Mendel [4]).

Type-1 fuzzy sets are unable to handle higher degrees of uncertainty. Moreover, some datasets are very changeable over time. This variability causes MFs to change and include more than one single-valued membership degree for each point in a universe of discourse. Because type-1 MFs assign only one single-valued membership degree to each point they are incapable of modeling systems using those kinds of uncertain data sets. Hence, type-2 fuzzy sets which assign another membership degree to the primary membership values are used in such systems. Figure 6 shows type-1 and type-2 fuzzy MFs (interval type-2 fuzzy MF and general type-2 fuzzy MF). Comparing both MFs in Fig. 6 indicates that type-2 fuzzy MFs are capable of including more information and handling variable data sets using two MFs.

A type-2 fuzzy set $\tilde{A}$ is represented by a type-2 membership function $\mu_{\tilde{A}}$ as follows.

$$\mu_{\tilde{A}} : X \rightarrow [0, 1] \tag{31}$$

Mendel [32] defines the type-2 fuzzy set as it is shown in (32).

$$\tilde{A} = \{((x, u), \mu_{\tilde{A}}(x, u)) | \forall x \in X,\ \forall u \in J_x \subseteq [0, 1]\} \tag{32}$$

where $0 \le \mu_{\tilde{A}}(x, u) \le 1$, J_x is a primary MF in the interval [0, 1], and u is the primary membership value. Another form of representing type-2 fuzzy set $\tilde{A}$ is as follows.

$$\tilde{A} = \int\limits_{x \in X} \int\limits_{u \in Jx} \mu_{\tilde{A}}(x, u)/(x, u) = \int\limits_{x \in X} \int\limits_{u \in Jx} f_x(u)/(x, u) \tag{33}$$

where $\iint$ shows the union of admissible x and u, $f_x(u)$ is the secondary MF [6].

Since type-2 fuzzy sets are presented as the union of secondary MFs ($f_x(u)$), type-2 fuzzy MFs are also represented with the secondary MFs for continuous and discrete universe of discourse as shown in (34) and (35) respectively.

$$\mu_{\tilde{A}}(x, u) = \int\limits_{u \in Jx} f_x(u)/u \tag{34}$$

$$\mu_{\tilde{A}}(x, u) = \sum_{u \in Jx} f_x(u)/u \tag{35}$$

However, due to the various degrees of uncertainty, different kinds of type-2 fuzzy sets are required. There are two kinds of type-2 fuzzy sets: (I) interval type-2 fuzzy sets (IT2 FSs); (II) general type-2 fuzzy sets (GT2 FSs). In IT2 FSs, MFs are interval instead of a single value. They are special type of GT2 FSs whose secondary membership values for the entire members of the primary domain are one. GT2 FSs are extensions of IT2 FSs, and they are bivariate interval valued [0,1] supporting more degrees of freedom [5]. However, IT2 FSs are less computationally expensive than GT2 FSs.

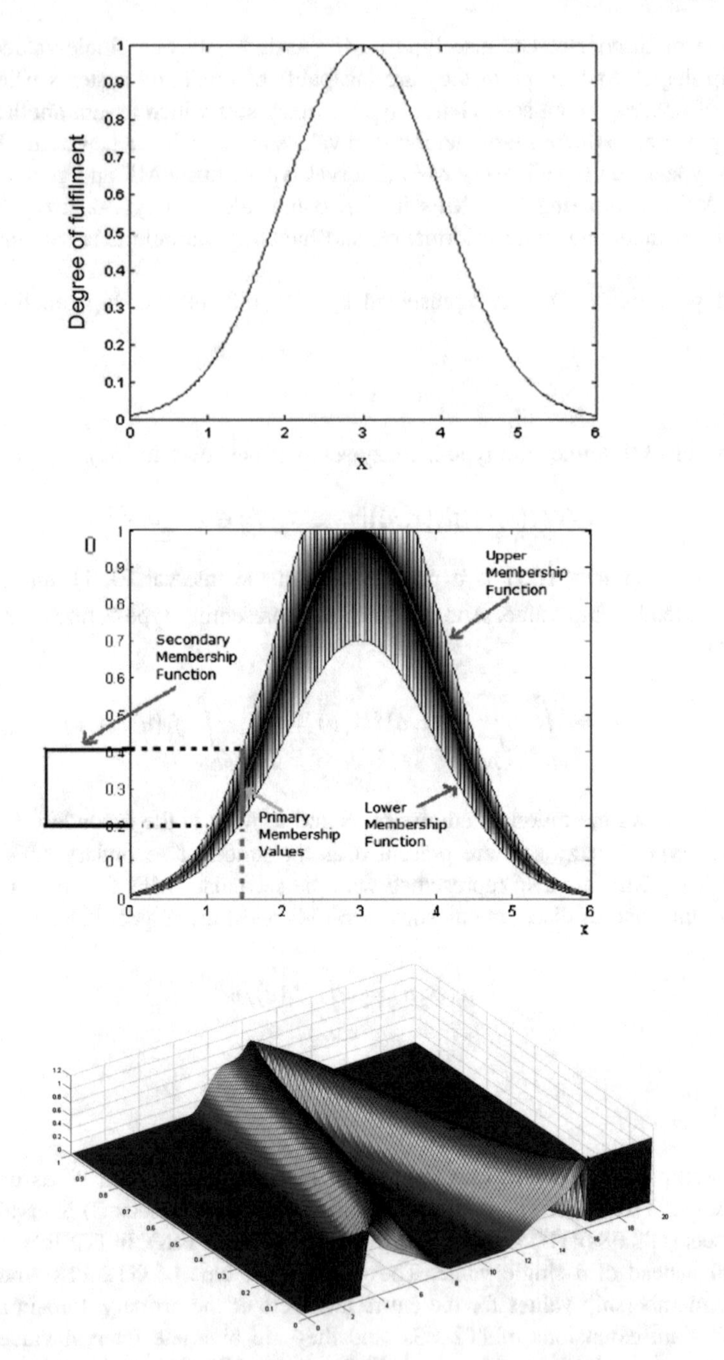

Fig. 6 Type1 fuzzy MF, interval type-2 fuzzy MF [6], and general type-2 fuzzy MF [31]

7.1 Interval Type-2 Fuzzy Sets

IT2 FSs have been proposed by Karnik et al. [33]. Since they are less complex and computationally expensive than GT2 FSs, researchers are most willing to use them instead of GT2 FSs. However, GT2 FSs are more capable of representing wider range of information and being used in face of variable data sets. MFs of interval type-2 fuzzy sets are defined by the following equations.

$$\mu_{\tilde{A}}(x)\colon X \to 1/u, \quad u \in J_x, J_x \subseteq [0,1] \tag{36}$$

$$f_x(u) = \forall x \in X, \quad u \in J_x, J_x \subseteq [0,1] \tag{37}$$

Equation (37) indicates that at each value of x, all the secondary membership values are equal to 1 as it is shown in Fig. 6. There is more than one MF for each value of x. The MFs includes the lower and upper MFs. The upper and lower MFs of IT2 FSs are defined by (38) and (39).

$$\mu_{\tilde{A}}^{U}(X) = \max_{u \in J_x}(u), \ \forall x \in X \tag{38}$$

$$\mu_{\tilde{A}}^{U}(X) = \min_{u \in J_x}(u), \ \forall x \in X \tag{39}$$

where $\mu_{\tilde{A}}^{U}(x)$ is the upper MF and $\mu_{\tilde{A}}^{L}(x)$ is the lower MF of the IT2 FS. Therefore, the IT2 FS is defined using upper and lower MFs by the following equation.

$$\mu_{\tilde{A}}(x)\colon X \to 1/u, \quad u \in \left[\mu_{\tilde{A}}^{L}(x), \mu_{\tilde{A}}^{U}(x)\right] \tag{40}$$

7.2 General Type-2 Fuzzy Sets

Using a type-2 fuzzy MF ($\mu_{\tilde{A}}(x, y)$), a GT2 FS $\tilde{A}$ is represented by the following equation [32].

$$\tilde{A} = \int_{x \in X} \int_{u \in J_x} \mu_{\tilde{A}}(x, u)/(x, u), J_x \subseteq [0, 1] \tag{41}$$

where $x \in X$ is the primary variable, $u \in J_x$ is the primary MF, J_x indicates an interval between the lower and upper MFs, and $\mu_{\tilde{A}}(x, u)$ is the secondary MF. GT2 FSs are expressed in two different ways: the vertical slice and the wavy slice [34].

Let $x = x_0$ be a specific point in the universe of discourse X; then, vertical slice $\mu_{\tilde{A}}(x', u)$ of the fuzzy MF $\mu_{\tilde{A}}(x, u)$ is obtained. In other words, a secondary MF $\mu_{\tilde{A}}(x = x', u)$ is denoted by vertical slices at each point for $x' \in x$ and $\forall u \in J_{x'} \subseteq [0, 1]$ as follows.

$$\mu_{\tilde{A}}(x = x', u) \equiv \int_{u \in J_{x'}} f_{x'}(u)/u, J_x \subseteq [0, 1] \tag{42}$$

where $f_{x'}(u)$ is the secondary MF and $f_{x'}(u) \subseteq [0, 1]$. Discretizing the entire universe of discourse into N samples, then the GT2 FS $\tilde{A}$ is denoted by aggregating the whole vertical slices as it is shown in (43) [35].

$$\tilde{A} = \sum_{i=1}^{N} \left[\int_{u \in J_{xi}} f_{x_i}(u)/u \right] \Big/ x_i \tag{43}$$

$\tilde{A}$ is also denoted by the union of all its embedded T2 FSs as follows.

$$\tilde{A} = \bigcup_{\forall \tilde{A}_e} \tilde{A}_e \tag{44}$$

where $\tilde{A}_e$ is represented as follows.

$$\tilde{A}_e = \int_{x \in X} u/x, \quad u \in J_x \tag{45}$$

The embedded T1 FS $\tilde{A}_e$, corresponding to an embedded T2 FS includes the primary MF of $\tilde{A}_e$. The centroid of $\tilde{A}_e$, i.e., $C_{\tilde{A}}$, is the union of the centroids of its entire embedded IT2 FSs. Its MF $C_{\tilde{A}}(\xi)$ ($\forall \xi \in X \subset R$), is defined as follows [36].

$$C_{\tilde{A}}(\xi) = \bigcup_{\forall \tilde{A}_e} \left\{ \min_{(x,u) \in A_e} f_x(u) \Big/ \xi = \frac{\int_{-\infty}^{+\infty} x u_{A_e}(x) dx}{\int_{-\infty}^{+\infty} u_{A_e}(x) dx} \right\} \tag{46}$$

The centroid of the GT2 FS $\tilde{A}$ is calculated by aggregating centroids of each α-plane ($C_{\tilde{A}}$) [35]. Hence, the center of the GT2 FS $\tilde{A}$ is presented by (47).

$$C_{\tilde{A}}(\xi) = \bigcup_{\alpha \in [0,1]} \alpha / C_{\tilde{A}\alpha}(\xi) (\forall \xi \in X) \tag{47}$$

where ${}^{\alpha}/C_{\tilde{A}}(\xi)$ is the centroid of $\tilde{A}_\alpha$.

$$C_{\tilde{A}_\alpha}(\xi) = \bigcup_{\forall \tilde{A}_e(\alpha)} \left[\xi = \frac{\int_{-\infty}^{+\infty} x u_{A_e(\alpha)}(x) dx}{\int_{-\infty}^{+\infty} u_{A_e(\alpha)}(x) dx} \right] = \left[C_l(\tilde{A}_\alpha), C_r(\tilde{A}_\alpha) \right] \tag{48}$$

where $C_l(\tilde{A}_\alpha)$ and $C_r(\tilde{A}_\alpha)$ are calculated by KM [37] and EKM [38] algorithms. $f_x(u)$ is considered as the combination of two increasing and decreasing functions called $g_x(u)$ and $h_x(u)$.

$$f_{x(u)} = \begin{cases} g_x(u) & u \in [S_L(x|0), S_L(x|1)] \\ h_x(u) & u \in [S_R(x|1), S_R(x|0)] \\ 1, & u \in [S_L(x|1), S_R(x|1)] \\ 0, & \text{Otherwise} \end{cases} \tag{49}$$

The upper and lower MF of GT2 FS with the secondary membership value of α are defined as $S_R(x|\alpha)$ and $S_L(x|\alpha)$ respectively. The slopes of the secondary MFs at α for $g_x(u)$ and $h_x(u)$ are equal to $g'_x(u)$ and $h'_x(u)$. Then in order to calculate upper and lower MFs at the plane "$\alpha + TS$" the following equations are required.

$$\begin{aligned} S_R(x|\alpha + Ts) &= S_R(x|\alpha) + \frac{T_S}{h'_x(u)} \\ S_L(x|\alpha + Ts) &= S_L(x|\alpha) + \frac{T_S}{g'_x(u)} \end{aligned} \tag{50}$$

8 Type-2 Fuzzy Systems

An interval type-2 fuzzy rule based system is presented in the following equation (Celikyilmaz and Turksen [6]).

$$\tilde{R}_i\text{:}IF\ \underset{j=1}{\overset{nv}{AND}} \left(x_j \in X_j\ isr\ \tilde{A}_{ij} \right) THEN\ y \in Y\ isr\ \tilde{B}_i \tag{51}$$

where,

- c is the number of rules in system model,
- x_j is jth input variable, $j = 1, \ldots, nv$, "nv" is total number of input variables,
- X_j is the domain of x_j,
- $\tilde{A}_{ij}$ is the linguistic label associated with jth input variable in the ith rule represented by a type-2 membership function, $\tilde{\mu}_i(x_j) \in X_j \rightarrow f_{x_j}(u)/u$, $u \in J_{x_i}, J_{x_i} \subseteq [0, 1]$.

- y is the output variable,
- Y is the domain of y,
- B_i is the linguistic label associated with output variable in the ith rule with type-2 membership, $\tilde{\mu}_i(y) \in Y \rightarrow f_y(w)/w, w \in J_y, J_y \subseteq [0, 1]$.
- AND is the logical connective used to aggregate membership values of input variables for a given observation in order to find the degree of fire of each rule,
- THEN ($\rightarrow$, $\Rightarrow$) is the logical IMPLICATION connective,
- ALSO is the logical connective used to aggregate model outputs of fuzzy rules, '*isr*' is introduced by Zadeh [1] and it represents that the definition or assignment is not crisp, it is fuzzy.

In order to find the result of the interval type-2 fuzzy rule based system (51), two inference methods have been proposed in literature. The first one is to considerupper and lower membership functions of interval type-2 fuzzy sets as lower and upper points of an interval (Liang and Mendel [4]).

Let $\mu_{\tilde{A}}^U(x)$ be the upper MF, and $\mu_{\tilde{A}}^L(x)$ be the lower MF. Equation (52) shows the inference mechanism proposed by Liang and Mendel [4].

$$\mu_{\tilde{A}}(x) : x \rightarrow {}^{1}/_{u}, u \in [\mu_{\tilde{A}}^L(x), \mu_{\tilde{A}}^U(x)] \tag{52}$$

Membership values of antecedents are aggregated for each rule as it is shown in (53).

$$\tilde{\mu}_i^L(x) = T_{j=1}^{nv}(\tilde{\mu}_i^L(x_j)), \tilde{\mu}_i^U(x) = T_{j=1}^{nv}(\tilde{\mu}_i^U(x_j)) \tag{53}$$

where, T denotes the T-norm connective. In (10), $\tilde{\mu}_i^{*L}(y)$ and $\tilde{\mu}_i^{*U}(y)$ are upper and lower memberships of output fuzzy set $\tilde{\mu}_i^*(y)$.

$$\tilde{\mu}_i^*(y) : Y \rightarrow {}^{1}/_{w}, w \in [\tilde{\mu}_i^{*L}(y), \tilde{\mu}_i^{*U}(y)] \tag{54}$$

The aggregated antecedents and consequents for lower and upper MFs are presented in (55) and (56).

$$\tilde{\mu}_i^{*L}(y) = T(\tilde{\mu}_i^L(x), \tilde{\mu}_i^L(y)) \tag{55}$$

$$\tilde{\mu}_i^{*U}(y) = T(\tilde{\mu}_i^U(x), \tilde{\mu}_i^U(y)) \tag{56}$$

Then, all rules are aggregated using (57) and (58).

$$\tilde{\mu}^{*L}(y) = S_{i=1}^{c^*}(\tilde{\mu}_i^L(y)) \tag{57}$$

$$\tilde{\mu}^{*U}(y) = S_{i=1}^{c^*}(\tilde{\mu}_i^U(y)) \tag{58}$$

where "S" is a s-norm or t-conorm operator, and "c" is the number of rules. At the end of the inference process, a crisp output is required. The following equation has been presented by Liang and Mendel [4] for calculating the crisp output (y^*).

$$(y^*) = \left[y^{*L} + y^{*U}\right]/2 \quad (59)$$

where, y^{*L} and y^{*U} are lower and upper bounds, i.e., $y^* \in \left[y^{*L}, y^{*U}\right]$.

The second inference method has been proposed by Turksen [39] using fuzzy disjunctive normal forms (FDNF) and fuzzy conjuctive normal forms (FCNF). Turksen's inference process with a rule set is represented as follows [39].

$$\mu_B^*(y) = [\bigvee_{x \in X} \mu_A'(x) T[\mu_{FDNF}(A \rightarrow B)(x,y)] \bigvee_{x \in X} \mu_A'(x) T[\mu_{FCNF}(A \rightarrow B)(x,y)]], \forall \mathrm{y} \in \mathrm{Y} \quad (60)$$

where "T" is used to indicate t-norm between two MFs. $\mu_{FDNF}(A \rightarrow B)(x,y)$ and $\mu_{FCNF}(A \rightarrow B)(x,y)$ are two boundaries of type-2 fuzzy systems based on fuzzy normal forms. $\mu_{\mathrm{B}}^*(y)$ is an inference result, and $\mu_A'(x)$ is the observed membership value.

$$\begin{cases} A\left(x_{D,i,r}^*(r)\right) = \mu(x_{FDNF(A(i,r))}^*(r)), \\ A\left(x_{C,i,r}^*(r)\right) = \mu(x_{FCNF(A(i,r))}^*(r)) \end{cases} \quad (61)$$

Equation (61) is the membership value of the FDNF of the left-hand side of the ith rule which has "r" input variables evaluated at $x_{(r)}^* = (x_1^*, x_2^*, \ldots, x_r^*)$. $A_{D,i,r}(x^*(r))$ is the membership value of the FDNF in the left-hand side of the ith rule which has "r" input variables evaluated at $x_{(r)}^* = (x_1^*, x_2^*, \ldots, x_r^*)$. $A\left(x_{C,i,r}^*(r)\right)$ is the membership value of the FCNF [39].

$A\left(x_{D,i,r}^*(r)\right)$ is computed recursively as:

$$\begin{cases} A_{D,i,\rho}(x^*(\rho)) = \left[A_{D,i,\rho-1}(x^*(\rho-1)) T A_{i,\rho}\left(x_\rho^*\right)\right] S \left[A_{D,i,\rho-1}(x^*(\rho-1)) T A_{i,\rho}\left(x_\rho^*\right)\right] \\ For\ \rho = 2, 3, 4, \ldots, r \end{cases} \quad (62)$$

Such that $X^*(2) = (X_1^*, X_2^*)$ and $A(i,2) = A_{i1}\ \mathrm{AND}\ A_{i2}$. "$S$" in the definition of "$A_{D,i,\rho}(x^*(\rho))$" indicates S-norm of two MFs.

$A_{C,i,r}(x^*(r))$ is computed recursively as [39]:

$$\begin{aligned} A_{C,i,\rho}(x^*(\rho)) = {} & [A(x^*_{C,i,\rho-1}(\rho-1)SA_{i\rho}\left(x^*_\rho\right)]T[A(x^*_{C,i,\rho-1}(\rho-1)SA_{i\rho}\left(x^*_\rho\right)] \\ & T[A(x^*_{C,i,\rho-1}(\rho-1)Sn(A_{i\rho}\left(x^*_\rho\right))]T[A(x^*_{C,i,\rho-1}(\rho-1)Sn(A_{i\rho}\left(x^*_\rho\right))] \\ & T[n(A(x^*_{C,i,\rho-1}(\rho-1))S(A_{i\rho}\left(x^*_\rho\right))]T\left[n\left(A\left(x^*_{C,i,\rho-1}(\rho-1)\right)S\left(A_{i\rho}\left(x^*_\rho\right)\right)\right], \\ & \rho = 2,3,4,\ldots,r \end{aligned} \tag{63}$$

In (64), $B^*(y)$ is the final output of the model, which is obtained by combining the final output of FDNF and FCNF.

$$B^*(y) = \beta B^*_D(y) + (1-\beta)B^*_C(y), \quad \forall y \in Y, 0 \le \beta \le 1, \tag{64}$$

where $\beta^*_D(y)$ is the final result for FDNF and $\beta^*_C(y)$ is the final result for FCNF obtained after completing the inference process.

9 Applications of Type-2 Fuzzy Systems

In this section, some of the applications of type-2 fuzzy systems are briefly reviewed. These applications include but are not limited to clustering, classification, pattern recognition, scheduling, forecasting, and supply chain management (SCM).

9.1 Type-2 Fuzzy Systems in Clustering, Classification, and Pattern Recognition

In this subsection, a brief review on the first categories of type-2 fuzzy applications including clustering, classification, and pattern recognition techniques is provided. Type-2 fuzzy systems have been used in clustering and classification models for managing real world problems which encounter higher degrees of uncertainty.

Data clustering is one of the main methods of structure identification. It is an unsupervised technique for putting similar data in a cluster [40]. Bezdek presented fuzzy c-means clustering (FCM) model. For structure identification of type-2 fuzzy systems different methods have been introduced in the literature. For more information please refer to [41–47]. In order to design a rule based system, Aliev et al. [48] proposed a type-2 FCM clustering model. Rhee and Choi [42] introduced three methods for interval type-2 fuzzy membership function generation: (I) Histogram based method; (II) Heuristics methods; (III) Interval type-2 fuzzy C-means (IT2 FCM). Fazel Zarandi, et al. [49] proposed a new interval type-2 fuzzy clustering method for functional systems called interval type-2 fuzzy c-regression clustering model (IT2 FCRM).

Type-2 fuzzy systems are also applied in classification problems. Melin and Castillo [50] reviewed the literature of clustering, classification, and pattern recognition. Tables 4 and 5 show the results of their work. Sharma and Bajaj [51, 52] proposed an interval type-2 fuzzy system for vehicle classification. They assigned each vehicle to its relevant class by investigating its characteristics such as the wheel base, ground clearance, and body length [50]. Tan et al. [53] proposed a type-2 fuzzy system for ECG arrhythmic beat classification. Three classes of ECG signals, called the normal sinusrhythm (NSR), ventricular fibrillation (VF) and ventricular tachycardia (VT) have been investigated in that research.

9.2 Type-2 Fuzzy Systems in Scheduling, Forecasting, and SCM

In this subsection, a brief review on the second categories of type-2 fuzzy applications including scheduling, forecasting, and SCM is provided.

Fazel Zarandi and Gamasaee [82] proposed a type-2 fuzzy hybrid expert system, which uses a combination of Mamdani and Sugeno methods, for tardiness forecasting in scheduling of steel continuous casting process. In that study, tardiness variables are represented by interval type-2 fuzzy MFs, and interval type-2 FDNF and FCNF proposed by Turksen [39] are used in the inference engine. The results of the method proposed by Fazel Zarandi and Gamasaee [82] show that using type-2 fuzzy system leads to the more accurate forecasted outputs in comparison to type-1 fuzzy system and other methods in literature.

Karnik and Mendel [83] applied type-2 fuzzy systems for forecasting of time-series. They showed that noisy data causes type-1 fuzzy system to be incapable of modeling the problem efficiently, so type-2 fuzzy systems are required to handle noisy data. Fazel Zarandi et al. [84] presented a type-2 fuzzy hybrid expert system for forecasting the amount of reagents in desulphurization process of a steel manufacturing company. The results of that model indicate that type-2 fuzzy system has less forecasting error in comparison to type-1 fuzzy system. Another research that uses type-2 fuzzy systems in time series prediction has been conducted by Gaxiola et al. [85]. They presented a generalized type-2 fuzzy weight adjustment for back propagation neural networks in time series prediction.

Pramanik et al. [86] used type-2 fuzzy systems for modeling and solving a fixed-charge transportation problem in a two-stage supply chain network. Since unit transportation costs, fixed charges, availabilities, and demands are imprecise, they have been indicated by Gaussian type-2 fuzzy numbers. The problem has been categorized in class of profit maximization problems, and retailers' demands are satisfied by selecting distribution centers. For solving the problem, type-2 fuzziness has been eliminated using generalized credibility measures. Fazel Zarandi and

Table 4 Type-2 fuzzy systems in clustering and classification [50]

Author(s) (pub. year)	References	Domain of the problem	Comparison with type-1	Why type-2 is required for the problem?
Sharma and Bajaj (2009 and 2010)	[51, 52]	Vehicle classification	Yes	Uncertainty and imperfection of data
Pimenta and Camargo (2010)	[54]	Classification	Yes	Imprecision in classification
Chumklin et al. (2010)	[55]	Cancer detection	Yes	Uncertainty in medical classification
Sanz et al. (2010)	[56]	Classification	No	Imprecision in classification
Wu and Mendel (2010)	[57]	Vehicle classification	Yes	Uncertainty in information
Phong and Thien (2009)	[58]	Arrhythmia classification	No	Uncertainty in medical classification
Aliev et al. (2011)	[48]	Clustering	Yes	Uncertainty in clustering
Abiyev et al. (2011)	[59]	Clustering	No	Uncertainty in clustering
Zheng et al. (2010)	[60]	Classification	No	Uncertainty in classification
Abiyev and Kaynak (2010)	[61]	Clustering	No	Uncertainty in clustering
Zhengetal. (2010)	[62]	Clustering	No	Uncertainty in clustering
Albarracin and Melgarejo (2010)	[63]	Signal clustering	Yes	Uncertainty in clustering
Ozkan and Turksen (2010)	[64]	Clustering	No	Uncertainty in clustering
Pedrycz (2010)	[65]	Clustering	Yes	Uncertainty in granulation
Juang et al. (2009)	[66]	Clustering	Yes	Uncertainty in clustering
Türkşen (2009)	[67]	Clustering	Yes	Uncertainty in clustering
Ren et al. (2010)	[68]	Clustering	No	Uncertainty in clustering
Qun et al. (2010)	[69]	Clustering	Yes	Uncertainty in clustering

Gamasaee [87] proposed a type-2 fuzzy system model for reducing bullwhip effect in supply chains. They concentrated on demand forecasting techniques where all demands, lead times, and orders have been represented by type-2 fuzzy sets. Using interval type-2 fuzzy c-regression clustering technique, demand data have been assigned to appropriate clusters, and structure of type-2 fuzzy expert system has

Table 5 Type-2 fuzzy systems in pattern recognition [50]

Author(s) (pub. year)	Ref. no.	Domain of the problem	Comparison with type-1	Why type-2 is required for the problem?
Melin (2010)	[70]	Edge detection	Yes	Uncertainty in edge detection
Lopez et al. (2010)	[71]	Finger print recognition	Yes	Uncertainty in finger print recognition
Li and Zhang (2010)	[72]	Pattern recognition	No	Uncertainty in pattern recognition
Own (2009)	[73]	Medical diagnosis	No	Uncertainty in diagnosis
Mendoza et al. (2009)	[74]	Face recognition	Yes	Uncertainty in face recognition
Kim et al. (2009)	[75]	Pattern recognition	No	Uncertainty in pattern recognition
Hidalgo et al. (2009)	[76]	Multimodal recognition	Yes	Uncertainty in multimodal pattern recognition
Lopez et al. (2008)	[77]	Finger print recognition	Yes	Uncertainty in finger print recognition
Ozkan and Turksen (2004)	[78]	Pattern recognition	Yes	Uncertainty in pattern recognition
Mitchell (2005)	[79]	Pattern recognition	No	Uncertainty in pattern recognition
Madasu et al. (2008)	[80]	Edge detection	Yes	Uncertainty in edge detection
Tizhoosh (2005)	[81]	Image thresholding	No	Uncertainty in image thresholding

been identified. Then, an interval type-2 fuzzy hybrid expert system was developed for demand forecasting. An interval type-2 fuzzy ordering policy was designed to determine orders in supply chain. Results of that model demonstrated that using type-2 fuzzy system leads to more bullwhip reduction and better forecasted results in comparison to type-1 fuzzy systems.

10 Type-n Fuzzy Systems

There are not too much works in the literature in type-3 and higher level fuzzy sets and systems. Transformation from full type-2 fuzzy sets and systems into higher level ones has serious modeling, representing, and computational complexity problems. First, we should find the real cases with for example type-3 fuzzy syntax and semantic. It is very difficult to show these cases in geometric configuration. Also, modeling type-n fuzzy needs lots of nonlinear variables, with high correlations with each other. Finally, the computational complexity of type-n fuzzy is too much. This fact needs some high level heuristic algorithms to manage it. So, there are rooms of potential scientific and applied works that should be done in the future to resolve these serious problems. We started it from three years ago and hope we can have some preliminary outputs in the near future.

11 Conclusions

In this chapter, the motivation for using fuzzy systems and the mathematical concepts of type-1 to type-n fuzzy sets, logic, and systems were discussed. This chapter has reviewed the most useful knowledge for researchers on several areas of fuzzy logic such as type-1 fuzzy sets, mathematical operations on type-1 fuzzy sets, fuzzy logic, approximate reasoning, fuzzy systems including mamdani and sugeno inference systems, interval type-2 fuzzy sets, general type-2 fuzzy sets, and type-2 fuzzy systems. Moreover, applications of those fuzzy approaches in solving real world problems were reviewed. Thus, this chapter is a reference for researchers to become more familiar with theory and applications of fuzzy logic.

References

1. Zadeh, L.A.: Fuzzy sets. Inf. Control, **8**, 338–353 (1965)
2. Zadeh, L.A.: Is there a need for fuzzy logic. Inf. Sci. **178**, 2751–2779 (2008)
3. Zadeh, L.A.: The concept of a linguistic variable and its application to approximate reasoning-I. Inf. Sci. **8**, 199–249 (1975)
4. Liang, Q., Mendel, J.M.: Interval type 2 fuzzy logic systems: theory and design. IEEE Trans. Fuzzy Syst. **8**, 535–550 (2000)
5. Aisbett, J., Rickard, J.T., Morgenthaler, D.G.: Type-2 fuzzy sets as functions on spaces. IEEE Trans. Fuzzy Syst. **18**(4), 841–844 (2010)
6. Celikyilmaz, A., Türksen, I.B.: Modeling uncertainty with fuzzy logic with recent theory and applications. Springer-Verlag, Berlin Heidelberg (2009)
7. Klir, G.J., Yuan, B.: Fuzzy sets and fuzzy logic theory and applications. Prentice Hall (1995)
8. Dombi, J.: A general class of fuzzy operators, the De Morgan class of fuzzy operators and fuzziness measures induced by fuzzy operators. Fuzzy Sets Syst. **8**, 149–163 (1982)
9. Frank, M. J.: On the simultaneous associativity of $F(x, y)$ and $x + y - F(x, y)$. Aequationes Mathe. **19**, 194–226 (1979)
10. Schweizer, B., Sklar, A.: Associative functions and abstract semi groups. Publ. Math. Debrecen **10**, 69–81 (1963)
11. Schweizer, B., Sklar, A.: Associative functions and statistical triangle inequalities. Publ. Math. Debrecen **8**, 69–81 (1961)
12. Schweizer, B., Sklar, A.: Statistical metric spaces. Pac. J. Math. **10**, 313–334 (1960)
13. Schweizer, B., Sklar, A.: Probabilistic Metric Spaces. North-Holland, New York (1983)
14. Weber, S.: A general concept of fuzzy connectives, negations and implications based on t-norms and t-conorms. Fuzzy Sets Syst. **13**, 247–271 (1984)
15. Yager, R.R.: On a general class of fuzzy connectives. Fuzzy Sets Syst. **4**, 235–242 (1980)
16. Dubois, D., Prade, H.: Fuzzy Sets and Systems: Theory and Applications. Academic Press, New York

17. Yu, Y.D.: Triangular norms and TNF-sigma-algebras. Fuzzy Sets Syst. **16**, 251–264 (1985)
18. Turksen, I.B.: An Ontological and Epistemological Perspective of Fuzzy Theory. Elsevier, The Netherlands (2006)
19. Zadeh, L.A.: Calculus of fuzzy restrictions, Fuzzy sets and their applications to cognitive decision processes, pp. 1–40. Academic Press, London (1975)
20. Zadeh, L. A.: Outline of a new approach to the analysis of complex systems and decision processes. IEEE Trans. Syst. Man Cybern. **1**, 28–44 (1973)
21. Gaines, B.R.: Foundations of fuzzy reasoning. Int. J. Man Mach. Stud. **8**, 623–668 (1976)
22. Goguen, J. A.: The logic of inexact concepts. Synthese **19**, 325–373 (1968–1969)
23. Łukasiewicz, J.: O logic etrójwartościowej (in Polish). Ruchfilozoficzny **5**, 170–171 (1920). English translation: On three-valued logic. In: Borkowski, L. (ed.) Selected Works by Jan Łukasiewicz. North–Holland, Amsterdam, pp. 87–88 (1970)
24. Smets, P., Magrez, P.: Implication in fuzzy logic. Int. J. Approximate Reasoning **1**, 327–347 (1987)
25. Reichenbach, H.: Wahrscheinlichkeitslehre: eine Untersuchungüber die logischen und mathematischen Grundlagen der Wahrscheinlichkeitsrechnung (1935)
26. Reichenbach, H.: The Theory of Probability, an Inquiry into the Logical and Mathematical Foundations of the Calculus of Probability. University of California Press (1949)
27. Willmott, R.: Two fuzzier implication operators in theory of fuzzy power sets. Fuzzy Sets Syst. **4**, 31–36 (1980)
28. Wu, W.M.: Fuzzy reasoning and fuzzy relational equations. Fuzzy Sets Syst. **20**, 67–78 (1986)
29. Klir, G. J.: Multivalued logic versus modal logics: alternative frameworks for uncertainty modeling, In: Wang, P.P. (ed.) Advances in Fuzzy Theory and Technology. Duke Univ., Durham, NC
30. Driankov, D., Saffiotti, A.: Fuzzy Logic Techniques for Autonomous Vehicle Navigation. Springer, Berlin, Heidelberg, GmbH (2001)
31. Shim, E.A., Rhee, F., C.-H.: General type-2 fuzzy membership function design and its application to neural networks. In: 2011 IEEE International Conference on Fuzzy Systems. Taipei, Taiwan, June 27–30, 2011
32. Mendel, J.M.: Uncertainty Rule-Based Fuzzy Logic Systems: Introduction and New Directions. Prentice Hall, Upper Saddle River (2001)
33. Karnik, N.N., Mendel, J.M., Liang, Q.: Type-2 fuzzy logic systems. IEEE Trans. Fuzzy Syst. **7** (6) (1999)
34. Mendel, J.M., John, R.I.: Type-2 fuzzy sets made simple. IEEE Trans. Fuzzy Syst. **10**(2), 117–127 (2002)
35. Liu, F.: An efficient centroid type reduction strategy for general type-2 fuzzy logic system. Inform. Sci. **178**, 2224–2236 (2008)
36. Zhai, D., Mendel, J.M.: Enhanced centroid-flow algorithm for computing the centroid of general type-2 fuzzy sets, IEEE Trans. Fuzzy Syst. **20**(5), 939–956 (2012)
37. Karnik, N.N., Mendel, J.M.: Centroid of a type-2 fuzzy set. Inform. Sci. **132**, 195–220 (2001)
38. Wu, D., Mendel, J.M.: Enhanced Karnik-Mendel algorithms. IEEE Trans. Fuzzy Syst. **17**(4), 923–934 (2009)
39. Türksen, I.B.: Type I and Type II fuzzy system modeling. Fuzzy Set. Syst. **106**, 11–34 (1999)
40. Oliveira, J.V., Pedrycz, W.: Advances in fuzzy clustering and its applications. Wiley (2007)
41. Fazel Zarandi, M.H., Turksen, I.B., Torabi Kasbi, O.: Type-2 fuzzy modeling for desulphurization of steel process. Expert Syst. Appl. **32**, 157–171 (2007)
42. Rhee, F., Choi, B.: Interval type-2 fuzzy membership function design and its application to radial basis function neural networks. In: Proceedings of the 2007 IEEE International Conference on Fuzzy Systems, pp. 2047–2052 (2007)
43. Rhee, F., Hwang, C.: A type-2 fuzzy C-means clustering algorithm, in: Proceedings of the 2001 Joint Conference IFSA/NAFIPS, 2001, pp. 1919–1926
44. Rhee, F., Hwang, C.: An interval type-2 fuzzy perceptron. In: Proceedings of the 2002 IEEE International Conference on Fuzzy Systems, pp. 1331–1335 (2002)

45. Rhee, F., Hwang, C.: An interval type-2 fuzzy K-nearest neighbor. In: Proceedings of the 2003 IEEE International Conference on Fuzzy Systems, pp. 802–807 (2003)
46. Choi, B. I., Rhee, F.C.: Interval type-2 fuzzy membership function generation methods for pattern recognition. Inf. Sci. **179**, 2102–2122 (2009)
47. Hwang, C., Rhee, F.: Uncertain fuzzy clustering: interval type-2 fuzzy approach to C-means. IEEE Trans. Fuzzy Syst. **15**, 107–120 (2007)
48. Aliev, A.R., Pedrycz, W., Guirimov, B.G., Aliev, R.R., Ilhan, U., Babagil, M.: Evolution optimization. Inf. Sci. **181**(9), 1591–1608 (2011)
49. Fazel Zarandi, M.H., Gamasaee, R., Turksen, I.B.: A type-2 fuzzy c-regression clustering algorithm for Takagi–Sugeno system identification and its application in the steel industry. Inf. Sci. **187**, 179–203 (2012)
50. Melin, P., Castillo, O.: A review on type-2 fuzzy logic applications in clustering, classification and pattern recognition. Appl. Soft Comput. **21**, 568–577 (2014)
51. Sharma, P., Bajaj, P.: Performance analysis of vehicle classification system using type-1 fuzzy, adaptive neuro-fuzzy and type-2 fuzzy inference system. In: Proceedings of the 2nd International Conference on Emerging Trends in Engineering and Technology. ICETET 2009, pp. 581–584, 2009 (art. no. 5395411)
52. Sharma, P., Bajaj, P.: Accuracy comparison of vehicle classification system using interval type-2 fuzzy inference system. In: Proceedings of the 3rd International Conference on Emerging Trends in Engineering and Technology, ICETET 2010, pp. 85–90 (2010)
53. Tan, W.W., Foo, C.L., Chua, T.W.: Type-2 fuzzy system for ECG arrhythmic classification. In: IEEE International Conference on Fuzzy Systems, 2007 (art. no. 4295478)
54. Pimenta, A.H.M., Camargo, H.A.: Interval type-2 fuzzy classifier design using genetic algorithms. In: 2010 IEEE World Congress on Computational Intelligence (WCCI), 2010 (art. no. 5584520)
55. Chumklin, S.: Auephanwiriyakul, S., Theera-Umpon, N.: Micro calcification detection in mammograms using interval type-2 fuzzy logic system with automatic membership function generation. In: 2010 IEEE World Congress on Computational Intelligence (WCCI), 2010 (art. no. 5584896)
56. Sanz, J., Fernandez, A., Bustince, H., Herrera, F.: A genetic algorithm for tuning fuzzy rule based classification systems with interval valued fuzzy sets. In: 2010 IEEE World Congress on Computational Intelligence (WCCI), 2010 (art.no. 5584097)
57. Wu, H., Mendel, J.M.: Classification of battlefield ground vehicles based on the acoustic emissions. Stud. Comput. Intell. **304**, 55–77 (2010)
58. Phong, P.A., Thien, K.Q.: Classification of cardiac arrhythmias using interval type2 TSK fuzzy system. In: Proceedings of the 1st International Conference Knowledge and Systems Engineering, pp. 1–6, 2009 (art. no. 5361742)
59. Abiyev, R.H., Kaynak, O., Alshanableh, T., Mamedov, F.: A type-2 neuro-fuzzy system based on clustering and gradient techniques applied to system identification and channel equalization. Appl. Soft Comput. J. **11**, 1396–1406 (2011)
60. Zeng, J., Liu, Z.-Q.: Type-2 fuzzy hidden Markov models to phoneme recognition. In: Proceedings of the International Conference on Pattern Recognition, vol. 1, pp. 192–195 (2004)
61. Abiyev, R.H., Kaynak, O.: Type-2 fuzzy neural structure for identification and control of time varying plants. IEEE Trans. Ind. Electron. **57**, 4147–4159 (2010)
62. Zheng, G., Xiao, J., Wang, J., Wei, Z.: A similarity measure between general type2 fuzzy sets and its application in clustering. In: Proceedings of the World Congress on Intelligent Control and Automation, pp. 6383–6387, 2010 (art. no.5554327)
63. Abiyev, R.H., Kaynak, O.: Type-2 fuzzy neural structure for identification and control of time varying plants. IEEE Trans. Ind. Electron. **57**, 4147–4159 (2010)
64. Ozkan, I., Turksen, B.: MiniMax ε-stable cluster validity index for type-2 fuzziness. In: Proceedings of the NAFIPS 2010 Conference, 2010 (art. no. 5548183)
65. Pedrycz, W.: Human centricity in computing with fuzzy sets: an interpretability quest for higher order granular constructs. J. Ambient Intell. Humaniz. Comput. **1**, 65–74 (2010)

66. Juang, C.-F., Huang, R.-B., Lin, Y.-Y.: A recurrent self-evolving interval type-2 fuzzy neural network for dynamic system processing. IEEE Trans. Fuzzy Syst. **17**, 1092–1105 (2009)
67. Türkşen, I.B.: Review of fuzzy system models with an emphasis on fuzzy functions. Trans. Inst. Meas. Control **31**, 7–31 (2009)
68. Ren, Q., Baron, L., Balazinski, M.: High order type-2 TSK fuzzy logic system. In: Proceedings of the NAFIPS 2010 Conference, 2008 (art. no. 4531215)
69. Qun, R., Baron, L., Balazinski, M.: Type-2 Takagi–Sugeno–Kang fuzzy logic modeling using subtractive clustering. In: Proceedings of the Annual Conference of the North American Fuzzy Information Processing Society—NAFIPS, 2006, pp. 120–125 (art. no. 4216787)
70. Melin, P., Interval type-2 fuzzy logic applications in image processing and pattern recognition. In: Proceedings of the 2010 IEEE International Conference on Granular Computing, GrC 2010, pp. 728–731 (2010)
71. Lopez, M., Melin, P., Castillo, O.: Comparative study of feature extraction methods of fuzzy logic type 1 and type-2 for pattern recognition system based on the mean pixels. Stud. Comput. Intell. **312**, 171–188 (2010)
72. Li, H., Zhang, X.: A hybrid learning algorithm based on additional momentum and self-adaptive learning rate. J. Comput. Inf. Syst. **6**, 1421–1429 (2010)
73. Own, C.-M.: Switching between type-2 fuzzy sets and intuitionistic fuzzy sets: an application in medical diagnosis. Appl. Intell. **31**, 283–291 (2009)
74. Mendoza, O., Melin, P., Castillo, O.: Interval type-2 fuzzy logic and modular neural networks for face recognition applications. Appl. Soft Comput. J. **9**, 1377–1387 (2009)
75. Kim, G.-S., Ahn, I.-S., Oh, S.-K.: The design of optimized type-2 fuzzy neural networks and its application. Trans. Korean Inst. Electr. Eng. **58**, 1615–1623 (2009)
76. Hidalgo, D., Castillo, O., Melin, P.: Type-1 and type-2 fuzzy inference systems as integration methods in modular neural networks for multimodal biometry and its optimization with genetic algorithms. Inf. Sci. **179**, 2123–2145 (2009)
77. Lopez, M., Melin, P., Castillo, O.: Optimization of response integration with fuzzy logic in ensemble neural networks using genetic algorithms. Stud. Comput. Intell. **154**, 129–150 (2008)
78. Ozkan, I., Türksen, I.B.: Entropy assessment for type-2 fuzziness. In: Proceedings of the IEEE International Conference on Fuzzy Systems, vol. 2, pp. 1111–1115 (2004)
79. Mitchell, H.B.: Pattern recognition using type-II fuzzy sets. Inf. Sci. **170**, 409–418 (2005)
80. Madasu, V.K., Hanmandlu, M., Vasikarla, S.: A novel approach for fuzzy edge detection using type II fuzzy sets. In: Proceedings of SPIE—The International Society for Optical Engineering, vol. 7075, 2008 (art. no. 70750I)
81. Tizhoosh, H.R.: Image thresholding using type II fuzzy sets. Pattern Recognit. **38**, 2363–2372 (2005)
82. Fazel Zarandi, M.H., Gamasaee, R.: Type-2 fuzzy hybrid expert system for prediction of tardiness in scheduling of steel continuous casting process. Soft. Comput. **16**, 1287–1302 (2012)
83. Karnik, N.N., Mendel, J.M.: Applications of type-2 fuzzy logic systems to forecasting of time-series. Inf. Sci. **120**, 89–111 (1999)
84. Fazel Zarandi, M.H., Gamasaee, R., Turksen, I.B.: A type-2 fuzzy expert system based on a hybrid inference method for steel industry. Int. J. Adv. Manuf. Technol. **71**, 857–885 (2014)
85. Gaxiola, F., Melin, P., Valdez, F., Castillo, O.: Generalized type-2 fuzzy weight adjustment for backpropagation neural networks in time series prediction. Inf. Sci. **325**, 159–174 (2015)
86. Pramanik, S., Jana, D.K., Mondal, S.K., Maiti, M.: A fixed-charge transportation problem in two-stage supply chain network in Gaussian type-2 fuzzy environments. Inf. Sci. **325**, 190–214 (2015)
87. Fazel Zarandi, M.H., Gamasaee, R.: A type-2 fuzzy system model for reducing bullwhip effects in supply chains and its application in steel manufacturing. Sci. Iranica **20**, 879–899 (2013)

Part III
Fuzzy Sets in Branches of Science

Fuzzy Sets in Earth and Space Sciences

Irem Otay and Cengiz Kahraman

Abstract Earth science refers to the field of science dealing with planet Earth while space science pertains several scientific disciplines studying the upper atmosphere, space, and celestial bodies rather than Earth. The fuzzy set theory is one of the tools that has been recently used in the earth and space sciences. In this chapter, we review and analyze the papers utilizing fuzzy logic in earth and space science problems from Scopus database. The graphical and tabular illustrations are presented for the subject areas, publication years and sources of the papers on earth and space sciences.

Keywords Earth science · Space science · Fuzzy set theory · Geology · Astronomy · Oceanography · Meteorology

1 Introduction

Earth science includes the disciplines *atmosphere, hydrosphere, lithosphere,* and *biosphere*. Earth scientists use tools from physics, chemistry, biology, chronology, and mathematics to understand how the Earth system works, and how it has evolved to its current state. The studies on earth science can be classified as follows: energy and mineral resources, protective activities for the environment of Earth, and dangerous natural disasters such as volcanoes, earthquakes and hurricanes.

Earth science can be divided into four sub-sciences. Geology handles the composition of Earth materials, Earth structures, Earth processes, and the organisms of the planet. Meteorology is the study of the atmosphere and how processes in the

I. Otay (✉)
Department of Managemet Engineering, Istanbul Technical University, 34367 Macka, Istanbul, Turkey
e-mail: iremotay@itu.edu.tr

C. Kahraman
Department of Industrial Engineering, Istanbul Technical University, 34367 Macka, Istanbul, Turkey

C. Kahraman et al. (eds.), *Fuzzy Logic in Its 50th Year*,
Studies in Fuzziness and Soft Computing 341,
DOI 10.1007/978-3-319-31093-0_7

atmosphere determine Earth's weather and climate. Oceanography studies the composition, movement, organisms and processes of oceans since the oceans are excellent energy sources and may cause climate changes. Astronomy is the study of the universe, which requires knowledge of Earth materials, processes and history to understand the planets in the universe. It is a natural science of celestial objects and the physics, chemistry, and evolution of such objects.

The space science covers the areas of science to which new knowledge can be contributed by means of space vehicles, i.e., sounding rockets, satellites, and lunar and planetary probes, either manned or unmanned. Space science makes extensive contributions to geophysics and astronomy. More accurate mapping of the earth's surface may offer a better basis for plotting important information, such as geological data, geographic locations, crops, forests, water resources, and land-use patterns. On the other hand, the measurements that make geodesy important to the scientists are also invaluable to the military navigation satellites. More accurate geodetic measurements can provide more information on the size and shape of the earth and give information about the distribution of mass in the earth's crust and stresses in the mantle.

The uncertainty, vagueness and lack of data in earth and space sciences complicate the problems that are hard to solve by classical mathematical approaches. For solving such complex problems, some scientists have intended to use the fuzzy set theory. Zadeh [39] introduced the fuzzy set theory to deal with the uncertainty due to imprecision and vagueness. A fuzzy set is a class of objects with a continuum of grades of membership. Fuzzy logic is a branch of mathematics that allows a computer to model the real world in the same way that people do. It provides a simple way to reason with vague, ambiguous, and imprecise input or knowledge [14]. It does not force the researchers to make sharp definitions about the membership of an element in a set but allows them to consider a degree of membership.

The rest of the chapter is organized as follows: In Sects. 2 and 3, earth and space sciences are briefly introduced. In Sect. 4, some graphical illustrations together with a literature review on fuzzy sets in earth and space sciences are presented. In Sect. 5, conclusions and the suggestions for further research are stated.

2 Earth Science

Earth Science is defined as the study of the Earth including its origin, its structure and all the changes it has gone through by considering the potential results of these changes to the past and the future [5]. Earth Science aims to understand the changes on Earth and its consequences on life for the planet. Earth science comprises of geology, meteorology, oceanography and astronomy [8].

Geology is the study of Earth by focusing on its solid materials and the processes which create the structure of Earth [42]. Geology is divided into two groups which are physical geology (the study of Earth's surface and materials) and historical geology (the origin of Earth and changes on it [20]).

Meteorology is basically the study covering all aspects of climate (long-term weather patterns), weather (short-term patterns) as well as atmosphere [42].

Oceanography, the study of oceans considering both living and non-living things, is divided into four groups which are physical oceanography, biological oceanography, chemical oceanography, and geological oceanography [23, 42].

Astronomy is the study of the nature and the movements of galaxies, planets, and stars. It briefly analyzes all objects and phenomenon outside from the atmosphere of Earth [29].

3 Space Science

Space science is defined as the study of the universe in which Earth moves. In other words, beyond the planet (galaxies, planets, and stars which are outside atmosphere of Earth) is studied in this area.

Early societies as Egyptians observed the movement of Sun, Moon and some of the planets in which only five of them can be seen with a naked eye. A combination of stargazing, myths, astrology and as well as religion, is called ancient astronomy [1].

For ages ago, people believed that Earth was located at the center of the universe (Earth-center cosmology), did not move and it had a shape of flat. But scientific inventions with the rapid advances in technology released by Galileo Galilei, Johannes Kepler and Sir Isaac Newton that Earth is not located in the center of the universe (heliocentric view of solar system) and the orbits of the planet is elliptical [42, 1].

With the invention of telescope around 16th century, scientist were able to see far into space. Since 20th century spacecrafts have been sent to many planets which are in the solar system and the first footprint were left on Moon's surface. Especially for two decades, lots of images from the universe have been collected from the Hubble Space Telescope and Cosmic Background Explorer [11]. There are a wide range of topics that have been studied in space science such as astronomy, astrophysics, geophysics, geography, geodesy, biology (molecular biology, space biology etc.), cosmochemistry, and space vehicles (design, construction a launch).

4 Fuzzy Sets in Earth and Space Sciences

The fuzzy set theory has been applied to almost all sciences including earth and space sciences. In the following, the works on earth and space sciences are classified with respect to the subjects such as geology, astronomy, oceanography, astrophysics, and meteorology.

By typing "*Fuzzy, Earth Science, Space Science, astronomy and astrophysics, geology, meteorology, oceanography*" on Scopus, all types of papers are sorted whether they include these words in their article title, abstract and keywords. The summarized data obtained from Scopus are listed in Table 1.

The subjects presented in Table 1 are analyzed in the following.

4.1 *Fuzzy Set Theory in Astronomy and Astrophysics*

Barry and Gathmann [3] described current space systems autonomy research and recent IntelliSTARTM advancements at the Space Systems Division in Rockwell International. The paper focused on the current application of fuzzy logic control, and an advancement designed to increase responsiveness with the architecture.

Shioiri and Ueno [31] proposed a new collision avoidance control law for aircrafts. Range of Closest Point of Approach (RCPA), representing risk in the future, and TCPA (Time to CPA), representing current risk, were used as risk functions. Fuzzy logic was introduced in order to achieve human maneuvering and to settle singular point and chattering problems. Avoidance strategy consisted of four main phases, maintaining course, avoidance, parallel flight and recovery phase, and three transition phases. The parallel flight phase and three transition phases were introduced to achieve moderate avoidance and smooth phase transition, respectively.

Wang et al. [37] combined the fixed thrust optimal-fuzzy combined formation control method and time-fuel optimal control method to form the optimal-fuzzy formation control. Optimal control was adopted when relative position errors reach a certain threshold, while fuzzy control worked for the small errors lower than that

Table 1 Summarized data obtained from Scopus

Search for	1990–1995	1996–2000	2001–2005	2006–2010	2011–2015	Total
Fuzzy Earth Science	2	11	21	47	53	134
Fuzzy Space Science	1	–	3	6	7	17
Fuzzy astronomy	1	7	7	24	13	52
Fuzzy astrophysics	–	4	3	12	4	23
Fuzzy geology	21	37	97	201	187	543
Fuzzy meteorology	5	25	52	109	139	330
Fuzzy oceanography	4	6	17	72	39	138
Total	34	90	200	471	442	1237

threshold. Valdés and Bonham-Carter [35] used an intelligent approach namely time dependent neural fuzzy network models to detect internal state changes in complex processes. The researchers highlighted the effectiveness of the proposed model applied to earth and planetary sciences using paleoclimate and solar data. Valdés and Bonham-Carter [35] suggested Grid and high throughput computing model composed of neuro-fuzzy networks and genetic algorithms for solving earth sciences and astrophysics problems. In the study, simulation experiments were done and for the application part paleoclimate and solar data were used. Shamir and Nemiroff [30] discussed modern astronomical pipelines by taking into account fuzzy-logic based decision-making. In their study, fuzzy-logic algorithms were proposed for the first time in the literature for astronomical purposes and tested with data from Night Sky Live sky survey. Furfaro et al. [9] suggested a fuzzy expert system acquiring past and present water/energy indicators as well as evaluating habitability. The proposed system was demonstrated for two hypothesized regions on Mars. The simulation results of the proposed model were given the same results provided by the human experts. Omar [24] proposed a systematic and simple procedure to develop an integrated fuzzy-based guidance law which consisted of three fuzzy logic controller (FLC). The parameters of all the fuzzy controllers, which included the distribution of the membership functions and the rules, were obtained simply by observing the function of each controller. These parameters were tuned by genetic algorithms by solving an optimization problem to minimize the interception time, missile acceleration commands, and miss distance. Barua and Khorasani [4] presented a hierarchical fault diagnosis framework and methodology that enabled a systematic utilization of fuzzy rule-based reasoning to enhance the level of autonomy achievable in fault diagnosis at ground stations. Fuzzy rule-based fault diagnosis schemes for satellite formation flight were developed and investigated at different levels in the hierarchy for a leader-follower architecture. Effectiveness of their proposed fault diagnosis methodology was demonstrated by utilizing synthetic formation flying data of five satellites that were configured in the leader-follower architecture. Moradi et al. [21] presented a method for three-dimensional attitude stabilization of a satellite. The pitch loop of the satellite was controlled by a momentum wheel; whereas the roll/yaw loops were stabilized by using two magnetic torques along their respective axes. An adjustable adaptive fuzzy system was proposed as the method to design the controller. More specifically, Glumov and Krutova [10] concentrated using fuzzy logic for designing control systems of space vehicles. Potyupkin [26] applied fuzzy measure for monitoring the technical state of space vehicles, realibility and tolerance monitoring. Kim et al. [15] solved the problem of spacecraft attitude control using an adaptive neuro-fuzzy inference system (ANFIS). ANFIS produced a control signal for one of the three axes of a spacecraft's body frame, so in total three ANFISs were constructed for 3-axis attitude control.

Figure 1 displays the distribution of the papers using the fuzzy set theory on astronomy with respect to years. It is remarkable that the studies on the fuzzy astronomy have a peak point in the year 2006.

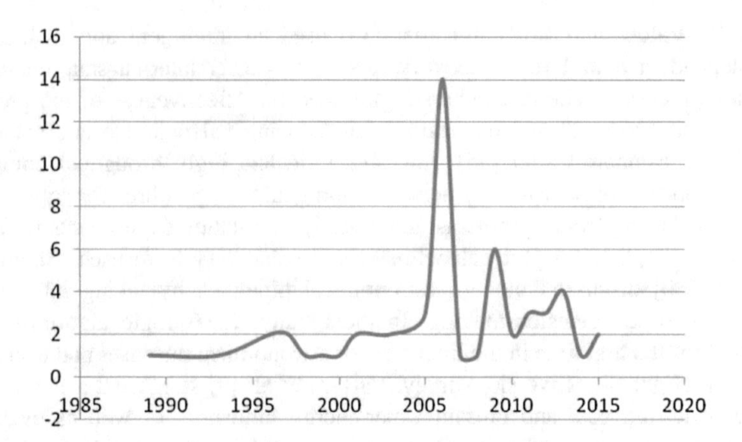

Fig. 1 Distribution of the papers using the fuzzy set theory on astronomy

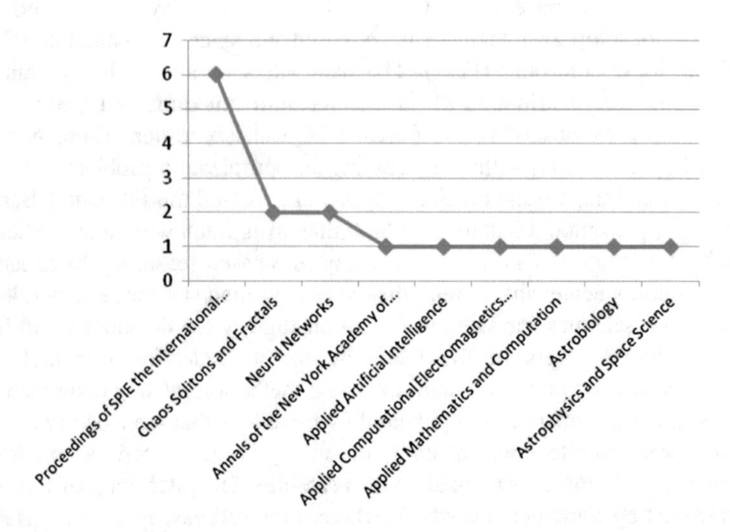

Fig. 2 Journals publishing fuzzy astronomy papers

Figure 2 shows the journals publishing astronomy papers by using the fuzzy set theory. According to the figure, most of the papers were published by Proceedings of SPIE—The International Society for Optical Engineering.

4.2 *Fuzzy Set Theory in Geology*

Kumar et al. [16] applied a fuzzy-MLP neural network for cross-borehole soil-geology interpolation offering many advantages in dealing with the nonlinearity inherent. The fuzzy-MLP neural network providing interpolation accuracy, takes advantage of both MLP neural networks and fuzzy set theory. Demicco [6] made an review on the application of fuzzy logic in geology. Lu et al. [19] concentrated on geological background and environmental engineering geology factors in Lueyang County. The paper proposed the quality assessment index system and the quality assessment standards. Using geographic information systems (GIS) technology and multiple stage fuzzy logic, environmental engineering geology quality of the country was evaluated. Lu et al. [18] used fuzzy mathematics and grey theory in order to assess and estimate the dangerous rank of mine geology disasters consisting of mine water, mine solid waste, apron and slide, ground collapse sink and underground fracture for different areas. Li et al. [17] developed an inference model based on the fuzzy logic technique in order to estimate the generation possibility of internal waves which has been one of the complex issue. The model providing satisfactory results highlighted that fuzzy based inference model was found as a useful tool for estimation of internal waves in ocean engineering. Sun and Li [33] developed a multidomain clustering inversion algorithm and used fuzzy c-means clustering technique for converting statistical petrophysical data into a deterministic geophysical inversion framework. They tested the algorithm with examples of synthetic and a field data.

In the literature, some of the fuzzy methods used in earthquake prediction are as fuzzy pattern recognition; fuzzy clustering analysis, fuzzy information retrieval, fuzzy similarity choice, fuzzy multi-factorial evaluation, fuzzy reasoning, fuzzy self-similarity analysis and fuzzy fractal dimension, gray fuzzy prediction, fuzzy neural network,fuzzy analyzing and processing software systems. For instance, Huang [13] studied on some topics such as fuzzy earthquake research and fuzzy earthquake engineering. The author also aimed to predict and analyze earthquake data using Hybrid Fuzzy Neural Networks with Information Diffusion Method.

Figure 3 illustrates the distribution of the papers using the fuzzy set theory on geology with respect to years. The figure indicates that the studies on the fuzzy geology have a peak point in the year 2009.

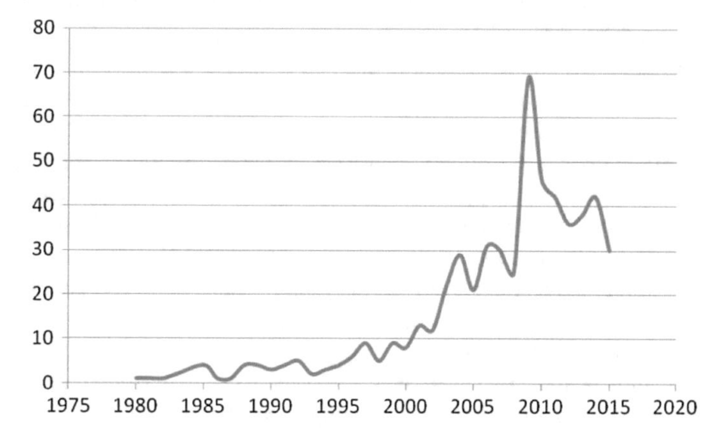

Fig. 3 Distribution of the papers using the fuzzy set theory on geology

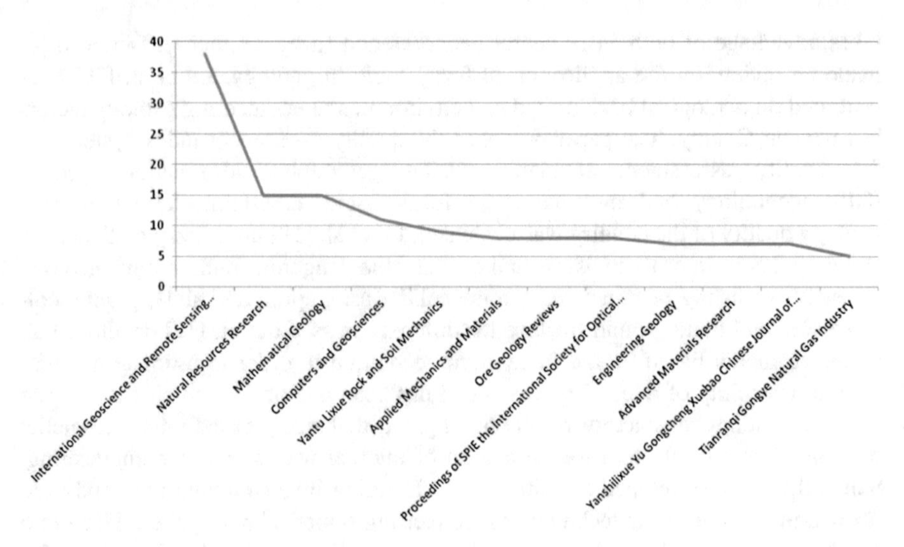

Fig. 4 Journals publishing fuzzy geology papers

Figure 4 displays the journals publishing geology papers by using the fuzzy set theory. Most of the papers were published by the proceedings of International Geoscience and Remote Sensing Symposium (IGARSS).

4.3 Fuzzy Set Theory in Meteorology

Amanullah et al. [2] concentrated on climatological changes, and analyzed statistical records of annual rainfall value from Uthamapalayam station between 1970 and 2000 and used the fuzzy set theory for the solution of the problem. Heidari et al. [12] employed an adaptive neural fuzzy system and studied climatic parameters for deriving the amount of monthly rainfall with acceptable accuracy for four cities of Semnan. In the study, fuzzy logic was used to establish a relationship among the observed and measured nonlinear meteorological phenomena. Srivastava et al. [32] aimed to forecast monthly rainfall in India by using Fuzzy-Ranking Algorithm (FRA) with some ocean-atmospheric predictor variables. Artificial Neural Network (ANN) technique was also applied for analyzing nonlinear relationship between inputs of the model and rainfall. In the study, a new approach based on fuzzy c-mean clustering was suggested. Rahoma et al. [27] proposed Adaptive Neuro-Fuzzy Inference System (ANFIS) integrating fuzzy logic and neural network techniques for forecasting meteorological parameters.

Figure 5 shows the distribution of the papers using the fuzzy set theory on meteorology with respect to years. It is noticeable that the studies on the fuzzy meteorology show an increasing trend starting from the year 1998.

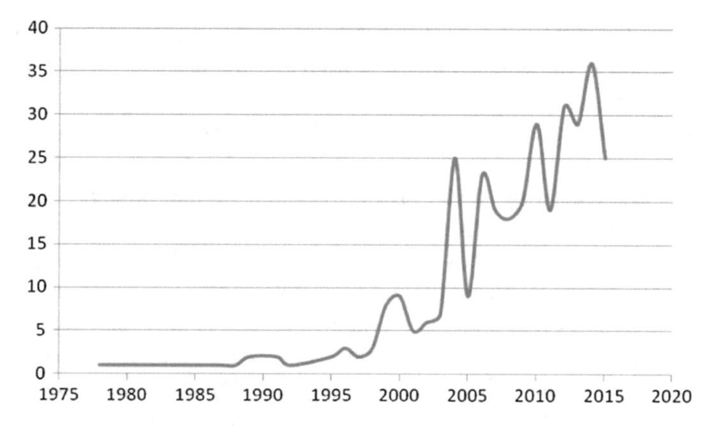

Fig. 5 Distribution of the papers using the fuzzy set theory on meteorology

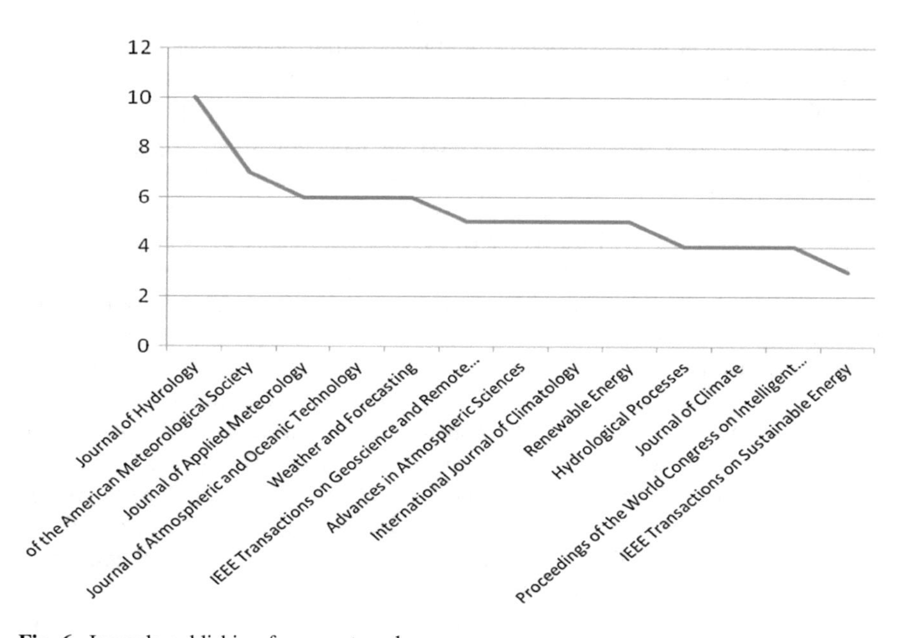

Fig. 6 Journals publishing fuzzy meteorology papers

Figure 6 illustrates the journals publishing meteorology papers by using the fuzzy set theory. Most of the papers were published by Journal of Hydrology.

4.4 *Fuzzy Set Theory in Oceanography*

Sylaios et al. [34] applied a Takagi-Sugeno-rule-based Fuzzy Inference System for forecasting the wind speed and direction as well as the lagged-wave characteristics in Operational Oceanography. In the study, the wind and wave dataset recorded in years 2000–2006 in the Aegean Sea was used and the proposed model performed good results between the predicted and the observed wave heights. Ding et al. [7] suggested intelligent data processing methods including fuzzy c-means algorithm based intelligent data cleaning method and the greedy clustering algorithm based data filtering and clustering method for the oceanic warning system. Sathiya et al. [28] employed neural networks and fuzzy logic for assessing the wave heights of the ocean for the coastal regions of Tamil Nadu. Zhang et al. [41] focused on the uncertain event detection method based on ontology reasoning and fuzzy logic as well as the uncertain event recognition method related to Bayesian network. These methods were used for updating uncertain environment ontology mode. Piedra-Fernandez et al. [25] studied on the development and validation of a content-based image retrieval system and used fuzzy logic and neurofuzzy systems. The researchers aimed to retrieve mesoscale oceanic structures from satellite images which were derived from the National Oceanic and Atmospheric Administration satellite. Morello et al. [22] presented a hybrid quality control system including a series of tests and an experimental fuzzy logic approach to assess data related to Australia's coastal ocean ecosystems. Zhang and Cao [40] presented a non-traditional fuzzy quantification method for modeling wave height. The proposed fuzzy model combining Poisson process and generalized Pareto distribution (GPD) model was used to identify the wave extremes in the time series data. Vidal-Fernandez et al. [36] suggested an object-based image analysis system. The system segmented and classified regions contained in sea-viewing field-of-view sensor and Moderate Resolution Imaging Spectro-radiometer -Aqua sensor satellite images into mesoscale oceanic structures. In the study, regions were labeled and classified by learning algorithms such as decision tree, Bayesian network, artificial neural network, and genetic algorithm. In addition, fuzzy descriptors were used in the study and the performance of the proposed system was tested with images from the Canary Islands and the North West African coast. Zhang and Cao (2015) suggested a non-traditional fuzzy quantification method integrating Poisson process and generalized Pareto distribution (GPD) model for modeling wave height. Wang et al. [38] implemented the fuzzy comprehensive evaluation model for evaluating operational ocean observing equipment in China.

Figure 7 demonstrates the distribution of the papers using the fuzzy set theory on oceanography with respect to years. The studies on the fuzzy oceanography have a peak point in the year 2009.

Figure 8 presents the journals publishing oceanography papers by using the fuzzy set theory. Most of the papers were published by Proceedings of the International Conference on Offshore Mechanics and Arctic Engineering—OMAE.

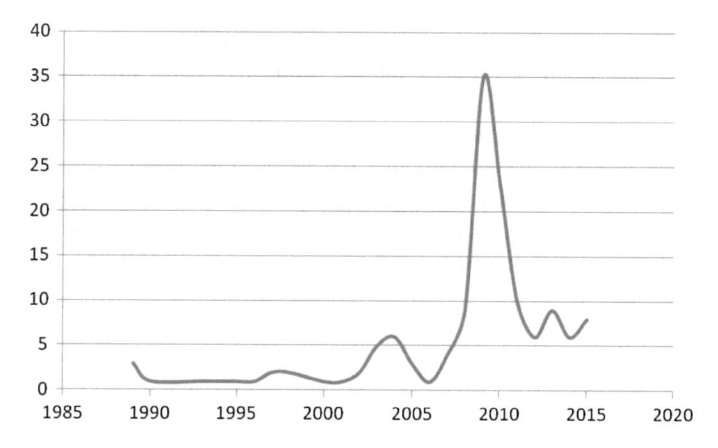

Fig. 7 Distribution of the papers using the fuzzy set theory on oceanography

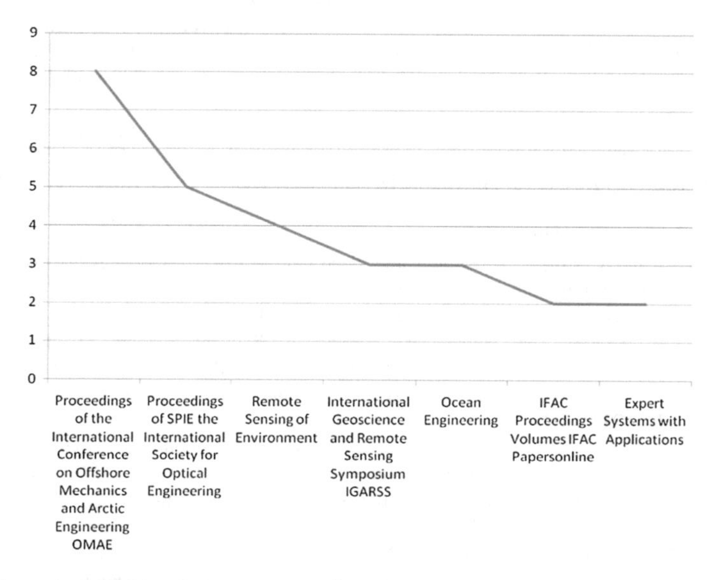

Fig. 8 Journals publishing fuzzy oceanography papers

5 Conclusion Remarks and Future Research Suggestions

Earth and space sciences are one of the sciences that uncertainty and incomplete information is often faced with. A geologist generally comes across with uncertain earth structures while an astronomer encounters vehicle control problems. The uncertainty in these kinds of problems can be captured by the tools of fuzzy set theory. Fuzzy logic control provides appropriate solutions for the complicated problems of not only earth sciences but also space sciences.

The literature review has demonstrated that there has been an increasing trend in the usage of the fuzzy set theory for the studies in earth and space sciences. Specifically, the year of 2000 is almost the starting point of this trend. It is expected that this trend will continue with a larger acceleration.

For further research, we suggest that the recently extended fuzzy sets such as intuitionistic fuzzy sets, hesitant fuzzy sets, type-2 fuzzy sets, fuzzy multisets, and non-stationary fuzzy sets to be employed in solving the complex problems of geology, astronomy, oceanography, astrophysics, and meteorology.

References

1. Angelo, J.A.: Encyclopedia of Space and Astronomy. Facts on File, Inc., NY-USA (2006)
2. Amanullah, S., Sundaram, A., Narayanan, E.S.L.: Fuzzy set theory application in meteorology —rainfall modelling. Indian J. Environ. Prot. **25**(8), 673–679 (2005)
3. Barry, J.M., Gathmann, T.P.: Application of fuzzy logic control to the IntelliSTARTM architecture. In: Proceedings of the 16th AAS Rocky Mountain Guidance and Control Conference on Advances in the Astronautical Sciences, vol. 81, pp. 423–430, Keystone-CO, USA (1993)
4. Barua, A., Khorasani, K.: Hierarchical fault diagnosis and fuzzy rule-based reasoning for satellites formation flight. IEEE Trans. Aerosp. Electron. Syst. **47**(4), 2435–2456 (2011)
5. Cullen, K.E.: Earth Science: The People Behind the Science. Chelsea House, NY-USA (2006)
6. Demicco, R.V.: Fuzzy Logic and Earth Science: An Overview, Fuzzy Logic in Geology. Academic Press, London (2003)
7. Ding, Y., Han, H., Liu, F.: Intelligent integrated data processing model for oceanic warning system. Knowl. Based Syst. **23**(1), 61–69 (2010)
8. Feather, R.M., Snyder, S.L. : Earth Science. McGraw-Hill, Glencoe (1999)
9. Furfaro, R., Dohm, J.M., Fink, W., Kargel, J., Schulze-Makuch, D., Fairen, A.G., Palmero-Rodriguez, A., Baker, V.R., Ferre, P.T., Hare, T.M., Tarbell, M.A., Miyamoto, H., Komatsu, G.: The search for life beyond Earth through fuzzy expert systems. Planet. Space Sci. **56**, 448–472 (2008)
10. Glumov, V.M., Krutova, I.N.: A fuzzy logic adaptation circuit for control systems of deformable space vehicles: its design. Autom. Remote C **63**(7), 1109–1122 (2002)
11. Guo, H., Wu, J.: Space Science & Technology in China: A Roadmap to 2050. Science Press Beijing, ISBN 978-7-03-025703-1,(2010)
12. Heidari, M., Nabavi, S.H., Shamshirband, S.: Application of an adaptive neural-fuzzy system to establish a relationship among nonlinear phenomena in meteorology to obtain monthly rainfall. In: ICSTE 2010 Proceedings of 2nd International Conference on Software Technology and Engineering, vol. 2, pp. 232–235 (2010)
13. Huang, C.: Fuzzy Logic and Earthquake Research, Fuzzy Logic in Geology. Academic Press, London (2003)
14. Kaya, İ., Kahraman, C.: Process capability analyses with fuzzy parameters. Expert Syst. Appl. **38**, 11918–11927 (2011)
15. Kim, S.-W., Park, S.-Y., Park, C.-D.: Preliminary test of adaptive neuro-fuzzy inference system controller for spacecraft attitude control. J. Astron. Space Sci. **29**(4), 389–395 (2002)
16. Kumar, J.K., Konno, M., Yasuda, N.: Subsurface soil-geology interpolation using fuzzy neural network. J. Geotech. Geoenviron. Eng. **126**(7), 632–639 (2000)
17. Li, J.-X., Zhang, R., Jin, B.-G., Wang, H.-Z.: Possibility estimation of generating internal waves in the northwest Pacific Ocean using the fuzzy logic technique. J. Marine Sci. Technol. **20**(2), 237–244 (2012)

18. Lu, D.-W., Wu, L.R., Li, Z.-X.: The evaluation of mine geology disasters based on fuzzy mathematics and grey theory. J. Coal Sci. Eng. **13**(4), 480–483 (2007)
19. Lu, Y.-D., Zhang, J., Li, M.-S.: GIS-based evaluation of urban environmental engineering geology by fuzzy logic: a case study of Lüeyang County, Shaanxi Province. J. Nat. Disasters **14**(2), 93–98 (2005)
20. Monroe, J.S., Wicander, R.: The Changing Earth: Exploring Geology and Evolution, 7th edn. Cengage Learning, CT-USA (2015)
21. Moradi, M., Esmaelzadeh, R., Ghasemi, A.: Adjustable adaptive fuzzy attitude control using nonlinear SISO structure of satellite dynamics. T. Jpn. Soc. Aeronaut. S **55**(5), 265–273 (2012)
22. Morello, E.B., Galibert, G., Smith, D., Ridgway, K.R., Howell, B., Slawinski, D., Timms, G. P., Evans, K., Lynch, T.P.: Quality control (QC) procedures for Australia's National Reference Station's sensor data-comparing semi-autonomous systems to an expert oceanographer. Methods Oceanogr. **9**, 17–33 (2014)
23. Nichols, C.R., Williams, R.G.: Encyclopedia of Marine Science. Facts on File, Inc., NY-USA (2009)
24. Omar, H.M.: Designing integrated fuzzy guidance law for aerodynamic homing missiles using genetic algorithms. T. Jpn. Soc. Aeronaut. S **53**(180), 99–104 (2010)
25. Piedra-Fernandez, J.A., Ortega, G., Wang, J.Z., Canton-Garbin, M.: Fuzzy content-based image retrieval for oceanic remote sensing. EEE Trans. Geosci. Remote Sens. **52**(9), 5422–5431 (2014)
26. Potyupkin, A.Y.: Application of a fuzzy measure in problems of monitoring the technical state of space vehicles. Meas. Tech. **45**(7), 689–695 (2002)
27. Rahoma, W.A., Rahoma, U.A., Hassan, A.H.: Application of neuro-fuzzy techniques for solar radiation. J. Comput. Sci. **7**(10), 1605–1611 (2011)
28. Sathiya, R.D., Vaithiyanathan, V., Suraj, M.S., Venkatraman, G.B., Sathivel, P.: Assessing the wave heights of the ocean using neural networks and fuzzy logic. In: Proceedings of 2013 International Conference on Emerging Trends in VLSI, Embedded System, Nano Electronics and Telecommunication System, ICEVENT 2013, Tiruvannamalai, Tamil Nadu, India (2013)
29. Schneider, P.: Extragalactic Astronomy and Cosmology. Springer-Verlag Berlin Heidelberg, (2015)
30. Shamir, L., Nemiroff, R.J.: Astronomical pipeline processing using fuzzy logic. Appl. Soft Comput. J. **8**(1), 79–87 (2008)
31. Shioiri, H., Ueno, S.: Three-dimensional collision avoidance control law for aircraft using risk function and fuzzy logic. T. Jpn. Soc. Aeronaut. S **47**(154), 253–261 (2004)
32. Srivastava, G., Panda, S.N., Mondal, P., Liu, J.: Forecasting of rainfall using ocean-atmospheric indices with a fuzzy neural technique. J. Hydrol. **395**(3–4), 190–198 (2010)
33. Sun, J., Li, Y.: Multidomain petrophysically constrained inversion and geology differentiation using guided fuzzy c-means clustering. Geophysics **80**(4) (2014)
34. Sylaios, G., Bouchette, F., Tsihrintzis, V.A., Denamiel, C.: A fuzzy inference system for wind-wave modelling. Ocean Eng. **36**(17–18), 1358–1365 (2009)
35. Valdés, J.J., Bonham-Carter, G.: Time dependent neural network models for detecting changes of state in complex processes: applications in earth sciences and astronomy. Neural Netw. **19** (2), 196–207 (2006)
36. Vidal-Fernandez, E., Piedra-Fernandez, J.A., Almendros-Jimenez, J.M., Canton-Garbin, M.: OBIA system for identifying mesoscale oceanic structures in SeaWiFS and MODIS-aqua images. IEEE J. Sel. Top. Appl. **8**(3), 1256–1265 (2015)
37. Wang, P., Zhao, X., Yang, D.: Fixed thrust optimal-fuzzy combined control for spacecraft formation flying. T. Jpn. Soc. Aeronaut. S **47**(155), 9–16 (2004)
38. Wang, Y., Li, Y., Liu, W., Gao, Y.: Assessing operational ocean observing equipment (OOOE) based on the fuzzy comprehensive evaluation method. Ocean Eng. **107**, 54–59 (2015)
39. Zadeh, L.: Fuzzy sets. Inf. Control **8**, 338–353 (1965)

40. Zhang Y.,Cao Y.: A fuzzy quantification approach of uncertainties in an extreme wave height modeling. Acta. Oceanologica. Sinica. **34**(3), 90–98 (2015)
41. Zhang, R., Yin, L., Gu, H.: Environment perception for AUV in uncertain ocean environment. Jisuanji Yanjiu yu Fazhan/Comput. Res. Dev. **50**(9), 1981–1991 (2013)
42. Zuchora-Walske, C.: Science Discovery Timelines, Key Discoveries in Earth and Space Science. Lerner Publishing Group, Inc., MN-USA (2015)

Fuzzy Sets and Fuzzy Logic in the Human Sciences

Michael Smithson

Abstract The development of fuzzy set theory and fuzzy logic provided an opportunity for the human sciences to incorporate a mathematical framework with attractive properties. The potential applications include using fuzzy set theory as a descriptive model of how people treat categorical concepts, employing it as a prescriptive framework for "rational" treatment of such concepts, and as a basis for analysing graded membership response data from experiments and surveys. However, half a century later this opportunity still has not been fully grasped. This chapter surveys the history of fuzzy set applications in the human sciences, and then elaborates the possible reasons why fuzzy set concepts have been relatively under-utilized therein.

Keywords Human sciences · Fuzzy sets · Fuzzy logic · Grade of membership · Fuzzy logical model of perception

1 Introduction

The development of fuzzy set theory and fuzzy logic provided an opportunity for the human sciences to incorporate a mathematical framework with attractive properties. These attractions have been expressed in largely overlapping ways by Smithson [42: 3–5], Ragin [35: 6–17], and Smithson and Verkuilen [45: 1–2]. They include the ability of fuzzy set theory and fuzzy logic to handle vagueness systematically, the simultaneously dimensional and categorical nature of many constructs in the human sciences, and the fact that many human science theories and hypotheses are expressed in quasi-logical or set-like terms. Given these properties, there would seem to be strong motivations for researchers in the human sciences to

---

M. Smithson (✉)
Research School of Psychology, The Australian National University, Canberra, Australia
e-mail: michael.smithson@gmail.com

C. Kahraman et al. (eds.), *Fuzzy Logic in Its 50th Year*,
Studies in Fuzziness and Soft Computing 341,
DOI 10.1007/978-3-319-31093-0_8

apply fuzzy sets and fuzzy logic, and initially there was some optimism both among fuzzy proponents and "pioneer" human scientists that these frameworks would be readily adopted throughout the human sciences.

However, half a century later this opportunity still has not been fully grasped. There are several possible reasons for this, some of which have persisted since the advent of fuzzy sets and others that have emerged in recent times. We will examine these reasons later, after a review of the history of fuzzy set applications in the human sciences. The primary focus of this chapter is on fuzzy sets rather than on fuzzy logic, chiefly because fuzzy set applications are much more common throughout the human sciences.

There are three kinds of potential applications of fuzzy sets throughout the human sciences. We will review these in this chapter, and it is important to distinguish among them. The first kind employs fuzzy set theory (FST) as a descriptive framework. Fuzzy set theory therefore is therefore evaluated on how well it corresponds to what humans do in dealing with categories and related phenomena. The second employs FST as prescriptive, i.e., how a "rational" agent would deal with categories. Human performance then is compared against FST as a "gold standard" for categorization judgements. The third kind employs FST as a data-analytic toolbox for testing hypotheses and theories (and thereby implicitly also as a prescriptive framework for dealing with categories). In this vein, researchers apply FST to operationalizing their concepts, measuring membership in categories, and testing hypotheses expressed in set-theoretic terms.

2 Fuzzy Set Theory as a Descriptive Framework

Most of these early applications treated FST as a descriptive framework. Early applications during the 1970s were predominately in psychology, although there also were applications in economics, anthropology, and medicine. Cognitive psychologists such as Rosch [39], Hersh and Caramazza [10], and Oden [27] conducted experimental studies comparing predictions from FST with human performance in categorization and related tasks. Oden and Massaro [28] published a fuzzy set theoretic account of perception. In economics, Ponsard [31, 32] was among the first to introduce fuzzy set concepts to the discipline in a comprehensive fashion. Several other papers appeared around this time, such as Biin and Whinston [2] on social choice and Grandmont [9] on preferences. Social anthropology had at least two early adopters in research on colour categories [14, 15]. Likewise, clinical medical research saw the development of a fuzzy set-based "grade of membership" model of types of disorders [60, 61].

By the early 1980s, several critiques of FST as a descriptive framework had appeared within psychology and related areas. The FST treatment of linguistic hedges was found not to always correspond with human intuition or behaviour [10, 15, 30, 42]. Osherson and Smith [29] observed that "guppy" is a better example of a "pet fish" than it is of a "pet" or a "fish", thereby contradicting the FST rules for membership in the conjunction of fuzzy sets. Researchers were divided on what could be implied from these differences, and seemed to fall into one of three camps. Critics of FST such as Osherson and Smith treated the differences as indicative that FST is both prescriptively and descriptively inadequate, a critique that they extended to all prototype theories of categorization. A minority of researchers, such as Hersh and Caramazza or Spies [50], identified some of the violations of FST as evidence of faulty reasoning or understanding on the part of the participants in the studies.

Smithson and Oden [48] pointed out a striking contrast between these verdicts and the nearly universal agreement that human violations of the conjunction rule in probability [56, 54] reflect inadequacies in human reasoning but do not cast any doubt on probability theory as a prescriptive framework. More recently, the quantum probability framework has been applied to account for these violations, thus potentially achieving both descriptive and prescriptive adequacy. No such reformation has been forthcoming for FST. Factors that may have produced this outcome include the facts that for some time FST lacked adequate axiomatic foundations, and had several alternative (and even competing) versions. Both of these may have resulted in doubts about the prescriptive adequacy of FST.

The third, and largest, camp of researchers has been prototype theorists such as Rosch, who may have had sympathy with the ideas behind FST but eschewed it and other formalisms. Journalists McNeill and Freiberger [25: 89–90] quoted Rosch as pointing out the parallels between her prototype-based framework and FST, but she never employed FST in her work. This remains the state of affairs today in most work on categorization and related topics. One can read this literature and identify occasional FST-like concepts and insights, but often not expressed in FST language. For example, the social psychological literature on social- and self-categorization has embraced the concept of prototypicality but not FST concepts such as simultaneous membership in multiple social identities. Capozza and Nanni [5] used FST to motivate theoretical ideas in this domain, and Smithson et al. [49] investigated factors influencing the extent to which people could allow simultaneously high membership in two national identities, but these are isolated instances.

Much the same can be said of other areas in psychology where FST has been used in nontrivial ways. Researchers of judgement and decision making under uncertainty might have found a role for FST, and a few did, but in a rather limited fashion. Early research on how people interpret verbal probability expressions (e.g., “likely”, “a small chance”, “improbable”) numerically found evidence of considerable variation among individuals (e.g., [41]) and even disagreements about their orderings. These findings were initially taken to imply unreliability in such judgements, but researchers applying FST to modelling these judgements demonstrated within-individual reliability and accounted for at least some of the inter-individual differences to fuzziness rather than unreliability (e.g., [37, 59]).

However, this successful application of FST did not generalize into applications elsewhere in the judgement and decision literature, even in research on decision making under “ambiguity”, which would have seemed a natural next step. There was a time when FST might have contributed to developments in subjective expected utility theory to account for human departures from it that arguably are not irrational (e.g., the development of rank-dependent expected utility theory and reformulation of Prospect Theory; see [33, 55]). In economics particularly, FST was trumped during the 1980s and 90s by Choquet capacities and related non-additive probability formalisms (see, e.g., the review by Billot [4]).

Billot's main point about the applications of FST in micro-economics is that they focused on extending standard theories to deal with various kinds of imprecision that heretofore had to be ignored or banished. For instance, FST applications to theories of preference generalized those theories somewhat but did not solve the fundamental problem of incomplete preferences. Moreover, they did not convince many economists that FST could do anything that probability could not.

In contrast to FST, the application of non-additive probabilities to micro-economics produced progress in areas where the discipline was stuck or in crisis. The primary problem this approach addressed is the representation of degrees of belief, and deposing probability as the sole or even best way of doing so. They also convincingly demonstrated that probability is insufficient to deal with problems in decision theory. Although there have been several generalizations of FST that can include these formalisms (and probability) as special cases, there are other generalized probability theories that can do so too. One has the sense of an opportunity that passed FST proponents by.

Nevertheless, FST has had some success in particular areas throughout the human sciences. In the area of organisational psychology, Hesketh and her colleagues (e.g., [1, 12, 11, 47]) employed FST for theoretical extensions and practical applications in the areas of work adjustment, personnel selection, person-environment fit, career choice, and occupational counselling.

Another success-story for FST is in the areas of speech and language perception, via elaborations and successful applications of the Oden-Massaro "fuzzy logical model of perception" (FLMP). Oden and Massaro [28] demonstrated that FLMP provided a good account of phoneme identification, and this initial success spawned applications in other areas such as letter recognition and lip-reading (e.g., [21]). FLMP has stood up against competing theories of language perception [24, 22].

3 Fuzzy Set Theory as a Prescriptive Framework and Data-Analytic Toolkit

Despite the fact that FST was given short shrift as a prescriptive standard for categorization against which to compare human performance, it has been rather uncritically adopted by researchers utilizing it to augment the data-analytic toolbox. Most researchers using FST for data analysis have not called its accounts of set intersection, union, dilation, or concentration into question and, in fact, have not even questioned the original Zadehian versions of these operations. Instead, most of the debates have centred on issues of measurement, inference, and the links between FST and traditional methods of data analysis (both quantitative and qualitative).

Let us begin with measurement. One of the limitations encountered by researchers across the human sciences is the problem of operationalizing and quantifying "degree of membership". As pointed out several times (e.g., [42: 77, 45, 57: 20–30]), this problem was neglected for some time in the FST literature. Much of the mathematical and engineering literature imposed degrees of membership "out of thin air", with little or no attention to grounding them in any systematic measurement framework. The measurement issue remains problematic, despite various attempts to remedy it.

Even within the FST community, there are several competing definitions of degree of membership [3, 45]. These may be translated into four camps. First, there is a "formalist" tradition that assigns membership via a (usually monotonic) transformation from a variable into the [0,1] interval. Second, a "probabilistic" tradition links degree of membership with the probability that a case gets assigned to the set on the basis of a subjective probability assignment, a poll, or some analogous sampling device. Third, a "decision-theoretic" camp assigns degrees of membership on the basis of some utility function describing the payoff of asserting that a case belongs in the set. Fourth is the "axiomatic measurement theory" camp (in the sense of the work of Krantz et al. [16]), which grounds membership assignments in an axiomatically grounded method of establishing that a membership scale behaves as a genuinely quantitative scale.

Even if researchers have agreed on how degree of membership is defined conceptually, what about quantification? Verbal labels, initially, seem straightforward for the same reason that constructing a Likert (agree-disagree) scale is straightforward. All that is required is a collection of verbal expressions for degrees of membership on which everyone agrees to a complete ordering (e.g., "not in the set", "slightly in the set", …, "fully in the set"). However, aside from assigning 0 to "not in the set" and 1 to "fully in the set", it is unclear what numbers should be assigned to the other verbal expressions. Moreover, quantified degrees of membership lack a behavioural interpretation analogous to the betting interpretation enjoyed by probability. Likewise, there is no apparent basis for a scoring rule such as the Brier score for probability judgements.

Now let us bring inference into the picture, although we will continue to refer to measurement because of the connections between these two issues. We will begin with inference in the sense employed by quantitative researchers, which focuses on statistical inference. An example where both degree of membership and statistical inference have been successfully dealt with is the "grade of membership" (GoM) model mentioned earlier, elaborated by Woodbury and colleagues over three decades. This approach is similar in some respects to latent class models, and throughout their book Manton et al. [18] compare the GoM and latent class models. It is among the most statistically sophisticated techniques associated with FST, and would at first glance appear to link FST with properly constructed statistical analyses. Indeed, the statistical methods are sound and the GoM has seen widespread application in clinical health, population health, and demography.

However, FST itself is not employed other than via the label "fuzzy" for the grade of membership. Instead, the GoM approach amounts to a hybrid classification and generalized linear model algorithm, and its machinery would remain undisturbed if the label "fuzzy" were dropped altogether. Also, it imposes restrictions on membership values that FST does not, such as requiring that they sum to 1 across a so-called "fuzzy" partition of clinical types. Thus, although it has been claimed by its authors that membership grades are not probabilities, they behave very much like probabilities in the GoM setup. Moreover, several latent class models have appeared that also employ grades of membership without reference to FST.

Perhaps most dismaying is the disconnection between virtually all of the FST discussions of methods for assigning membership and the long-established tradition of methods for assigning quantitative values to ordinal response scales. Probably the most widely used are Samejima's [40] "graded response model" and "partial credit" item response theory model [23], but numerous extensions and generalizations of these have been made since then, including variants of graded models (e.g., [52]), generalized partial credit models [26], and graded unfolding models [38]. All of these approaches are capable of dealing with degrees of membership, but of course they do not use FST concepts, and researchers employing FST concepts have largely ignored these psychometric approaches. A dialog between FST proponents and the psychometric community has yet to be broached.

In a separate line of work from the GoM, Smithson and Verkuilen [45] also targeted the integration of FST with statistical methods, oriented towards quantitative social science research. Their monograph includes statistical methods for assessing fuzziness, modelling fuzzy inclusion relations (see also [43]), and linking covariance with fuzzy intersection. They also point out the applicability of certain kinds of probability distribution functions for modelling fuzzy set membership distributions. Marchant [19, 20] and Verkuilen [57] demonstrated methods for converting standard measurement tasks including comparisons, subjective ratio scaling, and pairwise choices into fuzzy membership functions that adhere to axiomatic measurement theoretic criteria. Earlier, Crowther et al. [8] showed that the FLMP is equivalent to a Bradley-Terry-Luce model, but one where subjects provide interval ratings instead of pairwise choices.

In particular, the development of GLMs for doubly-bounded random variables using beta distributions to model them (e.g., [46, 58]) and the elaboration of GLMs for doubly- and interval-censored random variables (e.g., [44, Chap. 7]) arguably have paved the way for the statistical analysis of quantitative membership grades. However, as with the GoM, this work has not been linked with FST per se. Instead, it stands on its own as a technique for analysing any kind of doubly-bounded random variable. Perhaps for this reason, although the Smithson-Verkuilen 2006 paper has been cited more than 250 times in more than a dozen disciplines, it is largely ignored by researchers in the human sciences applying FST.

We now turn to developments regarding the utilization of FST for data analysis and inference that are more in the tradition of qualitative research. The early years of the 21st century saw a resurgence of interest in FST in sociology and political science, when Ragin linked it with his qualitative comparative analysis framework [35, 36]. Ragin's treatment ignored Smithson's earlier [42] book and covered far less material on FST than Smithson's monograph, but it proved considerably more popular in the domain of comparative politics. One explanation for this is that Ragin's presentation is embedded in a perspective that is more compatible with qualitative researchers' traditional assumptions and goals, whereas the treatments by Smithson [42] and Smithson and Verkuilen [45] are more oriented towards quantitative research traditions (see [17], for a catalogue of the differences between these two research traditions in political science).

Ragin's approach begins with a focus on issues of measurement, i.e., the construction of fuzzy set membership assignments and scales, embedded in his "qualitative comparative analysis" (QCA) framework [34, 35]. He presents this as a predominantly interpretive undertaking, with "qualitative anchors" for full set membership and non-membership, and "cross-over" (i.e., membership equalling ½). Ragin introduces the notion of "calibration" for justifying how these qualitative anchors are determined (e.g., the fuzzy set of "comfortable room temperature" for humans would require calibration of, say, the Celsius scale by human ratings of comfort). One criticism raised against this approach is that while it arguably generates an ordinal (and categorical) scale, numerical assignments to any of the grades of membership other than 0 or 1 are arbitrary. While some fuzzy set operations require only ordinal information (e.g., min and max), others require quantitative degrees of membership (e.g., concentration and dilation).

Research in the QCA tradition applies FST to what they term "case-based" analysis, usually on small samples and generally without using statistical inference. Instead, the focus is on identifying necessary and/or sufficient "causal" relationships between properties of cases (e.g., is a strong bourgeoisie necessary for a democratic social order to emerge?). This approach has been applied in numerous papers, in recent times more than 100 per year [53], and has spawned software resources in R and Stata. Here, then, is an example of a fairly wholehearted and sustained adoption of FST in the human sciences. QCA and its FST extension have not been without their critics, and the debates now span nearly 25 years.

As Thiem et al. [53] point out, these debates are littered with errors on both sides. Even within the QCA/FST camp, there are frequent lapses in standards of practice and sophistication of understanding in regard to FST. For example, almost no such research engages in any form of sensitivity analysis to assess robustness under variations in membership assignments. Necessary and/or sufficient "causal" relations without recognizing that skew in two variables can yield the appearance of an inclusion relation between two statistically independent fuzzy sets (as elaborated by Smithson and Verkuilen [45], but ignored by QCA/FST researchers). QCA/FST researchers also are unaware of logical quandaries that can arise in FST, especially when the data suffer from restricted range of membership in one or more sets [7]. However, to be fair, the same accusations of slipshod practices can be (and has been) said of researchers using conventional statistical methods in the human sciences.

A fundamental point of difference is whether the QCA/FST and traditional statistical modelling approaches are incommensurable or not. After all, if they are incommensurable then comparing their merits is largely pointless. On the one hand, Katz et al. [13] claim that "regression methods and fuzzy-set methods cannot test the same hypotheses" (p. 541) because they construe causality in different ways, and Thiem et al. [53] argue that QCA/FST and "regression" are incommensurably based on different algebras (Boolean vs. linear) and, therefore, FST operations such as conjunction cannot be duplicated by interaction terms in regression models. On the other hand, Clark et al. [6] argue that linear models with interaction terms are able to test hypotheses about necessity and sufficiency. Long before them, Smithson [42: Chap. 6] presented similar arguments and employed a more extensive variety of interaction terms than just the conventional product term, and more recently Smithson and Verkuilen [45: Chap. 6] draw a direct connection between fuzzy set conjunction and covariance (using the product operator for conjunction in FST). Nevertheless, the fact that the debates have continued is a sign of health.

4 Conclusion

FST still faces several challenges regarding its ability to contribute to the human sciences, and it would have to be said that much of its potential has not been realised even 50 years after its appearance. There are several possible reasons for this.

1. It was unclear from the start whether FST should play the role of a descriptive or prescriptive account for dealing with categories with blurry boundaries. While the basic notion of fuzziness was adopted early in the 1970s by researchers on categorization, the machinery of FST was ignored, and still is today. Researchers in categorization and related areas of cognitive psychology remain unconvinced that FST describes human categorization judgements. Where researchers have added FST to the methodological toolbox, they often have accepted its prescriptions uncritically and have employed a stripped-down version of FST.
2. In its original form, FST neglected the problem of measuring degrees of membership. That issue still is problematic today. There is no convincing behavioural interpretation of membership assignments, as in the bet-pricing interpretation of probability assignments. There is no "Dutch-Book" argument for identifying "calibrated" membership, nor a scoring-rule method to motivate judges to assign membership values in an honest fashion. Researchers employing FST in the human sciences have largely ignored well-established psychometric principles of measurement. While several well-founded grade-of-membership measurement techniques have long been available, these remain unexamined by researchers applying FST.

3. In its original form, FST was not linked with appropriate statistical techniques. That still is problematic to some extent, although the development of GLMs for doubly-bounded random variables has the potential to help.
4. There have been several alternative versions of FST operators (e.g., the t-norms), and it was not clear which of these should be preferred for which purposes. This problem has largely been swept under the carpet in most applications throughout the human sciences.
5. There was resistance to FST in the human sciences for similar reasons to resistance encountered in other disciplines (e.g., is fuzziness really just probability?).
6. There also was resistance in some areas of the human sciences because FST presented itself as forbiddingly mathematical, especially to researchers in the qualitative research traditions who stood to gain the most from FST. This issue has recurred even in largely sympathetic research communities such as QCA (e.g., [35]), and most of the applications therein are relatively unsophisticated versions of FST.
7. As time went on, various advances enabled the accomplishment of feats by statistical techniques and alternative generalizations of probability theory that previously might have been unique to FST. As demonstrated by the debates in the human sciences, many researchers believe that conventional statistical modelling techniques can do anything that FST can. Indeed, a stronger case could be made for such a claim than has been made in the published literature, e.g., via references to modern psychometric methods for dealing with graded membership. It is currently more difficult to demonstrate that FST has distinct advantages than it was in the past.
8. Often FST has been presented as an extension or generalization of an already established model or framework, rather than offering genuinely new results (e.g., the [4] article contrasting applications of FST with non-additive probabilities in economics).
9. At present, even where FST is being applied, a combination of conceptual confusion, lack of mathematical sophistication, and inattentive scholarship limits its utility. In sociology and political science, for instance, this has been a common criticism raised by sceptical quantitative researchers.
10. There are many areas in the human sciences where FST clearly could be helpful, but has never been considered (e.g., the diagnosis of psychological disorders, or actuarial risk categorizations). Perhaps the saddest evidence for its overall lack of impact is the recent paper by Stoklasa et al. [51] where the authors are repeating earlier efforts spanning the past 40 years, yet again pleading a case for the application of FST to various areas of psychology.

In a review of fuzzy set applications in psychology 16 years ago, Smithson and Oden [48] predicted that applications of FST might become more common but decreasingly distinct from other approaches and techniques. With the possible, but contested, exception of the FST extension of QCA in political science, that prediction seems to have been coming true. The psychometric and applied statistical literatures have seen a recent burgeoning of graded-response models, general linear models for doubly-bounded variables that model both location and dispersion, ordinal and quantile regression techniques that can deal with asymmetric fuzzy set inclusion-like relationships, and techniques such as taxometrics for determining whether to treat a construct as categorical or dimensional (a.k.a. "fuzzy"). This is not to say that FST has no place in the human sciences or that it has been made redundant. There are plenty of opportunities for FST to have wider application and greater impact than it has had thus far. Nonetheless, it is now more likely to assume the role of one among several alternative tools for a given task, rather than the only tool that will suffice.

References

1. Allinger, G.M., Feinzig, S.L., Janak, E.A.: Fuzzy sets and personnel selection: discussion and an application. J. Occup. Organ. Psychol. **66**, 163–169 (1993)
2. Biin, J.M., Whinston, A.B.: Fuzzy sets and social choice. J. Cybern. **3**, 28–33 (1973)
3. Bilgic, T., Turksen, I.B.: Measurement of membership functions: theoretical and empirical work. In: Dubois, D., Prade, H. (eds.) Handbook of Fuzzy Sets and Systems. Fundamentals of Fuzzy Sets, vol. 1. Kluwer, New York, (2000)
4. Billot, A.: From fuzzy set theory to non-additive probabilities: how have economists reacted? Fuzzy Sets Syst. **49**, 75–90 (1992)
5. Capozza, D., Nanni, R.: Differentiation processes for social stimuli with different degrees of category representativeness. Eur. J. Soc. Psychol. **156**, 399–412 (1986)
6. Clark, W.R., Gilligan, M.J., Golder, M.: A simple multivariate test for asymmetric hypotheses. Polit. Anal. **14**, 311–331 (2006)
7. Cooper, B., Glaesser, J.: Paradoxes and pitfalls in using fuzzy set QCA: illustrations from a critical review of a study of educational inequality. Sociol. Res. Online 16 (2011). http://www.socresonline.org.uk/16/3/8.html
8. Crowther, C.S., Batchelder, W.H., Hu, X.: A measurement-theoretic analysis of the fuzzy logic model of perception. Psychol. Rev. **102**, 396–408 (1995)
9. Grandmont, J.M.: Intermediate preferences and the majority rule. Econometrica **46**, 317–330 (1978)
10. Hersh, H.M., Caramazza, A.: A fuzzy set approach to modifiers and vagueness in natural language. J. Exp. Psychol. Gen. **105**, 254–276 (1976)
11. Hesketh, B., McLachlan, K., Gardner, D.: Work adjustment theory: an empirical test using a fuzzy rating scale. J. Vocat. Behav. **40**, 318–337 (1992)
12. Hesketh, B., Pryor, R.G., Gleitzman, M., Hesketh, T.: Practical applications and psychometric evaluation of a computerised fuzzy graphic rating scale. In: Zetenyi, T. (ed.) Fuzzy Sets in Psychology. Advances in Psychology, vol. 56. North-Holland, Amsterdam (1988)
13. Katz, A., vom Hau, M., Mahoney, J.: Explaining the great reversal in Spanish America. Sociol. Methods Res. **33**, 539–573 (2005)
14. Kay, C., McDaniel, C.: Color categories as fuzzy sets (Working Paper No. 44). University of California, Language Behavior Research Laboratory, Berkley (1975)
15. Kempton, W.: Category grading and taxonomic relations: a mug is a sort of a cup. Am. Ethnol. **5**, 44–65 (1978)
16. Krantz, D.H., Luce, R.D., Suppes, P., Tversky, A.: Foundations of Measurement. Academic Press, New York (1971)
17. Mahoney, J., Goertz, G.: A tale of two cultures: contrasting quantitative and qualitative research. Polit. Anal. **14**, 227–249 (2006)

18. Manton, K.G., Woodbury, M.A., Tolley, H.D.: Statistical Applications Using Fuzzy Sets. Wiley, New York (1994)
19. Marchant, T.: The measurement of membership by comparisons. Fuzzy Sets Syst. **148**, 157–177 (2004)
20. Marchant, T.: The measurement of membership by subjective ratio estimation. Fuzzy Sets Syst. **148**, 179–199 (2004)
21. Massaro, D.W.: Speech Perception by Eye and Ear: A Paradigm for Psychological Inquiry. Lawrence Erlbaum, Hillsdale (1987)
22. Massaro, D.W.: Testing between the TRACE model and the fuzzy logical model of speech perception. Cogn. Psychol. **21**, 398–421 (1989)
23. Masters, G.N.: A Rasch model for partial credit scoring. Psychometrika **47**, 149–174 (1982)
24. McDowell, B.D., Oden, G.C.: Categorical decision, rating judgments, and information preservation. Unpublished manuscript, University of Iowa (1995)
25. McNeill, D., Freiberger, P.: Fuzzy Logic. Simon and Schuster, New York (1993)
26. Muraki, E.: A generalized partial credit model: application of an EM algorithm. Appl. Psychol. Measure. **16**(2), 159–76 (1992)
27. Oden, G.C.: Fuzziness in semantic memory: choosing exemplars of subjective categories. Mem. Cognit. **5**, 198–204 (1977)
28. Oden, G.C., Massaro, D.W.: Integration of featural information in speech perception. Psychol. Rev. **85**, 172–191 (1978)
29. Osherson, D.W., Smith, E.E.: On the adequacy of prototype theory as a theory of concepts. Cognition **9**, 35–58 (1981)
30. Pipino, L.L., van Gigch, J.P., Tom, G.: Experiments in the representation and manipulation of labels of fuzzy sets. Behav. Sci. **26**, 216–228 (1981)
31. Ponsard, C.: L'imprécision et son traitement en analyse économique. Revue Économique Politique **1**, 17–37 (1975)
32. Ponsard, C.: Fuzzy mathematical models in economics. Fuzzy Sets Syst. **28**, 273–283 (1988)
33. Quiggin, J.: Generalized Expected Utility Theory: The Rank Dependent Model. Kluwer, Boston (1993)
34. Ragin, C.C.: The Comparative Method: Moving Beyond Qualitative and Quantitative Strategies. University of California Press, Berkeley (1987)
35. Ragin, C.C.: Fuzzy-Set Social Science. University of Chicago Press, Chicago (2000)
36. Ragin, C.C.: Redesigning Social Inquiry: Fuzzy Sets and Beyond. University of Chicago Press, Chicago (2008)
37. Reagan, R.T., Mosteller, F., Youtz, C.: Quantitative meanings of verbal probability expressions. J. Appl. Psychol. **74**, 433–442 (1989)
38. Roberts, J.S., Donoghue, J.R., Laughlin, J.E.: A general item response theory model for unfolding unidimensional polytomous responses. Appl. Psychol. Meas. **24**, 3–32 (2000)
39. Rosch, E.: Cognitive representations of semantic categories. J. Exp. Psychol. Gen. **104**, 192–233 (1975)
40. Samejima, F.: Estimation of latent ability using a response pattern of graded scores. Psychometrika Monogr. Suppl. **34**, 100–114 (1969)
41. Simpson, R.H.: The specific meanings of certain terms indicating differing degrees of frequency. Quart. J. Speech **30**, 328–330 (1944)
42. Smithson, M.: Fuzzy Set Analysis for the Behavioral and Social Sciences. Springer, New York (1987)
43. Smithson, M.: Fuzzy set inclusion: linking fuzzy set methods with mainstream techniques. Sociol. Methods Res. **33**, 431–461 (2005)
44. Smithson, M., Merkle, E.C.: Generalized Linear Models for Categorical and Continuous Limited Dependent Variables. Chapman and Hall, Boca Raton (2014)
45. Smithson, M., Verkuilen, J.: Fuzzy Set Theory: Applications in the Social Sciences. Quantitative Applications in the Social Sciences Series. Sage, Belmont (2006)
46. Smithson, M., Verkuilen, J.: A better lemon-squeezer? Maximum likelihood regression with beta-distributed dependent variables. Psychol. Methods **11**, 54–71 (2006)

47. Smithson, M., Hesketh, B.: Using fuzzy sets to extend Holland's theory of occupational interests. In: Reznik, L., Dimitrov, V., Kacprzyk, J. (eds.) Fuzzy System Design: Social and Engineering Applications. Studies in Fuzziness and Soft Computing, vol. 17. Physica-Verlag, Berlin (1998)
48. Smithson, M., Oden, C.G.: Fuzzy set theory and applications in psychology. In: Zimmermann, H.-J. (ed.) Practical Applications of Fuzzy Technologies. Kluwer, Norwell (1999)
49. Smithson, M., Sopena, A., Platow, M.: When is group membership zero-sum? Effects of ethnicity, threat, and social identity on dual national identity. PLoS ONE 1–18 (2015). doi:10.1371/journal.pone.0130539
50. Spies, M.: Syllogistic Inference under Uncertainty. Psychologie Verlags Union, Munich (1989)
51. Stoklasa, J., Talasek, T., Musilova, J.: Fuzzy approach: a new chapter in the methodology of psychology? Hum. Aff. **24**, 189–203 (2014)
52. Takane, Y., de Leeuw, J.: On the relationship between item response theory and factor analysis of discretized variables. Psychometrika **52**, 393–408 (1987)
53. Thiem, A., Baumgartner, M., Bol, D.: Still lost in translation! A correction of three misunderstandings between configurational comparativists and regressional analysts. Comp. Polit. Stud. 1–33 (2015). doi:10.1177/0010414014565892
54. Tversky, A., Kahneman, D.: Extensional versus intuitive reasoning: the conjunction fallacy in probability judgment. Psychol. Rev. **90**, 293–315 (1983)
55. Tversky, A., Kahneman, D.: Advances in prospect theory: cumulative representation of uncertainty. J. Risk Uncertain. **5**(4), 297–323 (1992)
56. Tversky, A., Kahneman, D.: Judgments by and of representativeness. In: Kahneman, D., Slovic, P., Tversky, A. (eds.) Judgment Under Uncertainty: Heuristics and Biases. Cambridge University Press, New York (1982)
57. Verkuilen, J.: Assigning membership in a fuzzy set analysis. Sociol. Methods Res. **33**, 462–496 (2005)
58. Verkuilen, J., Smithson, M.: Mixed and mixture regression models for continuous bounded responses using the beta distribution. J. Educ. Behav. Stat. **37**, 82–113 (2012)
59. Wallsten, T.S., Budescu, D.V., Rapoport, A., Zwick, R., Forsyth, B.H.: Measuring the vague meanings of probability terms. J. Exp. Psychol. Gen. **115**, 348–65 (1986)
60. Woodbury, M.A., Clive, J.: Clinical pure types as a fuzzy partition. J. Cybern. **4**, 111–121 (1974)
61. Woodbury, M.A., Clive, J., Garson, A.: Mathematical typology: a grade of membership technique for obtaining disease definition. Comput. Biomed. Res. **11**, 277–298 (1978)

Fuzzy Entropy Used for Predictive Analytics

Christer Carlsson, Markku Heikkilä and József Mezei

Abstract Process interruptions in (very) large production systems are difficult to deal with. Modern processes are highly automated; data is collected with sensor technology that forms a big data context and offers challenges to identify coming failures from the very large sets of data. The sensors collect huge amounts of data but the failure events are few and infrequent and hard to find (and even harder to predict). In this article, our goal is to develop models for predictive maintenance in a big data environment. The purpose of feature selection in the context of predictive maintenance is to identify a small set of process diagnostics that are sufficient to predict future failures. We apply interval-valued fuzzy sets and various entropy measures defined on them to perform feature selection on process diagnostics. We show how these models can be utilized as the basis of decision support systems in process industries to aid predictive maintenance.

Keywords Predictive analytics · Fuzzy entropy · Feature selection · Failure prediction

A preliminary version of this chapter was presented at the 2015 IEEE International Conference on Fuzzy Systems (FUZZ-IEEE 2015).

C. Carlsson (✉) · M. Heikkilä · J. Mezei
IAMSR, Åbo Akademi University, Turku, Finland
e-mail: christer.carlsson@abo.fi

M. Heikkilä
e-mail: maheikki@abo.fi

J. Mezei
e-mail: jmezei@abo.fi

C. Kahraman et al. (eds.), *Fuzzy Logic in Its 50th Year*,
Studies in Fuzziness and Soft Computing 341,
DOI 10.1007/978-3-319-31093-0_9

1 Introduction

The context we are addressing is one where we are required to predict if and when failures will occur and then to find some optimal set of actions to either prevent the failure from happening—or if this prevention is not successful, to minimize the consequences (in terms of time, cost, interruptions, quality problems, etc.) of the failures. We have quite some history of finding an optimal set of actions, more than 60 years of Operational Research and Management Science (cf. [14]) which now is more and more referred to as Analytics (cf. [8]). Throughout the history of building optimal action programs (cf. [4]), we have always come across the availability of data as one of the crucial requirements (cf. [15]) for building, testing, validating and implementing optimal action programs. It has been a recurring complaint that the (mathematical) models would be much more useful if some specific data needed could be collected and processed (cf. [15]). In the last few years we have got some new complications—we gather too much data and we cannot find key or relevant data quickly enough to deal with problems in real time.

Our specific context is the process industry and a Finnish multinational corporation that is specialized in automation systems for the process industry; we will work out models for predictive analytics for this context using the data that can be collected within the processes. In recent years—without going into much detail—it is often claimed that the task of getting data for optimal action programs is not a problem anymore; sensor technology has made it possible to fully monitor and collect data on all aspects of automated processes (cf. [37]). This is probably true, but what was not foreseen is the fact that we got too much data; we created a "big data" problem, which is described by the malady that we collect more data with sensor technology and monitoring computer systems than we can ever process with a reasonable use of time and resources.

We have to deal with a "big data" feature in our process industry context in order to predict if and when failures will occur. This added the need to find indications of a failure starting to shape up—also referred to as "weak signals" among the process engineers—in often very large data sets, and then to work out predictions of when the failure may or will occur.

The problem area is the monitoring and control of industrial processes that take place in a highly automated, technically very advanced centre where experienced and highly skilled personnel follows the process and is ready for interventions as soon as the processes start to deviate from predefined and pre-set standards (Fig. 1 shows a sketch of the combined data, information and knowledge elements for monitoring and control). These interventions are required to be (i) timely, (ii) correct and (iii) optimal in terms of cost and time so that any failure can be avoided.

The monitoring and control personnel has to deal with the following problems: to predict possibly coming component failures that build on deciding (i) what components could fail, (ii) where in the process this may happen and (iii) when a failure can be expected to happen. Typically this prediction is combined with a risk assessment of what specific failures will occur; we have previously worked out this

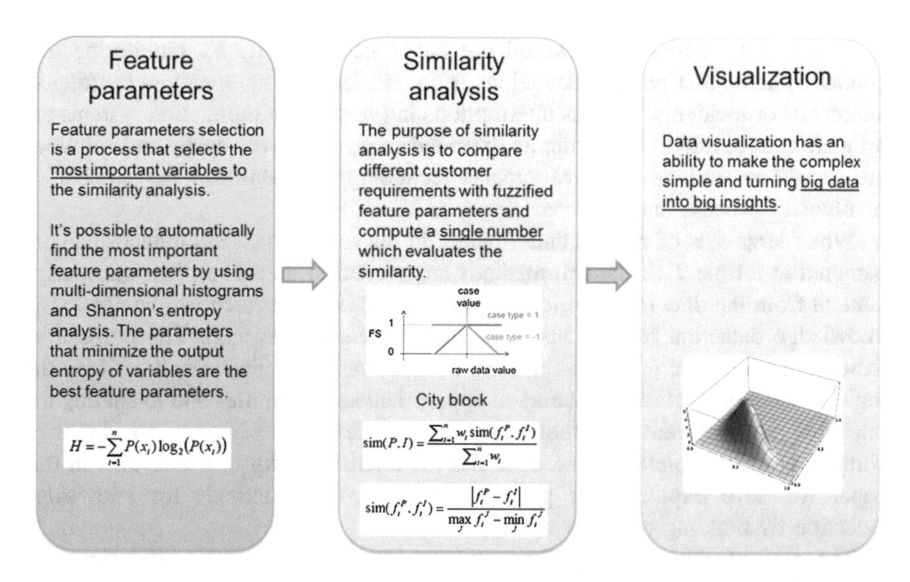

Fig. 1 Monitoring and control of industrial processes combine human and system intelligence (partner case description)

risk with Bayes predictive probabilities and Bayes predictive possibilities [5]. A different aspect is to work out the criticality of a failure: what is the impact on the process, what are the cost components and what are the estimated direct and indirect costs of a failure.

The predictions, the risk assessments and the criticality evaluations of a failure determine the choice of maintenance strategy: (i) corrective maintenance, or also known as run-to-failure (RTF) maintenance; (ii) preventive maintenance, that builds on inspections followed by repair or replacement according to some predetermined schedule; (iii) predictive maintenance, that is based on condition monitoring to identify the state of components and to predict possible failures, predict remaining time to failures and to decide on a maintenance program.

Most of the described situations can be considered as a big data context with all the challenges big data brings [11]. The big data challenges are (i) to find or develop algorithms that are fast and effective enough to produce indications in time for optimal maintenance decisions; (ii) to work out intelligent screening and selection tools that can offer effective advice to the personnel, and (iii) to work out decision support over mobile platforms, web-sites and cloud architectures that would be fast enough to support decisions that have to be made in near real-time.

Predictive modelling is basically to find out what is to happen next. Predictive models should unveil and measure patterns to identify risks and opportunities using transactional, demographic, web-based, historical, text, sensor, economic, and unstructured data. Predictive models typically should consider multiple factors and predict outcomes with a high level of accuracy.

Process interruptions are carried out either automatically by monitoring and control systems that react to deviations from standards or by operators reacting to anomalies or incidents. Process interruptions in (very) large production systems are difficult to trace and to deal with; an extended stop is also very costly and solutions are sought to find an effective support technology to minimize the number of involuntary process interruptions.

When large sets of control data (mainly in the form of process diagnostics) are gathered at a time T (instance), methods are needed to detect relevant knowledge content from the data (of historical events or incidents) that then can be applied to knowledge detection in new instances (new, upcoming events). The purpose of feature selection is to reduce the number of input variables (features) describing the logic of the incidents. By selecting an appropriate set of features and excluding the ones that are irrelevant the problem solving process can be improved in terms of computational complexity, time, cost and the interpretability of the results. In this paper we have explored the possibilities to develop methods for knowledge detection by building on fuzzy entropy.

The rest of the paper is structured as follows. Section 2 summarizes the state-of-the-art on fuzzy entropy, Big Data, and predictive analytics. Section 3 introduces standard concepts in feature selection and describes a model based on fuzzy entropy. In Sect. 4, we describe how feature selection method can be utilized as the basis for predictive analytics, with a discussion on a real-world application in monitoring and control, with recommendations on how to develop the algorithm for decision support in Sect. 5. Finally, summary and conclusions are provided in Sect. 6.

2 Fuzzy Entropy, Big Data and Predictive Analytics

In this section, we present preliminary definitions and summarize the relevant contributions related to our study in the context of (i) fuzzy entropy measures, (ii) big data and industrial analytics and (iii) predictive analytics.

2.1 Fuzzy Entropy

Zadeh [50] introduced fuzzy set theory and fuzzy logic as a tool to represent and manipulate imprecise expressions. The concept of fuzzy set extends the idea of a crisp set as it allows for elements to belong to the fuzzy set to some degree between 0 and 1; this degree is termed the membership value, $\mu(x)$. In the same article, Zadeh [50] introduced a generalization of Shannon's information entropy [41] using fuzzy membership values. This proposal includes the original definition as a special case but at the same time can model imprecision present in different aspects of the underlying problem:

$$H(X) = -\sum_{i=1}^{n} \mu_A(x_i) p(x_i) \log p(x_i),$$

where A is a fuzzy set, $P = \{p_1, p_2, \ldots, p_n\}$ is a probability distribution and $X = \{x_1, x_2, \ldots, x_n\}$ is the event space. If the membership value in the fuzzy entropy takes the form

$$\mu_A(x) = \begin{cases} 1, & \text{if } x \in X \\ 0, & \text{otherwise} \end{cases}$$

then we obtain the definition of entropy introduced by Shannon [41].

The common basis today for a well-defined entropy measure is the four De Luca–Termini axioms [29]:

1. $H(A) = 0$ if and only if $A \in X$ is a crisp set.
2. $H(A) = 1$ if and only if $\mu_A(x_i) = 0.5$ for every i.
3. $H(A) \leq H(B)$ if A is less fuzzy than B, i.e., if $\mu_A(x) \leq \mu_B(x)$ when $\mu_B(x) \leq 0.5$ and $\mu_A(x) \geq \mu_B(x)$ when $\mu_B(x) \geq 0.5$.
4. $H(A) = H(A^C)$ where $A^C = 1 - A$ denotes the complement of A.

According to these properties, the different fuzzy entropy measures quantify the "fuzziness" of fuzzy sets [22]. The first condition requires that the concept of fuzziness (entropy) disappears in the classical case, i.e. crisp sets are not fuzzy to any degree. According to the second condition, the fuzziness (entropy) takes its maximum when it is impossible to distinguish between a fuzzy set and its negation, i.e. every value in the support belongs (and does not belong) to the fuzzy set to the same degree. The other two axioms strengthen the imposed conditions of the first two required properties.

There are several existing definitions that satisfy the above axioms, in the following we list the most widely used ones:

- De Luca and Termini [9]:

$$H_{LT}(A) = -\sum_{i=1}^{n} [\mu_A(x_i) \log(\mu_A(x_i)) + (1 - \mu_A(x_i)) \log(1 - \mu_A(x_i))]$$

- Kosko [23]:

$$H_K(A) = \frac{\sum_{i=1}^{n} \min(\mu_A(x_i), \mu_{A^C}(x_i))}{\sum_{i=1}^{n} \max(\mu_A(x_i), \mu_{A^C}(x_i))}$$

- Li and Liu [27] defined entropy using the credibility measure [30]:

$$H_C(A) = \sum_{i=1}^{n}(-Cr(x_i)\ln(Cr(x_i) - (1 - Cr(x_i))\ln(1 - Cr(x_i))))$$

There are several other approaches termed fuzzy entropy that cannot be characterized by the above four conditions but try to capture the fuzziness of sets from different perspectives. The most widely used definition of this type was proposed by Yager [48], who argued that the measure of fuzziness should quantify the distinction between the fuzzy set and its complement.

Fuzzy entropy can be defined for different generalizations of traditional fuzzy sets, most importantly for interval-valued fuzzy sets.

Definition 1 [16] An interval-valued fuzzy set A defined on X is given by

$$A = \{(x, [\mu_A^L(x), \mu_A^U(x)])\}, \quad x \in X,$$

where $\mu_A^L(x), \mu_A^U(x) : X \to [0, 1]$; $\forall x \in X, \mu_A^L(x) \leq \mu_A^U(x)$, and the ordinary fuzzy sets $\mu_A^L(x)$ and $\mu_A^U(x)$ are called lower fuzzy set and upper fuzzy set about A, respectively.

As in the case of traditional fuzzy sets, there exist different definitions of entropy for interval-valued (or in general type-2) fuzzy sets [45]. The most widely used definitions are the following:

- Szmidt and Kacprzyk [43]:

$$H_{SK}(A) = \frac{1}{n}\sum_{i=1}^{n}\frac{1 - \max(1 - \mu_A^U(x_i), \mu_A^L(x_i))}{1 - \min(1 - \mu_A^U(x_i), \mu_A^L(x_i))}$$

- Burillo and Bustince [3]:

$$H_{BB}(A) = \sum_{i=1}^{n}(\mu_A^U(x_i) - \mu_A^L(x_i))$$

- Zeng and Li [51]:

$$H_{ZL}(A) = 1 - \frac{1}{n}\sum_{i=1}^{n}(\mu_A^U(x_i) + \mu_A^L(x_i) - 1)$$

Additionally, in recent years we have seen a large number of articles generalizing entropy to different families of fuzzy sets: hesitant fuzzy sets [47], intuitionistic fuzzy sets [28], vague sets [53] or intuitionistic and interval-valued soft sets [21]. While these definitions usually were proposed with descriptions of hypothetical applications, mainly in decision making problems, we have yet to see a practical implementation of any of these concepts.

Different fuzzy entropy measures have been applied extensively in the literature to solve various industrial problems, and several of these focus on fault detection and risk assessment. Ye [49] defines the fuzzy cross entropy of vague sets and utilizes it for fault diagnosis of turbines by focusing on identifying fault types and future trends in the turbine behaviour. Lee et al. [26] applies fuzzy entropy as an important tool in extracting features for fault detection of an induction motor. Wu et al. [46] perform risk assessment of engineering projects using fuzzy entropy weights, and they propose an inspection system to monitor and maintain safety on engineering projects.

One of the most important applications of fuzzy entropy is feature selection as a basis for classification and segmentation. Lee et al. [25] make use of fuzzy entropy to select relevant features with good separability for classification. Tao et al. [44] define a new fuzzy entropy measure and employ it for image segmentation. Parthalain et al. [38] combine fuzzy-rough set theory and fuzzy entropy and show that their procedure results in smaller feature subset sizes than other methods without significant loss of classification accuracy.

2.2 Big Data and Industrial Analytics

As the price of storage capacity to both store and transfer data have decreased drastically in recent years, both businesses and individuals are using this opportunity to store items perceived as valuable. This has led to the daily generation of quintillions of bytes of data [52]. For companies the potential value inherent in the vast amount of data stored is both an opportunity and a problem. It is an opportunity, because the data may have hidden, valuable information content. The big problem is how this hidden value could be uncovered, processed and utilized in daily business. This calls for new managerial approaches [32] as well as new business models.

For business utilization of big data, or information explosion as Beath et al. [1] call it, we should put the emphasis on business processes, product development and customer satisfaction. To do this, companies need, firstly, to identify the most important data items (about customers, sales orders, inventory items etc.), secondly, to define the work-flows that will use unstructured data and, thirdly, find potential business processes that benefit from the information explosion. The problem areas often identified here are (i) the increasing complexity of growing business into new areas that utilize data in different ways; (ii) the difficulty to utilize unstructured "metadata" generated in manual or automated processes, and (iii) the challenge to commit to an ongoing analysis of the business processes.

The key to utilizing big data is to understand its key features. The most common aspect is usually attributed to the mere volume of both the number of samples (vertical dimension) and the number of stored items, i.e. the features (horizontal dimension) of data sets. This "big dimensionality" [2] is usually called the first V ("volume") of the 5 big data attributes. The two most common attributes, in addition to volume, are "velocity" and "variety" (see e.g. [39]), but two additional features "value" and "veracity" have been identified as well [19, 54]. The first three Vs are defined as [39]:

- Constantly growing data that can easily grow into terabytes, or even petabytes (*volume*)
- Data used for time-sensitive processes (*velocity*)
- Data composed of various types of data: structured and unstructured (such as texts, sensor data, audio, video, log files, and so on) (*variety*)

The additional two V's are defined as follows:

- Data relevant for the context and useful for purposes of the user (*value*)
- Data that is reliable and verifiable (*veracity*)

As Hitzler and Janowicz [19] point out, the different aspects of big data have different significance for different areas: supercomputing researchers often find volume most interesting, social sciences are more interested in veracity and value, and the Semantic Web community is interested in the variety of data. For the velocity aspect for which data is used, Hitzler and Janowicz [19] refer to the development of sensor webs and the Internet of Things.

For the monitoring of pumps and valves (which is the real-world context for our work on predictive analytics), tagging enables the creation of a web of sensors which has the features of big data, since the volume is large with up to 500 variables stored *n* times per minute (where n may be >100). The speed required of data-driven processes, velocity, is a key issue and the number of pumps of various types that are used will add to the variety. Additionally the knowledge generated needs to be correct to build customer value (veracity).

Internet of Things (IoT) [6] is a fast growing area of study that works with ongoing processes where physical objects (products) are tagged with QR codes, barcodes and RFID tags that enable location-aware, person-centred and context relevant services. These objects are common in daily use and they are connected with mobile and wireless platforms. Such digital connectivity of data between objects, and from objects to services, is found not only in products but in production facilities as well. The Industrial Internet (II) works with key functions that enable modern automation which is used in production, logistics etc. Various sensor devices are used to monitor the functionality features of production facilities, machines and components.

What is typical for the IoT already today and even more for the II is that the amount of generated data is so large that it cannot be utilized directly. There are several reasons for this:

- The pace of development of analytical methods for utilization and operationalization of data has not been able to match the growth of data. Tools used for analytical purposes often date back to the 1990s or earlier [6].
- Data is collected in a number of databases that are of different types, they are run with different database systems and thus apply different logic based on local and

functional needs. The current data base management systems still in use were developed to serve the needs of unified corporate databases, mostly relational, that could retrieve data with SQL queries and get results from readily made statistical analyses. For data with a large number of locations such queries are a waste of resources. Instead, methods are needed to move queries to data and run them as parallel processes [36].
- Business models for the IoT still lag the developed technologies. Since it is common in the development work to have innovations to create demand both from existing services to products and from developed products to supporting services business models cannot be updated fast enough [34]. This means that the managerial understanding of innovations may exist on the conceptual level but practical implementations still require learning processes to take place.

There is a growing need for analytical methods, database management systems and business models to fill the growing gap between what the collected data enables and what the current production of products and services cannot accomplish. To fill this gap all of the above aspects need to be assessed. As Nieto-Santisteban et al. [36] point out: "It is a mistake to move large amounts of data to the query, when you can move the query to the data and exercise the query in parallel". Needless to say, there is not much trace of using fuzzy entropy as a basic principle for the business utilization of big data as theory and practice have not yet found the opportunity to form a common platform.

2.3 Predictive Analytics

In a report from the INFORMS 2011 Annual Meeting there was a statement: "There is no doubt—analytics are here" (as quoted in the OR/MS Today). This was after the INFORMS tried to ignore the new movement after it got started by Davenport's book in 2007. The movement coincides with the increased availability of data we have seen with advances in information and communication technology; with increasing data there is an increasing demand for analytic tools to build descriptive, predictive and prescriptive models. INFORMS now sees tremendous opportunity of synergy between analytics and Operations Research.

A customary way of building models to predict future failure events in industrial processes is to describe the logic behind the events in question (e.g. fatigue, decay or emergence of faults), to build a model to simulate the generation of the event results and to analyse data based on this model. This type of logic-driven models is common in industry because they usually generate good quality and verifiable results and can be shown to produce added value to the process.

The actual context we have worked with—one of pumps and valves in industrial processes—offered us a number of practical insights. For dynamic processes that are in constant change and for which several causes either decide or influence processes data-driven models are often insufficient. In such situations and especially if there is an abundance of data (such as hundreds of time series of sensor log data), it is hard to find a single model that would adjust to and be an adequate representation of the changing operational environment and the changing patterns of collected data. In such an environment advanced data analysis methods are needed and the process of analysis starts from data that is used to find decision models that are judged to be best suited for the problems at hand. The problems may change when the environment changes, for instance if a component in a device is changed, if a device is moved to another process, if a different maintenance program is performed, or if the data indicates some process-internal changes or changes of sub-processes.

Data and business analytics methods are today developed for both logic-driven and data-driven modelling; however, the importance of data-driven methods has grown with the growth of data storages and recorded sensor data. The use of predictive models often requires that data can be applied online for it to be useful. With the abundance of data collected from sensors it is necessary to pre-process it and to choose a representative data set with feature selection methods of the type described in this article. In order to carry out an analysis of a chosen set of features a number of approaches are possible, such as artificial neural networks, clustering analysis, decision tree analysis or multinomial statistical models.

Gandomi and Haider [13] point out that much work on big data implicitly assumes that we can work with predictive analytics and structured data, i.e. the problems with massive data are understood but it is assumed that analytics will offer remedies to come to terms with the large data sets; then it is further assumed that we will have structured data because work with analytics will be easier that way. Then it is ignored—something we found out in our industrial cases—that the largest component of big data is unstructured and available as audio, images, video and unstructured text. Predictive analytics relies on statistical methods and it now starts to be understood that big data will require new statistical methods. First, statistical significance which is used to generalize from small samples of a population to the entire population will not work for big data as very large samples will produce an abundance of statistically significant relationships that have no practical relevance. Second, conventional sampling methods for small samples do not scale up to big data. Third, big data suffers from heterogeneity, noise accumulation, spurious correlations and incidental endogeneity.

Diamantoulakis et al. [12] have worked out some of the problems encountered by big data analytics in dynamic energy management, which appear to neatly summarize some of the problems to be expected when tackling industrial processes. The first concern is that traditional computer techniques cannot handle fast data processing, which is required for real-time monitoring, dynamic energy management and power flow optimization; the remedy is to employ efficient high performance computing techniques (virtualization, in-memory computing). The second concern is that available communication resources are not enough for data acquisition in a centralized manner; the remedy is distributed data mining and dimensionality reduction to reduce the required resources. The third concern is the need for significantly increased data storage and computing resources which will require investments and generate rapidly increasing operating costs; the remedy is cloud computing and payment for resources as needed. The fourth concern is the demand response for which the algorithms rely on sensor data with correlations, trends and patterns but should include user reactions to real-time prices and stochastic parameters, everything in a big data setting; the remedy is predictive analytics and machine learning techniques that can facilitate real-time decision making.

Janke et al. [20] found some surprising benefits of predictive analytics and big data in emergency care. They note that many large emergency departments have more than 100,000 patient visits per year that through electronic medical records of all procedures carried out in modern emergency care will create (very) large amounts of real-world observational data. Associations in big data can drive prediction-based, personalized improvements in care quality; large observational data sets have been found to be very usable for the development of predictive models, and this is now starting to be used in emergency care as big data analytics is gaining acceptance. An interesting observation made in the big data environment is that advances in technology will reduce the need to apply simple decision tools to patients with complex disease. Parallels are often not well-founded, but we have come across the same type of observations in the process industry where a lack of analysis tools has encouraged the operators in the control room to use ad hoc methods and tacit heuristic rules (also called rules of thumb).

We found out in this section that predictive analytics offers some means to deal with failure prediction in industrial processes despite the challenges offered by big data. Our proposal is to work out predictive models that make use of fuzzy entropy, an approach that has not been much used but which we have found to offer some benefits.

3 Feature Selection with Fuzzy Entropy

We will now work out how fuzzy entropy measures can be used as a basis for predictive analytics. The basic data set used as the motivation consists of time series of several process diagnostic variables measured by a system of sensors. We propose a two-stage process for failure prediction:

- First, we choose a subset of the recorded diagnostic variables that contain the information essential to predict a failure
- Second, the identified features are used to predict failures (or in other words, classify a given state as critical or non-critical)

The main motivation behind this two-step process is the availability of big data collected on sensor-monitored valves that are key components of industrial pumps. The number of diagnostic variables can be in the range of hundreds for an individual valve, and the failure prediction has to be performed in real time, or with a few seconds delay in the worst-case scenario. For this reason, we need to identify a subset of variables that contain the most information regarding the state of the system.

An initial observation guided by the requirements of the engineers is that a decision support system should perform feature selection or extraction, but not component extraction. An important issue when using feature selection—and subsequent classification—relates to the interpretability of the results. If for example, Principal Component Analysis (PCA) is used, original variables are decomposed into new constructs by a specific combination method, which makes the results difficult if not impossible to interpret in many cases. A feature selection method can identify a set of original variables that can represent the whole data set to an acceptable level. In our context, this can provide important help for the engineers to identify which part of the process most likely will be responsible for the potential failure.

3.1 Feature Selection Methods

In real-world problems, a complex process can be described and analyzed through a large number of variables. Some of these variables contain essential information about the underlying process while others are irrelevant when we try to understand the behavior of the process and estimate future events. For this reason, in many cases it is an essential task to identify the "important" features of the problem. In the age of big data, this task is even more crucial than before.

In our application case (described in Sect. 4) the feature selection serves as a first stage of a prediction algorithm to estimate whether the system is in a state that indicates a high possibility of failure. In other words, the goal is to select a subset of features such that the prediction based on this smaller subset of features is as close as possible to a prediction based on the original set of features [31].

According to Dash and Liu [7], there are four main issues to be considered when choosing the appropriate feature selection method. The generation procedure refers to the method of selecting the candidate feature subsets. A traditional approach is to apply a greedy selection procedure [24], which means that we start from an empty set of features and iteratively add the best candidate from the remaining set of features.

The second issue is the choice of an evaluation function. The function evaluates the candidate feature subsets based on different attributes. There are numerous choices available also for the evaluation function, for example distance measures [10], or correlation measures [18]. Another important approach makes use of different types of information measures [26], most frequently Shannon's entropy. In the different methods based on entropy, a feature is preferred in a given step if the information gain (i.e. reduction of entropy) is higher by choosing that feature over any other. In practice, this translates to reduced uncertainty concerning the outcome variable; in this paper, fuzzy entropy will be used as the basis for feature selection.

The last two issues concern the stopping criterion of the algorithm and the validation process [17]. The stopping criterion is usually dependent on the approach used for determining the evaluation function. In case of an information theory based evaluation measure, the stopping criterion specifies a critical improvement value: we select new features until the contribution of a selected new feature to the information gain is lower than a predefined value. The validation process usually tests the predictive validity (or classification precision) of the identified feature subset. In many practical problems, there can be several feature combinations that provide equally good performance in predicting the outcome; experienced process managers and engineers can provide additional help in identifying alternative solutions.

3.2 *Fuzzy Entropy for Feature Selection*

In this section, we will describe an algorithm for feature selection based on fuzzy entropy introduced in [33]. An important step in applying entropy is the discretization of the continuous range of variables. As a consequence, there will always be numerous observations which lie close to the borders between two neighbouring intervals; as it was pointed out by Shie and Chen [42], this can result in situations for which an incorrect choice in the discretization process can decrease the performance of the chosen feature set. To overcome this problem, fuzzy entropy can be applied; as a result, observations close to borders between intervals will be classified as belonging to both intervals to some extent and consequently increasing the information extracted from the chosen features.

Mezei et al. [33] propose to use type-2 fuzzy sets and fuzzy entropy in the feature selection process. The approach makes use of the concept of interval-valued fuzzy numbers. A fuzzy number is a normal, convex fuzzy set with the referential set X as the set of real numbers [4]. An interval-valued fuzzy number can be described by two membership functions: upper and lower membership values representing a second level of imprecision present in the process of defining the membership values for belonging to the fuzzy sets. By applying this tool, we can further improve feature selection methods by considering the observations close to the borders of intervals. We apply trapezoidal-shaped interval-valued fuzzy numbers, where the upper and lower fuzzy numbers are ordinary trapezoidal fuzzy numbers represented by the following membership function:

$$\mu_A(x) = \begin{cases} 1 - \frac{a-x}{\alpha} & \text{if } a - \alpha \leq x \leq a \\ 1 - \frac{x-b}{\beta} & \text{if } b \leq x \leq b + \beta \\ 1 & \text{if } a \leq x \leq b \\ 0 & \text{otherwise} \end{cases}$$

3.3 Fuzzy Entropy Models

To describe the algorithm, we need the following notations:

- $C_1, \ldots, C_m$ denote the possible values of the outcome variable. In the simplest case, we have two classes, which translates to a binary classification problem.
- $F_1, \ldots, F_l$ are the variables describing the system/object that is modelled, i.e. the set of possible features. We want to choose a subset that contains sufficient information to determine the class membership in the outcome variable.
- $A^i_j, \ldots, A^i_{j_i}$ are the interval valued fuzzy numbers describing feature i. The optimal number of fuzzy numbers has to be determined individually for every feature.

To determine the information contained in a feature concerning the outcome variable, we use the match degree based on the fuzzy entropy for every outcome class j as

$$D_j = \sum_{i=1}^{n} \frac{1 - \max\left(1 - \mu^U_A(x_i), \mu^L_A(x_i)\right)}{1 - \min\left(1 - \mu^U_A(x_i), \mu^L_A(x_i)\right)}.$$

The fuzzy entropy for one fuzzy set describing a feature can be calculated as

$$H_j(A) = -D_j \log D_j$$

while the fuzzy entropy of the feature with a given set of fuzzy numbers is

$$H(A) = \sum_{j=1}^{m} H_j(A).$$

In the feature selection algorithm, we first derive the optimal number of fuzzy numbers for every feature. Additionally, we also calculate the overall entropy of the feature with respect to the outcome variable (failure indicator in our case). This helps us to identify the feature with the optimal entropy value. In the subsequent steps of the algorithm, new features are added to the feature set by calculating and minimizing the joint entropy of a subset of features including the already chosen features and the ones which have not been chosen yet.

The joint entropy can be calculated using the match degree for two fuzzy sets as in [33]; for higher numbers it can be extended in a similar way:

$$D^h_j(s,t) = \sum_{i=2}^{m} \frac{\min(A,B)}{\max(A,B)},$$

where

$$A = 1 - \max\left(1 - \mu_{A_s^{U_1}}(x_i), \mu_{A_s^{L_1}}(x_i)\right),$$
$$B = 1 - \max\left(1 - \mu_{A_t^{U_h}}(x_i), \mu_{A_t^{L_h}}(x_i)\right)$$

The entropy of the pair (A_s^1, A_t^h) is

$$H\left(A_s^1, A_t^h\right) = \sum_{j=1}^{m} -D_j^{1,h}(s,t) \log D_j^{1,h}(s,t),$$

and finally the overall entropy

$$H(F_1, F_h) = \sum_{s=1}^{k_1} \sum_{t=1}^{k_h} H\left(A_s^1, A_t^h\right).$$

We continue to include features as long as the improvement in the gained information is higher than a predefined value δ.

4 Case Study: Failure Prediction Using Fuzzy Entropy

In this section, we will describe an application of the fuzzy entropy-based feature selection methodology for failure prediction. In the industrial processes we have been working on, maintaining continuous, uninterrupted operations is of essential importance; consequently processes are monitored constantly in real-time. The occurrence of a failure usually implies stopping the operation of the whole process resulting in significant losses in terms of production output. For this reason, as we concluded previously, predictive maintenance offers a better strategy to minimize losses. In the following we specifically focus on the case of data describing the working behaviour of pumps and valves in the industrial process we have been working on.

As we learned from expert engineers, the operational states of valves usually can be classified into three main cases:

- normal: the process is fully operational, there is no need for any type of intervention
- failure: the process has to be shut down and the cause of the failure identified
- gradual transition from normal operation to failure: the behaviour is still normal, or almost normal, but there are signs indicating a possible failure

In our analysis, we will consider data generated using a simulation model developed at our partner company. The model has been improved continuously based on the understanding of the real behaviour of valves, and provides a means to obtain sufficiently detailed data to understand how failures develop and how they can be predicted. In general, behaviour of different industrial processes can be predicted and is frequently described using simulation models [35]. Simulation studies are performed outside the real-world system, but on a model of the system that is specifically created to obtain an understanding of the relationships among some system characteristics and the operational behaviour of the system. If the simulation model resembles the real process to a sufficient degree, we can formulate conclusions that are valid for the system. The major reasons for developing a model, as opposed to analysing the real system, include the costs (usually significant) of experimenting with the system, unavailability of the "real" system, and the goal of achieving a deeper understanding of the relationships between the elements of the system.

In practice, there are hundreds of diagnostic measures to be observed, stored and analysed. Before predicting the failure of the system, it can be beneficial both from computational complexity and interpretability perspectives to reduce the number of measures to be included in the prediction models. This selection is performed using a fuzzy entropy-based feature selection algorithm. In the simulated data, three different fault types are present, as it is assumed that different characteristics are associated and can be used to predict different problems. Additionally, to control for the type of pump (our partner company manufactures tens of thousands of different versions of pumps), different valves sizes and different tuning parameters are present in the data.

According to the above description of the algorithm, we will use the following notations:

- C_1, C_2, C_3 denote the 3 possible values of the outcome variable. In our case, as our main goal is failure prediction of valves, the outcome variable is the actual state of a valve (or the underlying process). Additionally to the two straightforward classes, normal behaviour and failure, we include a transition state as the third class.
- $F_1, \ldots, F_{16}$ are the diagnostic variables describing the valve, i.e. the set of possible features. We want to choose a subset that contains sufficient information to determine the class membership in the outcome variable. The diagnostic measures include as an example "stiction", "steady state deviation" or "valve reversals per hour".

The simulated data is generated with a simplified model of a pipeline process that consists of a pump and a control valve that generates a flow through the pipeline. There are three simulated faults: (i) air leakage; (ii) faulty H-clip; (iii) friction.

The three faults are simulated for different valve sizes and different controller tuning parameters. The different combinations resulted in 32 valves in total, with 120 days of each valve simulated. With the associated 16 diagnostic variables, we performed the analysis on 61,440 data values. To test the effectiveness of the fuzzy entropy-based feature selection model, we first identified the optimal feature subset for the three fault types separately. We evaluated the selected feature subsets by comparing their prediction performance to the model using all the 16 diagnostic variables, using a multinomial logistic regression analysis. Additionally, we tested the three different fuzzy entropy measures discussed in Sect. 2.1.

In the following, we will discuss the results of the feature selection process and the subsequent classification modelling. Table 1 lists the results of the analysis. The models were implemented using the R statistical software [40]. The running time of the models increases with the complexity of the underlying data representation; transforming the original values into interval-valued fuzzy memberships can be time consuming compared to type-1 fuzzy sets, or in the crisp case, compared to no transformation at all. As one can find, the number of selected features can significantly vary among methods; this behaviour can be explained by the different structure of the entropy measures and consequently the way in which their value improves when increasing the number of features in the selected feature sets. In the analysis, we used the same threshold for limiting the number of selected features (at least 5 % improvement in the overall entropy); different threshold specifications for different entropy measures may result in more similar outputs. The predictive performance in general shows high accuracy with different varying performance depending on the fault types. We can observe that there is no single approach that performs better than the others. However there is at least one method for every fault type with high accuracy. According to this, we can conclude that fuzzy entropy based feature selection can be a useful alternative to traditional methods, when the specific attributes of the underlying problem are considered properly and the fuzzy entropy measure is chosen based on the problem attributes.

In Table 2, the chosen features (process diagnostics) are listed for three entropy measures. As we can observe, different fault types can be explained with different sets of diagnostic measures; while some faults are more straightforward to recognize and closely related to a specific feature of the process (*Faulty H-clip—Valve travel*), some other fault types are more complex and require combining knowledge from different diagnostics. From the perspective of the entropy measures, one can observe that the most important features chosen are always the same independently from the entropy used. However, there can be differences regarding less important, but still selected features, and regarding the number of features selected.

Table 1 Comparison of the results of the different approaches (number of selected features/classification performance)

Entropy/Fault	Fault 1	Fault 2	Fault 3
H	3 (90 %)	2 (96 %)	4 (87 %)
H_{LT}	4 (92 %)	1 (93 %)	5 (88 %)
H_{SK}	5 (93 %)	1 (93 %)	6 (85 %)
H_{BB}	2 (88 %)	3 (96 %)	8 (82 %)
H_{ZL}	3 (90 %)	3 (96 %)	6 (90 %)

Table 2 Detailed list of process diagnostics chosen by three different entropy measure for predicting/explaining the three considered fault types

Fault/Entropy	H	H_{LT}	H_{SK}
Air leakage	DZ min Valve reversals Setpoint travel	DZ min Valve reversals Setpoint travel DZ max	DZ min Setpoint travel DZ max Setpoint ratio PWM ratio
Faulty H-clip	Valve travel Setpoint travel	Valve travel	Valve travel
Friction	Stiction Setpoint travel Valve ratio DZ min	Stiction Setpoint travel Valve travel Valve ratio PWM ratio	Stiction Setpoint travel Valve travel Valve ratio DZ min Travel ratio

5 Discussion—Decision Support for Monitoring and Control

The fuzzy entropy used for feature selection is one of the methods we have developed and tested as part of a project with the process industry. Our task is to work out some solutions for predictive maintenance, i.e. we need to identify failures, (i) what may fail, (ii) where in the process and (iii) when can a failure be expected to happen. The context is a plant where the process is regulated by hundreds of pumps, some of which represent a critical path for the working of the plant and have back-up pumps built for parallel flows to replace the primary flow in case of failures; other pumps are stand-alone for various reasons—cost-effectiveness, low risk, (assumed) fail-safe constructs, etc. The pumps have components that determine how well the pump will carry out the tasks for which it has been constructed; the components are monitored and controlled by sensors that continuously collect data on the performance of the components.

The support process we need for the monitor and control personnel builds on predictive analytics and modelling; the feature selection with fuzzy entropy is only part of the process as modern analytics offers support in different ways [8]: (i) descriptive techniques and models for reviewing and examining data set(s); (ii) diagnostic techniques and models to determine what has happened and why; (iii) predictive techniques and models that analyze current and historical data to determine what is most likely to (or not to) happen; (iv) prescriptive techniques and models for computationally developing and analyzing alternatives that can become courses of action and that may discover the unexpected; (v) decision techniques and models for visualizing information to facilitate human decision-making.

The data that is collected from the sensors is first stored in a simple flat file format; cf. Fig. 2, on which sorting, classifying and clustering operations can be carried out with tools that scale for big data. This sensor data is collected on a

ISO-8859-1

Tag name (file name)	hartTag	longTag	description	message	performanceLevel	maxTurningAngle	direction	positionSensorRotation	positionerFailAction	valveActingType	valveType	actuatorType	actType	deadAngle	deadAngleCompensation	setpointCutOffClose
41PV1101	41PV1101	ND9000	ND9000	ND9000	3	59	0	1	1	2		2		0		
01DV1412	01DV1412	ND9000	ND9000	1101-PDV-1412	5	24	1	1	2	4		2		0		
02FV1204	02FV1204	ND9000	ND9000	ND9000	5	30	0	1	1	2		2		0		
02FV1205	02FV1205	ND9000	ND9000	ND9000	5	45	0	1	1	2		2		0		
04FV1212	04FV1212	ND9000	ND9000	1104-FV-1212	5	47	1	1	2	4		2		0		
05DV1511	05DV1511	ND9000	05PDV1511	ND9000	5	24	1	1	2	4		2		0		
05FV1601	05FV1601	ND9000	05FCV1601	ND9000	5	47	0	1	1	4		2		0		
05LV1602	05LV1602	ND9000	05LCV1602	ND9000	5	46	0	1	1	4		2		0		
05QV1701	05QV1701	ND9000	ND9000	ND9000	3	57	0	1	1	2		2		0		
05QV1702	05QV1702	ND9000	ND9000	ND9000	5	44	0	1	1	2		2		0		
05TV1603	05TV1603	ND9000	05TCV1603	ND9000	5	29	0	1	1	4		2		0		
07FV1207	07FV1207	ND9000	07FCV1207	ND9000	5	56	1	1	2	4		2		0		

Fig. 2 Data collected on components and pumps

regular basis for every valve and the spreadsheet file contains more than 500 features. Previously, process engineers used this data file by manually highlighting suspicious cases without the support of any real advanced analytics tools. We use fuzzy entropy for feature selection in the early stages when we are in the process of learning about what failures occur in what part of the process and how frequently. Later, when we have collected a more solid set of observations we can decide on feature parameters and identify key variables and combine this with some heuristics to reduce "big data" to some meaningful data sets. The fuzzy entropy measures are shown in Fig. 3 and demonstrate the type of insight we can get from processing the large data sets; we can identify the features of failure events and build on that when predicting future failures. As it is illustrated on this visual representation, we can identify features (horizontal axis) with improvements in the fuzzy entropy (vertical axis).

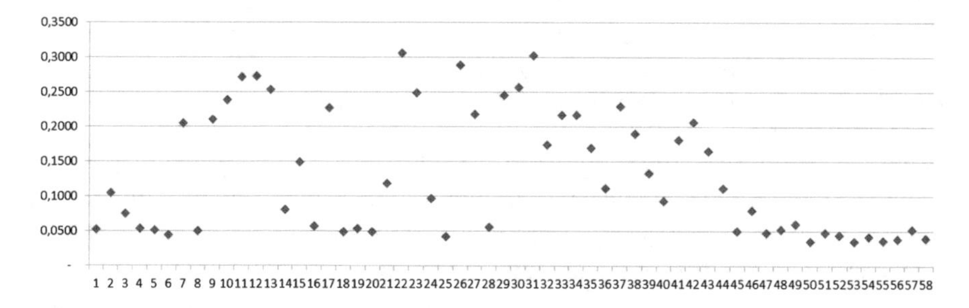

Fig. 3 Entropy measures approximate entropy

As a result of choosing the optimal number of fuzzy subsets for every feature, we will also obtain the overall entropy of the features with respect to the outcome variable. The feature with the optimal entropy value will be selected as the best describing variable of the model and used in the consequent steps. In the algorithm, in every step one new feature is added to the set of selected features: the choice is determined by calculating the joint fuzzy entropy of a subset of features with respect to the outcome variable.

The feature selection set is used to identify similar incident cases that can be used in case-based reasoning implementations to significantly speed up the problem-solving processes. Additionally to this, we can find out how likely the incidents are to occur at various instances within a certain time frame $(T, T+t)$.

We can use Bayesian models for the prediction of future failure events; probabilistic Bayesian modelling when we get robust and large time series; possibilistic Bayesian modelling for imprecise data and for short and incomplete time series (cf. [5]). We have found that when (for proper risk assessment and management) knowledge related to risky events is needed, this knowledge can be derived by combining probabilistic Bayesian models and feature selection with fuzzy entropy.

The online decision support that is needed for the monitoring and control personnel is first producing predictions on possible failures with the information of what, where and when. This is then combined with a "quick fix" support function that shows how similar problems have been dealt with previously by selecting similar cases from a knowledge base (cf. [5]).

6 Conclusions

In this chapter, we proposed a systematic approach for applying predictive analytics for predictive maintenance decision support. In a process industry case we were able to identify three decisive conditions: (i) we can assume that sensor systems will produce "big data" that we have to process in order to work out meaningful strategies for predictive maintenance; (ii) data can be imprecise, have errors, have missing elements, have outliers, etc.; (iii) processing cannot take much time, it should be in almost real time and there will be a trade-off between precision and relevance. A decision rule could be to find sufficiently good precision in such a time that costly problems can be avoided.

There are further lessons to be learned from the material as it is common experience in the process industry that if predictive maintenance can be carried out just-in-time before a failure the maintenance cost is minimized. This is part of building an optimal maintenance policy for the components, for modules built around the components, for devices (pumps in our case) and eventually for the process. Thus the fuzzy entropy, which is part of the process, can have further importance as a key part of maintenance decisions.

A different extension of the proposed approach would be to work out maintenance as a customer service with its own business models or as a supported program carried out by process or equipment owners. The business models build on finding a good balance between three central factors: the cost of risk, the cost of maintenance policy and the revenue of an optimal maintenance policy.

Acknowledgment This research has been funded through the TEKES strategic research project Data to Intelligence [D2I], project number: 340/12.

References

1. Beath, C., Becerra-Fernandez, I., Ross, J., Short, J.: Finding value in the information explosion. MIT Sloan Manage. Rev. **53**(4), 18 (2012)
2. Bolón-Canedo, V., Sánchez-Maroño, N., Alonso-Betanzos, A.: Recent advances and emerging challenges of feature selection in the context of big data. Knowl. Based Syst. **86**, 33–45 (2015)
3. Burillo, P., Bustince, H.: Entropy on intuitionistic fuzzy sets and on interval-valued fuzzy sets. Fuzzy Sets Syst. **78**(3), 305–316 (1996)
4. Carlsson, C., Fullér, R.: Fuzzy Reasoning in Decision Making and Optimization, vol. 82. Springer Science & Business Media (2002)
5. Carlsson, C., Heikkilä, M., Mezei, J.: Possibilistic Bayes modelling for predictive analytics. In: 2014 IEEE 15th International Symposium on Computational Intelligence and Informatics (CINTI), pp. 15–20. IEEE (2014)
6. Chen, H., Chiang, R.H., Storey, V.C.: Business intelligence and analytics: from big data to big impact. MIS Q. **36**(4), 1165–1188 (2012)
7. Dash, M., Liu, H.: Feature selection for classification. Intel. Data Anal. **1**(3), 131–156 (1997)
8. Davenport, T.H., Harris, J.G.: Competing on Analytics: The New Science of Winning. Harvard Business Press (2007)
9. De Luca, A., Termini, S.: A definition of a nonprobabilistic entropy in the setting of fuzzy sets theory. Inf. Control **20**(4), 301–312 (1972)
10. De Mántaras, R.L.: A distance-based attribute selection measure for decision tree induction. Mach. Learn. **6**(1), 81–92 (1991)
11. Demirkan, H., Delen, D.: Leveraging the capabilities of service-oriented decision support systems: putting analytics and big data in cloud. Dec. Supp. Syst. **55**(1), 412–421 (2013)
12. Diamantoulakis, P.D., Kapinas, V.M., Karagiannidis, G.K.: Big data analytics for dynamic energy management in smart grids. Big Data Res. **2**(3), 94–101 (2015)
13. Gandomi, A., Haider, M.: Beyond the hype: big data concepts, methods, and analytics. Int. J. Inf. Manage. **35**(2), 137–144 (2015)
14. Gass, S.I., Harris, C.M. (eds.): Encyclopedia of Operations Research and Management Science. Kluwer Academic Publishers, Dordrecht (1996)
15. Gil-Aluja, J.: Fuzzy Sets in the Management of Uncertainty. Springer Science & Business Media (2004)
16. Gorzałczany, M.B.: A method of inference in approximate reasoning based on interval-valued fuzzy sets. Fuzzy Sets Syst **21**(1), 1–17 (1987)
17. Guyon, I., Elisseeff, A.: An introduction to variable and feature selection. J. Mach. Learn. Res. **3**, 1157–1182 (2003)
18. Hall, M.A.: Correlation-based feature selection for machine learning. Ph.D. thesis, The University of Waikato (1999)
19. Hitzler, P., Janowicz, K.: Linked data, big data, and the 4th paradigm. Seman. Web **4**(3), 233–235 (2013)
20. Janke, A.T., Overbeek, D.L., Kocher, K.E., Levy, P.D.: Exploring the potential of predictive analytics and big data in emergency care. Ann. Emerg. Med. (2015)

21. Jiang, Y., Tang, Y., Liu, H., Chen, Z.: Entropy on intuitionistic fuzzy soft sets and on interval-valued fuzzy soft sets. Inf. Sci. **240**, 95–114 (2013)
22. Knopfmacher, J.: On measures of fuzziness. J. Math. Anal. Appl. **49**(3), 529–534 (1975)
23. Kosko, B.: Fuzzy entropy and conditioning. Inf. Sci. **40**(2), 165–174 (1986)
24. Kwak, N., Choi, C.-H.: Input feature selection for classification problems. IEEE Trans. Neural Netw. **13**(1), 143–159 (2002)
25. Lee, H.-M., Chen, C.-M., Chen, J.-M., Jou, Y.-L.: An efficient fuzzy classifier with feature selection based on fuzzy entropy. IEEE Trans. Syst. Man Cybern. Part B Cybern. **31**(3), 426–432 (2001)
26. Lee, S.-H., Kim, S., Kim, J.-M., Choi, C., Kim, J., Lee, S., Oh, Y.: Extraction of induction motor fault characteristics in frequency domain and fuzzy entropy. In: 2005 IEEE International Conference on Electric Machines and Drives, pp. 35–40. IEEE (2005)
27. Li, P., Liu, B.: Entropy of credibility distributions for fuzzy variables. IEEE Trans. Fuzzy Syst. **1**(16), 123–129 (2008)
28. Li, J., Deng, G., Li, H., Zeng, W.: The relationship between similarity measure and entropy of intuitionistic fuzzy sets. Inf. Sci. **188**, 314–321 (2012)
29. Liu, X.: Entropy, distance measure and similarity measure of fuzzy sets and their relations. Fuzzy Sets Syst. **52**(3), 305–318 (1992)
30. Liu, B.: Uncertainty Theory. Springer, Berlin (2007)
31. Liu, H., Yu, L.: Toward integrating feature selection algorithms for classification and clustering. IEEE Knowl. Data Eng. **17**(4), 491–502 (2005)
32. McAfee, A., Brynjolfsson, E.: Big data: the management revolution. Harvard Bus. Rev. **90** (10), 60–68 (2012)
33. Mezei, J., Morente-Molinera, J.A., Carlsson, C.: Feature selection with fuzzy entropy to find similar cases. In: Advance Trends in Soft Computing, pp. 383–390. Springer, Berlin (2014)
34. Morabito, V.: Big data driven business models. In: Big Data and Analytics, pp. 65–80. Springer, Berlin (2015)
35. Muir, B.M., Moray, N.: Trust in automation. Part ii. Experimental studies of trust and human intervention in a process control simulation. Ergonomics **39**(3), 429–460 (1996)
36. Nieto-Santisteban, M.A., Szalay, A.S., Thakar, A.R., O'Mullane, W.J., Gray, J., Annis, J.: When database systems meet the grid. arXiv preprint cs/0502018 (2005)
37. Papermaking—Parts 1–3. Paperi ja Puu Oy, Jyväskylä (2007)
38. Parthalain, N., Jensen, R., Shen, Q.: Fuzzy entropy-assisted fuzzy-rough feature selection. In: 2006 IEEE International Conference on Fuzzy Systems, pp. 423–430. IEEE (2006)
39. Qin, S.J.: Process data analytics in the era of big data. AIChE J. **60**(9), 3092–3100 (2014)
40. R Core Team: R: A Language and Environment for Statistical Computing. R Foundation for Statistical Computing, Vienna, Austria (2014)
41. Shannon, C.E.: A mathematical theory of communication. Bell Syst. Tech. J. **27**(3), 379–423 (1948)
42. Shie, J.-D., Chen, S.-M.: Feature subset selection based on fuzzy entropy measures for handling classification problems. Appl. Intell. **28**(1), 69–82 (2008)
43. Szmidt, E., Kacprzyk, J.: Entropy for intuitionistic fuzzy sets. Fuzzy Set. Syst. **118**(3), 467–477 (2001)
44. Tao, W.-B., Tian, J.-W., Liu, J.: Image segmentation by three-level thresholding based on maximum fuzzy entropy and genetic algorithm. Patt. Rec. Lett. **24**(16), 3069–3078 (2003)
45. Wu, D., Mendel, J.M.: Uncertainty measures for interval type-2 fuzzy sets. Inf. Sci. **177**(23), 5378–5393 (2007)
46. Wu, X., Gu, H., Hu, X., Dong, Y.: Application of the fuzzy entropy weight in risk assessment of the engineering project. In: Fifth International Conference on Information Assurance and Security, vol. 1, pp. 145–148. IEEE (2009)
47. Xu, Z., Xia, M.: Hesitant fuzzy entropy and cross-entropy and their use in multiattribute decision-making. Int. J. Intell. Syst. **27**(9), 799–822 (2012)
48. Yager, R.R.: A procedure for ordering fuzzy subsets of the unit interval. Inf. Sci. **24**(2), 143–161 (1981)

49. Ye, J.: Fault diagnosis of turbine based on fuzzy cross entropy of vague sets. Expert Syst. Appl. **36**(4), 8103–8106 (2009)
50. Zadeh, L.A.: Fuzzy sets. Inform. Cont. **8**(3), 338–353 (1965)
51. Zeng, W., Li, H.: Relationship between similarity measure and entropy of interval valued fuzzy sets. Fuzzy Sets Syst. **157**(11), 1477–1484 (2006)
52. Zhai, Y., Ong, Y.-S., Tsang, I.W.: The emerging "big dimensionality". IEEE Comput. Intell. Mag. **9**(3), 14–26 (2014)
53. Zhang, Q.-S., Jiang, S.-Y.: A note on information entropy measures for vague sets and its applications. Inf. Sci. **178**(21), 4184–4191 (2008)
54. Zikopoulos, P., Eaton, C.: Understanding big data: analytics for enterprise class hadoop and streaming data. McGraw-Hill Osborne Media, New York (2011)

Fuzzy Sets in Agriculture

Elpiniki I. Papageorgiou, Konstantinos Kokkinos and Zoumpoulia Dikopoulou

Abstract Agricultural modeling and management are complex conceptual processes, where a large number of variables are taken into consideration and interact for system analysis and decision making. Most of the processes in the agricultural sector include the uncertainty, ambiguity, incomplete information and human intuition characteristics. These processes are not only constrained by their environment (e.g., market, climate, seasons, consumer choices), but they are also highly influenced by human factors (stakeholders' perceptions). Fuzzy sets are able to manage and represent uncertainty, assure that the incomplete information is valued and provide solutions to issues which are crucial in agriculture like fertilization, land degradation, soil erosion and climate variability during planting material selection in physiological analysis. Fuzzy sets have gained constantly increasing research interest in the last twenty years and have found great applicability in the agricultural domain, helping farmers to take right decisions for their cultivated.

Keywords Agriculture · Fuzzy sets · Irrigation · Fuzzy cognitive maps · Crop simulation

E.I. Papageorgiou (✉)
Department of Computer Engineering, Technological Educational Institute (TEI) of Central Greece, 3rd Km Old National Road Lamia-Athens, 35100 Lamia, Greece
e-mail: epapageorgiou@teiste.gr; epapageorgiou@iti.gr

K. Kokkinos
Centre for Research and Technology Hellas, 6th Km Charilaou-Thermi Road, 57001 Thessaloniki, Greece
e-mail: kokkinos@iti.gr; kokkinos@uth.gr

E.I. Papageorgiou · Z. Dikopoulou
Faculty of Business Economics, Hasselt University, Agoralaan—Building D, 3590 Diepenbeek, Belgium
e-mail: zoumpolia.dikopoulou@student.uhasselt.be

C. Kahraman et al. (eds.), *Fuzzy Logic in Its 50th Year*,
Studies in Fuzziness and Soft Computing 341,
DOI 10.1007/978-3-319-31093-0_10

1 Introduction

Nowadays, at almost every continent, agriculture is facing new major challenges. Over the next 50 years, global population is expected to be increased to more than nine billion people, which indicates that the total food production should be increased by 70 % to face and prevent massive famine. The arable land will "cover" only 10 % of this growth, which still remains a limited resource, while the remaining 90 % should be resulted from the intensification of current production [1]. The modern agriculture is developing ways to increase the productivity and efficiency on all kinds of agricultural production. These novel agricultural methods have developed into a highly competitive and globalized industry to address factors such as: climate changes, resource depletion, climate intensification (floods and droughts) and dramatic political shifts, in order to ensure healthy economy and sustainable production [2].

Based on the literature, new solutions for all aspects of agricultural production required to resolve problems involving the complexity, uncertainty, ambiguity and probabilistic in nature. There is a need for some issues to be improved, such as: better and predictable planning of crops, precision in agriculture, optimized application of resources, support the effective and collaborative processes by using modern technology, autonomous solutions (completely or partially) for tedious work, viable long-term growth of useful knowledge [3]. Fuzzy sets (FS), as an intelligent methodology to handle ambiguity and fuzziness have already applied in the agricultural domain too. They have been mainly used to cope with a variety of complex processes in agriculture, for modeling, management and decision support. Fuzzy modeling has substituted the analytical approach that has proved inefficient to develop optimal solutions in a number of agricultural systems [4, 5]. Even though the application of FS in agriculture has a relatively short life, intensive research projects based on FS and their extensions which are integrated with humanity, are ongoing reaching a new phase of evolution.

On the whole, fuzzy information is included in human thinking and reasoning, which flows from inexact human concepts. FS theory can deal with the vagueness of qualitative values by the capturing and the reflection of the human knowledge [6]. With the formation of the theory of fuzzy sets, Lotfi Zadeh introduced the fuzzy logic (FL) term which copes with the approximate reasoning. FL is an approach to reasoning that declares that, the variables may have a truth value that ranges in degree between 0 and 1, instead of having a precise and definite value [6]. The simulation of human reasoning processes is accomplished in a specific way, using fuzzy rules, which are performing computations between linguistic variables, instead of the classical numeric variables [7]. The application of fuzzy if-then rules also improves the interpretability of the results and provides more insight into the decision making process [8].

Based on the FS theory and FL, fuzzy inference systems (FIS), can provide insights into management behavior complex agricultural systems. Imperfect information, uncertainty and other issues create conditions where expert knowledge is needed and which fuzzy sets and systems can imitate. FS and the inference methods have found large applicability in agriculture in the recent years [9].

The popularity of FS stems from the fact that they offer a series of advantages including flexibility and adaptation to practically any problem domain, capabilities to handle efficiently complex issues in environments full of uncertainty and ambiguity, ability to execute with incomplete information, relatively simple and comprehensible modeling philosophy which is very close to human reasoning.

More often, the ambiguity in decision making activities in the agricultural domain, is compounded by imprecision and intuition. The capabilities of FS on handling complexity in agricultural processes that include uncertainty and ambiguity, can help farmers to take the right decisions for cultivation activities such as: sowing, fertilizer management, irrigation management, integrated pest management, storage etc. to claim higher crop productions [5].

In this line of research, the present chapter aims at contributing to a refinement of review studies by applying FS, FL and fuzzy cognitive mapping (FCM) to the exploration of agriculture modeling and management. In the next section we provide a literature review on the latest related research which is divided into three subsections namely: (a) FL models, (b) FIS and (c) FCM's. We only concentrate to research published since 2011 as the aforementioned methodologies became popular. In Sect. 3 we describe rigorously the basic mathematical framework that surrounds the three techniques. Then, two representative case studies on agriculture are presented. The motivation of these studies emanates from within a wider research endeavour on the applications of FS and FIS in agriculture. Furthermore, FCM's have been explored in this domain to provide better understanding on the relationships between agricultural factors involving complexity and uncertainty, and achieve prediction and classification capabilities in crop and yield grading in some agricultural cases. We finally conclude on the value of concepts and methods and the applications of FS, FIS and FCM in the field of agricultural engineering. The future of the development and applications of these methodologies in agricultural engineering is discussed, especially in the soil and crop quality context for crop management and in decision support in precision agriculture.

2 Literature Review

Modern agriculture faces great challenges due to the fact of the constantly increased production demand and the competition introduced by its transformation into a globalized industry. In order for the farmers to meet the production increase or at least the production sustainability criterion, fuzzy systems unite the accumulated expertise of individual disciplines such as FL, FIS and FCM to provide innovative yet near optimal methodologies in various aspects related to agriculture. The vast majority of this research deals with: (a) the study of soil and water regimes related to crop growth, (b) the crop simulation models and human decision simulation models, (c) the irrigation, the water conservation and the drought tolerant models

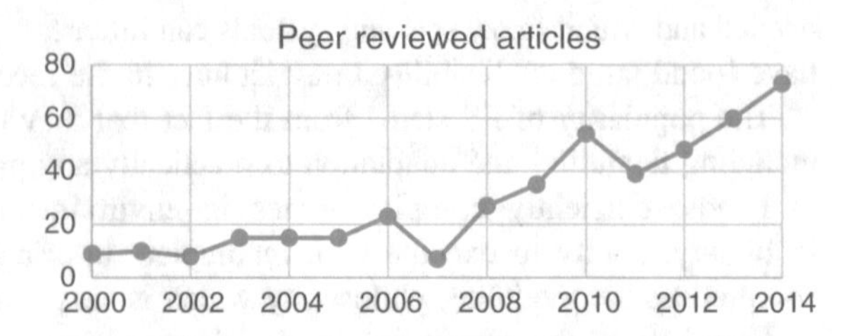

Fig. 1 Number of peer-reviewed articles in Scopus after 2001

and (d) the prediction and the handling of environmental and climate parameters related to precision agriculture.

The interest for FS in agriculture is manifested by an increasingly active literature, as reported in Fig. 1. Making a literature review in scopus with keywords “fuzzy sets” or “fuzzy logic” or “fuzzy inference system” and “agriculture”, a number of peer reviewed articles were presented.

We categorize the state of the art section into three subsections relatively to the major FS disciplines and for each one of these, we provide sub-categories of related works to the aforementioned agricultural studies.

2.1 *Fuzzy Logic Models*

2.1.1 Crop Simulation Models

Creating Fuzzy Logic Models

Bosma et al. [10] describe a methodology for the development of fuzzy logic models (FLMs), simulating strategic decision making in which human drivers and motives are the key variables. They further propose a 10 steps modeling approach that includes determination of the membership functions and the fuzzy rules. This methodology is applied on farmers’ choices for components in a hybrid/mixed agriculture–aquaculture farming system. Their system was tested and evaluated in the floodplain and in the hills of the Mekong Delta, Vietnam (VMD). The FS has hierarchical structure composed of five subsets: (a) the primary production factors, (b) the product opportunities, (c) the product options, (d) the farmers’ reference frames and (e) the final output layer. This transparent structure allows stakeholders’ participation to strengthen the ability of the modeler to empathize with the experts’ reasoning. Towards the same goal, research in [11, 12] developed a method for assessing the overall performance of biogas plants focusing on four assessment aspects: (a) biogas production, (b) biogas utilization, (c) environmental impact and (d) socioeconomic efficiency. The method involves the use of imprecise and uncertain data via FS theory and fuzzy mathematics pre-assessing biogas plant development in their planning phase and proving that biogas utilization is the aspect with the largest potential for performance improvement, by increasing the external heat utilization.

Using FLMs and Image Processing Techniques

An important parameter for planning and management of irrigation is the information on the land usage under command. Murmu and Biswas [13] claim that image classification of satellite data is an excellent technique in creating area crop thematic maps and thus estimating crop water requirements for irrigation planning. They used various fuzzy classifiers and found that FL takes advantage of simple rules resulting in equal or even less time consuming results than the other conventional methods. They also showed that the fuzzy c-means classifier proved to be a promising tool to build models for multivariate studies by identifying proper parameters and making use of the interpretation of satellite data. The same technique is also used in [14] where the FIS coupled with an image processing technique in order to develop a decision-support system for qualitative grading of milled rice. The quality indices used were namely, the degree of milling (DOM) and the percentage of broken kernels (PBK) and they were broken into five classes. The classifier was fed with images and the two quality indices and produced one output variable (Quality), all in the form of triangle membership functions. A similar application of fuzzy classification of remotely sensed data in agriculture is shown in [15] but on a different aspect. In this work, a subtractive based FIS is implemented to estimate potato crop parameters like biomass, leaf area index, plant height and soil moisture. Input to the fuzzy network included the scattering coefficients for HH- and VV-polarizations using microwave remote sensing. The measurements of the above parameters were sampled in various growth stages and were used as the target variables during the training and validation of the network. As for the subtractive clustering and more specifically for the cluster centers, all the data points were considered as candidates resulting on values of crop/soil parameters to be much closer to the experimental values.

2.1.2 Estimation and Evaluation of Croplands in Conjunction to Climate

Emergy is a type of available energy that is consumed in direct and indirect transformations needed to make a product or a service. Note that agricultural lands are dynamic systems combining ecology and social economics. For that reason, assessment of the health of ecosystems is achieved via emergy analysis based on principles of systems ecology and energy. Li and Yan [16] presented an integrated evaluation model based on emergy analysis and FL to evaluate the health of agricultural lands in certain regions of North China. They mapped the emergy index to a fuzzy index in order to create a composite index for the evaluation analysis. A similar tool for the mapping and quantification of crops was used in [17] to verify the accuracies of indicator kriging and fuzzy classification as supervised classifiers in identifying sugarcane and citrus crops in Brazil. The classification results were evaluated using the kappa index indicating that the fuzzy classifier performed the best. Also from the point of view of the soil structure, it has been shown by

Abbaspour-Gilandeh and Sedghi [18] that correct tillage operations lead to prevention from soil degradation and help maintain and improve the physical, chemical, and biological characteristics. The soil fragmentation during tillage operation for seedbed preparation was described by a fuzzy model. The model inputs included soil moisture content, tractor forward speed and soil sampling depth and consisted of 50 rules. The evaluation of the model was done using the root mean square error, the relative error, and the coefficient of determination and proved to be of high accuracy. As for the evaluation of the available water resources and more specifically for the most influential reference evapotranspiration parameters for a land under study, an adaptive neuro-fuzzy inference system (ANFIS) was used in [19]. The inputs to the model were weather datasets for seven meteorological parameters and the results obtained indicated that, among the input variables, sunshine hours, actual vapor pressure and minimum air temperature, are the most influential parameters for evapotranspiration estimation.

2.2 *Fuzzy Inference Systems*

2.2.1 Crop Simulation Models

The authors in [20] created a FIS aiming in maximizing the soya-bean production in India. The system enabled farmers and researchers to benefit from an intelligent knowledge base created to capture the variety of field conditions. The development approach used was modular with most important steps to be: (a) the detection of nutrient levels in soil, (b) the farm size calculation, (c) the sowing period and (d) the sowing method and rain fall. A similar method was used in [21] for a GIS based land suitability assessment for tobacco production using an Analytic Hierarchy Process, (AHP) and FS in Shandong province of China. The assessment used 20 factors as suitability parameters with climatic condition, soil type and nutrient characters, and topography data to be the most important. This technique succeeded to integrate the FS model, the AHP method, and the GIS thematic maps to create an overall land suitability map. The results showed that 29.82 % of the total area was highly suitable for tobacco production and 17.74 % was unsuitable. Additionally, the use of a FIS becomes valuable when specific cases under studying suffer from input data imperfection and incompleteness. Coulon-Leroy et al., in [22] show that, imperfect knowledge to model complex ergonomic features can be used to develop a FIS under the condition that all relationships of data are identified. Their aim was the vine-vigor evaluation and how this process links to the shoot growth and leaf areas observed on vine plots in order to assess the interactions between environmental factors, agricultural practices and vine growth.

2.2.2 Irrigation and Drought Evaluation

Since agriculture is the major water consumer, water conservation has become an increasingly important issue in countries that face changes both in climate and in the agricultural practices. Most of the solutions to overcome this issue include statistical DSS's for water preservation which primarily use complex mathematical formulations that require large data sets to perform analysis. The introduction of a fuzzy model into these systems overcomes the above limitations. Such a model was introduced by [23] for the estimation of reductions in the pollutant loads based on selected strategies (e.g., storms, water management ponds, vegetated filter strips). This model employs an export coefficient approach and accounts for the number of animals to estimate the pollutant loads generated by different land usages (e.g., agriculture, forests, highways, livestock, and pasture land). Also, water quality index is used for the assessment of water quality once these pollutant loads are discharged into the receiving waters. Several other such systems have been published also.

Most of the FS in this category are parts of a bigger DSS aiming in managing water resources and providing near optimal irrigation schedules [24]. Most recently Giusti and Marsili-Libelli [25] developed a fuzzy DSS for irrigation and water conservation in agriculture which is based on the IRRINET model [26], and it is composed by: (a) a predictive soil moisture model, calibrated with data produced by the IRRINET model using inputs from its agro-meteorological database, (b) the irrigation inference system, consisting of a FIS to decide the timing and amount of irrigation on the basis of the crop phenophase, previous irrigations, and the prescribed soil moisture thresholds and (c) the irrigation performance index, consisting of the sum of the past irrigations. Similar to this work is the research by Binte Zinnat and Abdullah [27] with the design of an automated FS for monitoring and controlling the shading and the irrigation process. This system incorporates an integrated web system developed by Trono et al. [28] with their FS and provides information about different rain rates of different rain events. These rainfall data are considered as input to the FS and using a knowledge base it generates decisions to maintain the shading and irrigation process in the crop field for different rain events.

All such described above systems can help governments to identify future water supply problems in order to plan mitigation measures. Water reuse has a potential as a suggested solution but the major challenge is to identify the variables that determine the reclamation level. A FIS proposed by Almeida et al. [29] provides a conceptual approach based on multi-criteria decision making. This approach relates the water reuse to environmental factors such as drought, water exploitation index, water use, population density and the wastewater treatment rate resulting warnings about future water supply.

2.2.3 Land Evaluation

Most of these systems concentrate on the concepts of precision agriculture and suitability of lands for specific productions. Such FIS's usually integrate genetic algorithms to introduce machine learning into fuzzy evaluation for the suitability of agricultural lands relatively to the specifics of the agricultural products under study, and construct the so-called GA-optimized fuzzy inference models [30]. Additionally, these models can adjust the original criteria for land evaluation by genetic learning and reduce the subjective uncertainty in an ordinary FIS. Related works include a research by Mawale and Chavan [31] that invented a FIS to predict soil productivity and soil fertility using expert knowledge under the sustainability hypothesis and the research by Papadopoulos et al. [32] that provides a FIS for solving the nitrogen equation under agricultural conditions and is based on knowledge elicitation and FL methodologies. For the first work, the authors developed a FLM using a "mamdani"-FIS. The inference rules were framed using expert knowledge in the form of IF…THEN structures for the prediction of soil productivity. For the later work, the authors provided a system composed of two parts; a knowledge base and an analytical modular part which simulates nitrogen balance. This analytical part is built in a four level structure consisting of eleven FS's with the goal to evaluate characteristics such as soil, weather and farming practices.

2.3 Fuzzy Cognitive Maps

FCMs form a field of intelligent modeling and computing that has gained constantly increasing research interest in the last ten years. FCMs constitute graphical models in the form of directed graphs consisting of two basic elements: *Nodes*, which correspond to Cognitive Concepts bearing different states of activation depending on the knowledge they represent; *Arrows* denoting the causal effects that source nodes exercise on the receiving concept expressed through weights [33, 34]. It can be considered as an integration of multifold subjects, including neural network, graph theory and fuzzy logic. Because of its excellent representation and inference ability, FCM has spread widely to a variety of fields, including ecosystem modeling, business management, medical diagnosis, strategic planning etc. [35].

In the agricultural context, FCM techniques have been used mainly in management and decision support. They have been recently applied in precision agriculture for providing methods and tools for decision support like predicting yield in apples [36] and cotton [37, 38]. They have been successfully employed to model the relationships among the factors influencing the agricultural yield and analyze their cause effect relationships.

The first work on FCM exploitation in agriculture was accomplished by Papageorgiou et al. [37], proposing a modeling tool able to help the farmers to take informed decisions in precision agriculture. The same team of researchers [38], one year later, developed an FCM model for decision support in precision agriculture for cotton yield prediction. The proposed FCM model consists of nodes representing the main factors in cotton crop production such as soil texture, organic matter, electrical conductivity, pH level of the soil etc., as well as the level of cotton yield as the concept to be decided. Afterwards, Papageorgiou et al. [36] proposed a new FCM model trained using Non Linear Hebbian Learning (NHL) algorithm to predict apple fruit yield level of Central Greece apple farms. In a recent study [39], FCM were used to analyze the impact of the climatic variations and the soil parameters on coconut production and categorize the coconut production level for a given set of agro climatic conditions.

Halbrendt et al. [40] elaborated the differences in farmer and expert beliefs and the perceived impacts of conservation agriculture. An agricultural project in Nepal was used as a case study to show the possible relationships between trends in expert and rural farmer reasoning and predictions regarding the outcomes associated with development technology based on these beliefs. The outcome of the results was that conservation agriculture techniques should not be applied universally, development practitioners should engage in a two-way learning with local communities to benefit from locally situated knowledge.

Christen et al. [41] examined the potential of FCM as a tool to overcome the several barriers considered to the uptake of the prescribed environmentally beneficial farm management practices. The FCM methodology was investigated through the application in a Scottish case study on how environmental regulation affects farmers and farming practice and what factors are important for compliance or non-compliance with this regulation. The study compares the views of two different stakeholder groups on this matter using FCM network visualizations that were validated by interviews and a workshop session.

3 Methods

This section describes the basic mathematical framework of FS theory, the method of FL, the FIS as well its extension, the FCMs.

3.1 Fuzzy Sets

The traditional tools which are using to model, clarify, and compute are usually crisp, deterministic, and definite. A crisp set is divided only in two groups (referred also as dual logic), that is, yes-or-no type instead of more-or-less type and the corresponding statement can be true or false and nothing in between [42]. A FS deals with the concept of partial truth which is involved in human language, in human judgment, evaluation, and decisions [6, 7]. FS's are defined as an important tool for the representation of conceptual models that meet the needs of the expert and designer. Lotfi A. Zadeh, was the "father" of the theory of FS [6], proposing them as an extension of crisp set theory. The concept of FS theory is a membership function $\mu_{\widetilde{A}}(x)$, which represents the relationship of an element in a FS $\widetilde{A}$. In other words, element $\bar{x}$ is mapping to a "membership degree" which expressed in a continuous scale from one (full membership) to zero (full non-membership), $\mu_{\widetilde{A}}(x) \in [0, 1]$ and X is a collection of objects which generally denoted by x in Eq. (1).

$$\widetilde{A} = \{(x, \mu_{\widetilde{A}}(x)) | x \in X\} \quad (1)$$

When the universe of discourse, X, is discrete and finite or continuous and infinite, for the sake of convenience, a fuzzy set A is denoted in Eqs. (2) and (3), respectively.

$$\widetilde{A} = \{\mu_{\widetilde{A}}(x_1)/x_1 + \mu_{\widetilde{A}}(x_2)/x_2 + \cdots + \mu_{\widetilde{A}}(x_n)/x_n\} = \sum_i \mu_{\widetilde{A}}(x_i)/x_i \quad (2)$$

$$\tilde{A} = \left\{ \int \mu_{\tilde{A}}(x)/x \right\} \quad (3)$$

In above equations, the division slash, "/", is not equivalent to the operator of division but "acts" as a delimiter between the numerator and the denominator. Each term of the numerator is the membership value in set $\widetilde{A}$, which is associated with the element of the universe of the denominator. Furthermore, in Eq. (1), the plus sign, "+" and the summation symbol "∑", are not corresponding to the algebraic "sum" but as an aggregation operator. In Eq. (2), the integral symbol, "∫", is corresponding to continuous function-theoretic aggregation operator for continuous variables and not to an algebraic integral [43].

Moreover, the membership values of a crisp set are a subset of the interval [0, 1], can be considered as a special case of FS's. Therefore, the properties of crisp sets such as: commutativity, associativity, distributivity, idempotency, identity, transitivity and involution, are the same to those for FS's [43].

3.2 *Fuzzy Logic*

FL is multi-valued and handles the concept of partial truth. The statements of FL, can be represented with degrees of truthfulness and falseness. For example, the statement, "today is a rainy day", might be 100 % true if it rains all day, 80 % true if there are many dark clouds, 50 % true if it's hazy and 0 % true if there are no clouds. Furthermore, FL allows decision making with estimated values under vague or insufficient information. According to Zadeh [7], there is a correlation between complexity and unreliability: "The closer one looks at a real world problem, the fuzzier becomes its solution". FL uses three main operators, which equivalent to the in Boolean algebra operators: (a) union or OR, (b) intersection or AND, and (c) negation or NOT. The min and max operators are special cases of t-norms and t-conorms respectively [44]. Let A and B be fuzzy sets with A, B $\in$ X and x be an element in the X universe of discourse.

$$AND \rightarrow A \cap B = \mu_{A \cap B}(x) = \min\{\mu_A(x), \mu_B(x)\}$$
$$OR \rightarrow A \cup B = \mu_{A \cup B}(x) = \max\{\mu_A(x), \mu_B(x)\}$$
$$NOT \rightarrow A = \mu_A(x) = 1 - \mu_A$$

The membership function of the intersection of two FS's A and B with membership functions $\mu_A(x)$ and $\mu_B(x)$ respectively is defined as the minimum of the two individual membership functions. This is called the minimum criterion. The membership function of the Union of two FS's A and B with membership functions $\mu_A(x)$ and $\mu_B(x)$ respectively is defined as the maximum of the two individual membership functions. This is called the maximum criterion. The membership function of the complement of a FS A with membership function $\mu_A(x)$ is defined as the negation of the specified membership function. This is called the negation criterion [6].

The relationships among FS's are represented with fuzzy rules. Since the mapping between a number and its classification is crisp and distinct, then the fuzzy rule set is a representation of a relationship between the real values that can be classified by the FS in their antecedents and consequents [45]. The most common and widely used interpretation considers a fuzzy rule in the form "IF x is A THEN y is B", where A and B are all sets of X and Y. The general form of fuzzy rule is: IF ($input_1$ *is membership function*$_1$) and/or ($input_2$ *is membership function*$_2$) and/or … then ($output_n$ *is output membership function*$_n$).

3.3 Fuzzy Inference Systems

FIS are one of the most popular applications of FL and FS theory. FIS have been successfully applied in domains, such as automatic control, data classification, decision analysis, expert systems, and computer vision. Owing to its interdisciplinary nature, FIS are associated with other fuzzy domains, such as fuzzy-rule-based systems, fuzzy expert systems, fuzzy modeling, fuzzy associative memory, fuzzy logic controllers, and simply (and ambiguously) FS's [46]. FIS are particularly useful for modelling the relationship between variables in complex environments where human knowledge is available and where there is insufficient information to supply traditional mathematical models [8].

A FIS implements a nonlinear mapping from a particular input with a particular output utilizing FL. The process of FIS consists of three main components: fuzzification, knowledge base and defuzzification shown in Fig. 2, using membership functions, FL operators, and if-then rules. In the case of crisp inputs, the

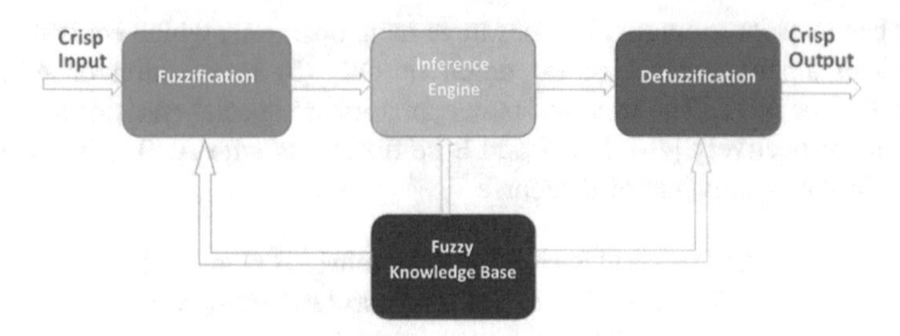

Fig. 2 The process of fuzzy inference system

following processes are performed: The fuzzification process is necessary to convert the "crisp" inputs into fuzzy inputs by membership functions and to determine the FS's of output variables. Thereafter, the fuzzy inference process is performed using a reasoning mechanism. The participating knowledge base consists of the rule base and the database. The database defines the membership functions of the FS's used in the fuzzy rules. In this step, the degree is determined to which each part of the antecedent has been satisfied for each rule. Therefore, in order to capture all of these combinations, numerous rules are composed and evaluated in parallel, using fuzzy reasoning "If-Then rules" [45, 47].

The rules should be combined in a way to make a decision. This is achieved by the process of aggregation, whereby the FS's that represent the outputs of each rule obtained and combined into a single FS. The aggregation is performed before the process of defuzzification in each output variable. The last process is the defuzzification which defuzzifies the aggregated FS of output values into a single output (crisp) value for decision making. There exist plenty of defuzzification methods. The most well-known defuzzification method is the CENTROID (center of area or center of gravity).

3.4 *Fuzzy Cognitive Maps*

FCM's consists of nodes, called "concepts" and weighted interconnections among them, called "weights". Concepts denote entities, states, variables, or characteristics of the system under investigation [34]. Using mathematical terms, a FCM can be defined using a 4-tuple $(\mathrm{C, W, A, f})$ where $\mathrm{C} = \{\mathrm{C_1, C_2, \ldots, C_M}\}$ is a set of M graph nodes (the nodes are also called "concepts" in FCM terminology), $\mathrm{W}:(\mathrm{C_i, C_j}) \rightarrow \mathrm{w_{ij}}$ is a function which associates a causal value $\mathrm{w_{ij}} \in [-1, 1]$ to each pair of nodes $(\mathrm{C_i, C_j})$. It denotes the weight of the directed edge from $\mathrm{C_i}$ to $\mathrm{C_j}$. Each connection in the FCM is determined by the experts using the fuzzy linguistics variables. The linguistics variables suggested by the experts of each interconnection are aggregated using the sum method and an overall linguistic weight is produced and then defuzzified using CoG (center of gravity) method [33]. Finally the numeric influence factor w_{ij} from concept 'i' to concept 'j' is obtained. The matrix $W_{M \times M}$ gathers the system causality which could be estimated by experts or automatically computed from historical data [35, 36].

Equally, $A{:}(C_i) \rightarrow A_i$ associates an activation value $A_i \in \mathbb{R}$ to each node C_i at each time t ($t = 1, 2, \ldots, T$). The activation values of concepts are updated following the rule given in Eq. (4).

$$A_i^{t+1} = f \sum_{j=1}^{M} w_{ji} A_j^t \quad (4)$$

The transformation or threshold function $f : \mathbb{R} \rightarrow [0, 1]$ is used to keep the activation value of concepts in the allowed range. The selection of the transfer function is crucial for the system behavior, and frequently depends on the problem requirements. The calculated activation vector through the propagation rule is iteratively repeated until a hidden pattern is observed (which is an ideal outcome), or a maximal number of cycles T is reached [36] which means that the map was unable to converge towards a fixed-point attractor. The transformation allows for qualitative comparisons between concepts. The most commonly used threshold functions are: the bivalent function, the trivalent function, and also the sigmoid variants [36].

An important characteristic of FCM is its learning capabilities. Learning in FCM involves updating the strengths of causal links between the concept nodes. A learning strategy is to improve the FCMs by fine-tuning its initial causal links or edge strengths by applying training algorithms similar to that of artificial neural networks [48].

4 Two Representative Case Studies in Agriculture

4.1 Case Study on Fuzzy Sets Application in Viticulture

Recently a FS model was proposed to model grape quality based on expert knowledge, with respect to precision viticulture principles. A set of linguistic rules were built to explain the relationship between grape total quality and individual grape variables. Finally a FIS was developed and validated to classify grape quality based on selected grape attributes in a commercial vineyard in Greece. The proposed FS model consists of four grape attributes as input variables to model grape quality: total soluble solids (TSS), titratable acidity (TA), total anthocyanins (Anth) and berry fresh weight (BW), and one output variable representing the Grape Total Quality (GTQ) [45].

To define the fuzzy sets of each one of the input parameters/grape attributes, the related literature and the knowledge from vineyard experts used in our study [47] was embraced. At first it was needed to define the threshold(s) for the four input parameters for Agiorgitiko variety belonging to red table wines and next based on the threshold(s) to determine the grape attributes classes.

Based on the literature for red wines, low sugar (i.e. TSS) is undesirable but extremely high values lead to excessive alcohol in the wine which can be detrimental for wine balance and consumer health [49]. At harvest, the sugar and acid concentration should be around 150–250 g/L, i.e. approximately 17–25 brix [50], and 6.5–8.5 g/L respectively [51]. In the case of Agiorgitiko which is a late maturing variety, a moderate TSS concentration at harvest of 21 brix was considered for the production of red table wines. This threshold leads to the definition of two classes of low (<21 brix) and high (>21 brix) sugar content in the must [52]. The two classes for TSS were chosen because sugar content of the must usually presents the lowest variation among various maturity indices in grapes [53]. For TA, three classes were chosen based on must acidity data reported for this variety [54]. These classes are low (<5 g/L), medium (5–7 g/L) and high (>7 g/L) respectively with values superior to 7 g/L generally considered as favourable for wine ageing.

The role of skin anthocyanins in grape and wine quality is determinant as shown in [55, 56]. For Agiorgitiko grapes, moderate levels of skin anthocyanins are present [54]. Due to the highest, among berry components, range of within-vineyard variation [55], four quality classes were chosen for this input parameter: low (<600 mg/L), medium (600–800 mg/L), high (800–1000 mg/L) and very high (>1000 mg/L). Grape berry weight also shows high diversity among Vitis vinifera cultivars, ranging for the most part of wine grapes between 1.5 and 3.5 g at harvest. For Agiorgitiko grapes, berries' average size is situated between 1.6 and 2.0 g [54]. In this study, three classes were chosen as follows: low (<1.6 g), medium (1.6–2.0 g) and high (>2.0 g).

The GTC as the output parameter of the system was corresponded to five FSs determining the grape quality classes with levels: very poor (0–0.28), poor (0.18–0.47), average (0.38–0.68), good (0.58–0.88) and excellent (0.78–1). Figure 3 shows the membership functions of the inputs and Fig. 4 shows the membership functions of the GTQ.

To construct the fuzzy model, a set of rules based on vineyard expert knowledge was provided. As a general guideline, best grape quality was associated with high values of TSS, TA and Anth and low values of BW, this means that small berries are associated with higher wine quality. An example of rule definition is: If TSS is 'high', TA is 'high', Anth is 'very high' and BW is 'low', then GTQ is 'Excellent'. Similarly other rules were developed, for example: If TSS is 'high', TA is 'medium', Anth is 'low' and BW is 'low', then GTQ is 'Average' or if TSS is 'low', TA is 'low', Anth is 'low' and BW is 'high', then GTQ is 'Very poor'. Table 1 shows an example of the fuzzy rules used to elaborate the FIS.

Following the fuzzification, inference and defuzzification process, a numerical value is calculated as the result of the proposed FIS. This numerical value defining the score of the grape quality grade (class) is interpreted through its membership degrees to the five different fuzzy sets (very poor, poor, average, good and excellent).

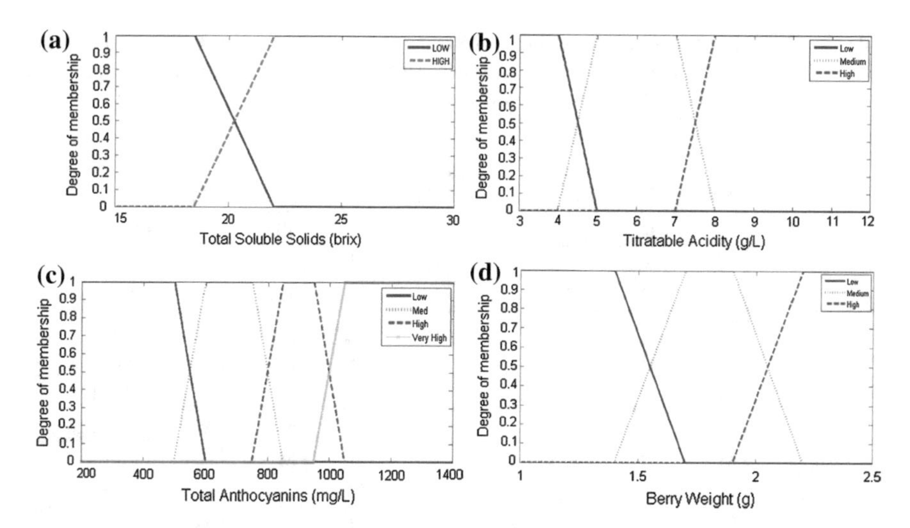

Fig. 3 Corresponding membership functions of: **a** total soluble solids (two FS—Low and High), **b** titratable acidity (three FS—Low, Medium and High), **c** total anthocyanins (Four FS—Low, Medium, High and Very high) and **d** berry fresh weight (three FS—Low, Medium and High)

The FIS was implemented using the Mamdani type fuzzy rule-base inference, and the CENTROID method for defuzzification. The proposed FIS was validated by 5 vineyard farmers and winemakers of the Greek wine industry with experience in 'Agiorgitiko' variety. These experts also assigned five categories of grape total quality (GTQ): very poor, poor, average, good, and excellent. The level of agreement between the GTQ proposed by the FIS and expert knowledge was calculated for three subsequent years. Furthermore, output fuzzy quality values were compared to soil and yield parameters to verify their spatial relevancy.

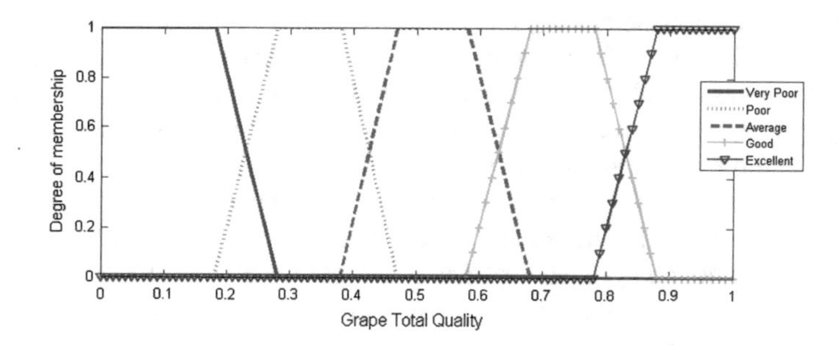

Fig. 4 Grape total quality membership functions corresponding to five FS (very poor, poor, average, good and excellent) that represent the five different categories of grape total quality

Table 1 Example of fuzzy rules relating grape composition parameters to Grape Total Quality (GTQ): total soluble solids (TSS, brix), titratable acidity (TA, g/L), total anthocyanin content (Anth, mg/L) and berry fresh weight (BW, g) [45]

Fuzzy inputs				Fuzzy output	Fuzzy inputs				Fuzzy output
TSS	TA	Anth	BW	GTQ	TSS	TA	Anth	BW	GTQ
H	H	VH	L	Excellent	L	H	VH	L	Good
H	H	VH	M	Excellent	L	H	VH	M	Good
H	H	VH	H	Good	L	H	VH	H	Average
H	H	H	L	Excellent	L	H	H	L	Good
H	H	H	H	Good	L	H	H	H	Average
H	H	M	L	Excellent	L	H	M	L	Good
H	H	M	M	Good	L	H	M	M	Average
H	H	M	H	Average	L	H	M	H	Poor
H	H	L	L	Good	L	H	L	L	Average
H	H	L	M	Average	L	H	L	M	Poor
H	H	L	H	Poor	L	H	L	H	Very poor
H	M	M	H	Poor	L	M	M	H	Very poor
H	M	L	L	Average	L	M	L	L	Poor
H	M	L	M	Poor	L	M	L	M	Very poor
H	M	L	H	V.Poor	L	M	L	H	Very poor
H	L	L	H	V.Poor	L	L	L	H	Very poor

For the conducted experiments in a number of 48 cells (parts) of the field, the overall agreement between FIS results and expert evaluation was 77.20 % for 2010, 81.83 % for 2011 and 82.35 % for 2012, which suggests that the FIS was able to model expert knowledge successfully for this wine grape variety and wine style. In general, the conducted evaluation process showed high agreement between GTQ and expert evaluation suggesting that the FIS was able to model expert knowledge successfully. Moreover, GTQ exhibited higher variability than the individual grape quality attributes in all years. Concerning the four input grape components, anthocyanins and berry weight seemed to be more important in determining GTQ than total soluble solids and titratable acidity [45].

As it can be shown, FIS robustness depends on the number and the quality of the fuzzy rules established. The adoption of FIS or other FL tools can provide an alternative way to analyze and combine the available data for more accurate decision making regarding the definition of optimum harvest time or the application of differential management strategies such as selective harvesting.

4.2 Case Study on FCM Application in Precision Agriculture

Precision agriculture is a methodology to maximize the profit, rationalize agricultural inputs and lower the environmental damage which is related to adjustment of agricultural practices with the site requirements [57]. Often the uncertainty and missing information leads to wrong conclusions. Thus, the application of FCMs has found a great extent because they are able to assure that the incomplete information is valued and to provide solutions to many management and optimization issues [58].

The development and design of the appropriate FCM for the modeling of a system in agriculture requires the contribution of human knowledge. Usually, knowledgeable experts, who are familiar with the FCM formalism, are required to develop FCM using an interactive procedure of presenting their knowledge on the operation and behavior of the system [35]. Experts are usually able to store in their mind the correlation among different characteristics, states, variables and events of the system and in this way they encode the dynamics of the system using fuzzy if-then rules.

We show a case study investigated in [39], where the FCM technique was used for analyzing the impacts of the climatic variations and the soil parameters on coconut production and categorize the coconut production level for a given set of agro climatic conditions. In this case study, experts from the specific agriculture domain were asked to determine the concepts that best describe the model of the system, since they know which factors are the key principles and functions of the system operation and behavior; therefore, they introduced the important concepts.

More explicitly, the experts constructed the FCM model for predicting coconut production level following the step by step approach described in [36]. In step I, they determined the concepts and the fuzzy membership functions. For this case study modeling the productivity level of coconut for yield management, twenty concepts were identified with the help of three agricultural domain experts representing various soil conditions as well as the seasonal and weather parameters. These are: electric conductivity, organic matters, bulk density of the soil, temperature, humidity, pest, age of plant, rainfall (RF), rainfall duration (RFD), soil moisture (SM), pH, Ca, Mn, Fe, Zn, Mg, Copper level in the soil (Cu), potassium (K), phosphorous (P), and nitrogen (N). The twentieth first concept is the Yield (called Decision Concept (DC)) and represents the coconut yield category. The fuzzy membership functions of each concept were defined by experts following their experience and the related literature. Based on the literature and the experts' knowledge, the FSs of the assigned concepts were defined (see Table 2). These FSs are in accordance with the range of values provided from the respective literature.

In step II, the same experts described the strength of influence of each concept to the other using fuzzy if-then rules among 20 factor concepts and yield from the term set T{influence} = {very low, low, medium, high and very high} following the FCM construction process as suggested in [35]. Each fuzzy rule infers a fuzzy

Table 2 Fuzzy sets of each concept and their corresponding numerical ranges

EC (mS/m)		OM (ppm)		BD (g/cm³)		Temp (°C)		Humidity (%)	
Range	Three fuzzy set	Range	Three fuzzy set	Range	Three fuzzy set	Range	Seven fuzzy set	Range	Six fuzzy set
<15	Low	<1.0	Low	0–1.3	Low	<18	Very very low	<20	Very very low
15–35	Medium	1.0–2.0	Medium	1.3–1.9	Medium	18–22	Very low	20–35	Very low
>35	High	>2.0	High	>1.9	High	22–26	Low	35–50	Low
						26–30	Medium	50–70	Medium
						30–34	High	70–85	High
						34–38	Very high	85–100	Very high
						>38	Very very high		

Pest (%)		AoP (years)		RF (mm)		RFD (months)		SM (%)	
Range	Six fuzzy set	Range	Six fuzzy set	Range	Five fuzzy set	Range	Five fuzzy set	Range	Five fuzzy set
<10	Very very low	<10	Very low	<500	Very low	0–2	Very low	<40	Very low
10–20	Very low	10–20	Low	500–1000	Low	2–4	Low	40–50	Low
20–30	Low	20–35	Medium	1000–2000	Medium	4–8	Medium	50–70	Medium
30–50	Medium	35–50	High	2000–3000	High	8–10	High	70–85	High
50–70	High	50–70	Very high	>3000	Very high	10–12	Very high	85–100	Very high
>70	Very high	>70	Very very high						

pH		Ca (ppm)		Mn (ppm)		Fe (ppm)		Zn (ppm)	
Range	Five fuzzy set	Range	Four fuzzy set	Range	Three fuzzy set	Range	Three fuzzy set	Range	Four fuzzy set
<4	Very low	<200	Very low	<100	Low	<10	Low	<2	Low
4–6	Low	200–800	Low	100–300	Medium	10–20	Medium	2–6	Medium
6–8	Medium	800–1800	Medium	>300	High	>20	High	6–10	High
8–10	High	>1800	High					>10	Very high
>10	Very high								

Mg (ppm)		Cu (ppm)		K (ppm)		P (ppm)		N (ppm)	
Range	Five fuzzy set	Range	Four fuzzy set	Range	Five fuzzy set	Range	Five fuzzy set	Range	Five fuzzy set
<50	Very low	<0.3	Very low	<50	Very low	0–5	Very low	<3	Very low
50–150	Low	0.3–0.6	Low	50–100	Low	5–15	Low	3–15	Low
150–300	Medium	0.6–1	Medium	100–250	Medium	15–25	Medium	15–30	Medium
300–500	High	>1	High	250–400	High	25–40	High	30–45	High
>500	Very high			>400	Very high	>40	Very high	>45	Very high

weight, which in sequence is translated to a numerical one used in the FCM reasoning process [35, 36]. The inference of the rule determines the strength of the connection between the two related concepts.

In the case of fuzzy rules for the specific problem of yield description in coconut, some examples, as derived from experts, are given:

IF a small change occurs in the value of concept “humidity”, THEN a small change in the value of “yield” (decision concept) is caused.

This means that: the influence from concept “humidity” to “yield” is low.

IF a medium change occurs in the value of concept “OM”, THEN a high change in the value of concept “yield” is caused.

This means that: the influence from concept “OM” to “yield” is high.

In step III, the inferred fuzzy linguistic weights (strengths of influences) were combined using the SUM method described by Zadeh [6]. Then, using the defuzzification method of centroid [6, 7], the linguistic weights with membership functions were transformed into a numerical weight w_{ji} which lies in the range [−1, 1].

The produced FCM model for modeling and predicting yield in apples including the initial calculated values of weights is shown in Fig. 5.

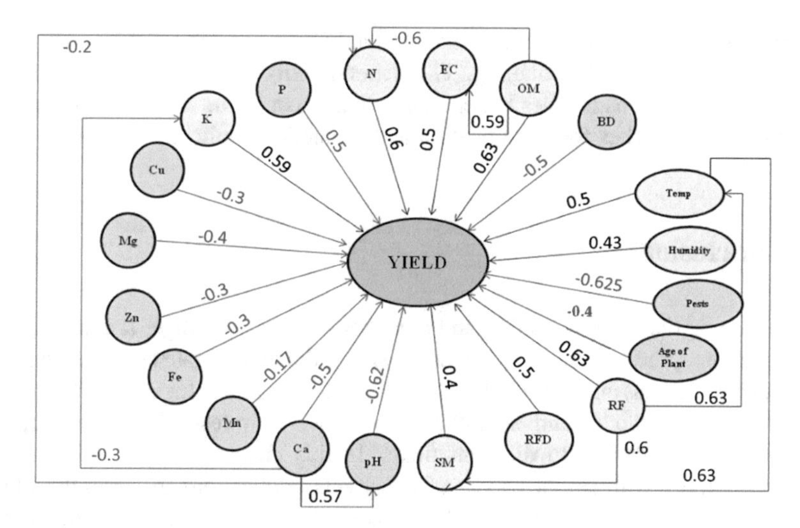

Fig. 5 FCM model for coconut yield management

After the FCM construction, an efficient classification algorithm for FCMs, the data driven NHL (DDNHL) [48] was used for training the proposed FCM model based on the available historic data and then to classify the yield category of the given cases [39]. Through this training, the initial weights of causal links assigned by the group of experts were optimized based on the training data set so as to produce more accurate classification results.

Fifty records were used for training and subsequently for classifying the coconut production into three classes' viz. *high* yield, *medium* yield and *low* yield. Forty nine records of the coconut yield cases were considered to be used for training and one remaining for testing, with a random selection of these cases for training and testing at each algorithm performance. During each algorithm run, the classification accuracy was calculated by the testing cases. The overall classification accuracy was estimated by the mean value of the calculated classification accuracies produced after a large number of experiments.

The classification accuracy obtained was 96 % which was indeed very promising comparing the model with other benchmark machine learning models, like decision trees, neural networks. Furthermore the FCM model was used for scenario analysis to study the effects of climate and weather variables on the coconut yield response. This study was performed to answer several "what-if" scenarios with respect to the variations in climatic conditions and how they affect the coconut yield.

Furthermore, yield prediction software was developed to provide the agronomists and farmers with a front-end decision support tool for estimating the yield level of coconut for the given set of weather and soil parameters. Modeling using the FCM technique makes it easy to study the effect of the various factors influencing the agricultural productivity in detail and answer several "what-if" questions more precisely. The modification of FCM weighted connections through learning algorithms is an efficient technique making the FCM a powerful and efficient tool to assist farmers in decision making. The dependability of the FCM model is well proven in literature as it gives better results than most of the other bench mark soft computing approaches for classification and prediction applications [36].

5 Conclusions

To sum up, agricultural modeling and management is a challenge. It claims the use of a suitable representation of the object of study, the appropriate characterization of the decision factors, and the reliable definition of criteria. As fuzzy sets is a method that embraces a model and an inference engine, their research is growing and application is spreading to different fields. In this chapter, we presented a short literature survey with recent papers of fuzzy sets applications in agriculture highlighting the way that the fuzzy sets have been applied in this complex domain, as well as two representative case studies of fuzzy sets and an enhanced methodology of them, fuzzy cognitive maps for crop management and decision support in precision agriculture. The main and extremely useful capabilities of fuzzy sets to manage and represent uncertainty, assure the researchers about the suitability of them in agricultural problems with inherent uncertainty and fuzziness. Further fuzzy sets assure that the incomplete information is valued and provide solutions to issues which are crucial in agriculture.

References

1. Food and Agriculture Organization: Investing in food security, p. 3. I/I1230E/1/11.09/1000, Italy (2009)
2. Popa, C.: Adoption of artificial intelligence in agriculture. Bull. UASVM Agric. Electron. **68** (1), 284–293 (2011)
3. Dengel, A.: Special issue on artificial intelligence in agriculture. Künstl Intell. **27**, 309–311 (2013). doi 10.1007/s13218-013-0275-y
4. Center, B., Verma, B.P.: Fuzzy logic for biological and agricultural systems. Artif. Intell. Rev. **12**, 213–225 (1998)
5. Dubey, S., Pandey, R.K., Gautam, S.S.: Literature review on fuzzy expert system in agriculture. Int. J. Soft Comput. Eng. (IJSCE) **2**(6) (2013)
6. Zadeh, L.A.: Fuzzy sets. Inf. Control **8**(3), 338–353 (1965)
7. Zadeh, L.A.: Outline of a new approach to the analysis of complex systems and decision, processes. IEEE Trans. Syst. Man Cybern. **3**, 28–44 (1973)
8. Azeem, M.F.: Fuzzy inference system—Theory and applications. InTech 518 (2012)
9. Huang, Y., Lan, Y., Thomson, S.J., Fang, A., Hoffmann, W.C., Lacey, R.E.: Development of soft computing and applications in agricultural and biological engineering. Comput. Electron. Agric. **71**, 107–127 (2010)
10. Bosma, R., van den Berg, J., Kaymak, U., Udo, H., Verreth, J.: A generic methodology for developing fuzzy decision models. Expert Syst. Appl. **39**, 1200–1210 (2012)

11. Djatkov, D., Effenberger, M., Lehner, A., Martinov, M., Tesic, M.: New method for assessing the performance of agricultural biogas plants. Renew. Energy **40**, 104–112 (2011)
12. Djatkov, D., Effenberger, M., Martinov, M.: Method for assessing and improving the efficiency of agricultural biogas plants based on fuzzy logic and expert systems. Appl. Energy **134**, 163–175 (2014)
13. Murmu, S., Biswas, S.: Application of fuzzy logic and neural network in crop classification: a review. Aquat. Procedia **4**, 1203–1210 (2015)
14. Zareiforoush, H., Minaei, S., Alizadeh, M.R., Banakar, A.: A hybrid intelligent approach based on computer vision and fuzzy logic for quality measurement of milled rice. Measurement **66**, 26–34 (2015)
15. Pandey, A., Prasad, R., Singh, V.P., Jha, S.K., Shukla, K.K.: Crop parameters estimation by fuzzy inference system using X-band scatterometer data. Adv. Space Res. **51**, 905–911 (2013)
16. Li, Q., Yan, J.: Assessing the health of agricultural land with emergy analysis and fuzzy logic in the major grain-producing region. Catena **99**, 9–17 (2012)
17. da Silva, A.F., Barbosa, A.P., Zimback, C.R.L., Landim, P.M.B., Soares, A.: Estimation of croplands using indicator kriging and fuzzy classification. Comput. Electron. Agric. **111**, 1–11 (2015)
18. Abbaspour-Gilandeh, Y., Sedghi, R.: Predicting soil fragmentation during tillage operation using fuzzy logic approach. J. Terrramech. **57**, 61–69 (2015)
19. Petkovic, D., Gocic, M., Trajkovic, S., Shamshirband, S., Motamedi, S., Hashim, R., Bonakdari, H.: Determination of the most influential weather parameters on reference evapotranspiration by adaptive neuro-fuzzy methodology. Comput. Electron. Agric. **114**, 277–284 (2015)
20. Prakash, C., Thakur, G.S.M.: Fuzzy based agriculture expert system for Soya-bean. In: International Conference on Computing Sciences WILKES100-ICCS2013, Jalandhar, Punjab, India (2013)
21. Zhang, J., Su, Y., Wu, J., Liang, H.: GIS based land suitability assessment for tobacco production using AHP and fuzzy set in Shandong province of China. Comput. Electron. Agric. **114**, 202–211 (2015)
22. Coulon-Leroy, C., Charnomordic, B., Thiollet-Scholtus, M., Guillaume, S.: Imperfect knowledge and data-based approach to model a complex agronomic feature—Application to vine vigor. Comput. Electron. Agric. **99**, 135–145 (2013)
23. Ceballos, M.R., Gorricho, J.L., Gamboa, O.P., Huerta, M.K., Rivas, D., Rodas, M.E.: Fuzzy system of irrigation applied to the growth of Habanero Pepper (Capsicum chinense Jacq.) under protected conditions in Yucatan, Mexico. Int. J. Distrib. Sens. Netw. **2015**, 124–137 (2015). doi:10.1155/2015/123543
24. Islam, N., Sadiq, R., Rodriguez, M.J., Francisque, A.: Evaluation of source water protection strategies: a fuzzy-based model. J. Environ. Manage. **121**, 191–201 (2013)
25. Giusti, E., Marsili-Libelli, S.: A Fuzzy Decision Support System for irrigation and water conservation in agriculture. Environ. Model Softw. **63**, 73–86 (2015)
26. Rossi, F., Nardino, M., Mannini, P., Genovesi, R.: IRRINET Emilia Romagna: online decision support on irrigation. Online agrometeological applications with decision support on the farm level. Cost Action **718**, 99–102 (2004)
27. Binte Zinnat, S., Abdullah, D.: Design of a fuzzy logic based automated shading and irrigation system. In: 17th International Conference on Computer and Information Technology, 22–23 Dec 2014, Daffodil International University, Dhaka, Bangladesh (2014)
28. Trono, E.M., Guico, M.L., Labuguen, R., Navarro, A., Libatique, N.G., Tangonan, G.: Design and development of an integrated web-based system for tropical rainfall monitoring. Procedia Environ. Sci. **20**, 305–314 (2014)
29. Almeida, G., Vieira, J., Marques, A.S., Kiperstok, A., Cardoso, A.: Estimating the potential water reuse based on fuzzy reasoning. J. Environ. Manage. **128**, 883–892 (2013)
30. Liu, Y., Jiao, L., Liu, Y., He, J.: A self-adapting fuzzy inference system for the evaluation of agricultural land. Environ. Model Softw. **40**, 226–234 (2013)

31. Mawale, M.V., Chavan, V.: Fuzzy Inference System for productivity and fertility of soil. Int. J. Eng. Dev. Res. **2**(3), 2321–9939 (2014)
32. Papadopoulos, A., Kalivas, D., Hatzichristos, T.: Decision support system for nitrogen fertilization using fuzzy theory. Comput. Electron. Agric. **78**, 130–139 (2011)
33. Kosko, B.: Fuzzy cognitive maps. Int. J. Man-Machine Stud. **24**, 65–75 (1986)
34. Kosko, B.: Neural Networks and Fuzzy Systems. Prentice-Hall, Englewood Cliffs (1992)
35. Papageorgiou, E.I. (ed.): Fuzzy Cognitive Maps for Applied Sciences and Engineering—From Fundamentals to Extensions and Learning Algorithms, Intelligent Systems Reference Library 54. Springer, Berlin (2014)
36. Papageorgiou, E.I., Aggelopoulou, K., Gemptos, T., Nanos, G.: Yield prediction in apples related to precision agriculture using fuzzy cognitive map learning approach. In: Computers and Electronics in Agriculture, vol. 91, pp. 19–29, December 2012 (2013)
37. Papageorgiou, E.I., Markinos, A., Gemptos, T.: Application of fuzzy cognitive maps for cotton yield management in precision farming. Expert Syst. Appl. **36**, 12399–12413 (2009)
38. Papageorgiou, E.I., Markinos, A., Gemptos, T.: Fuzzy cognitive map based approach for predicting yield in cotton crop production as a basis for decision support system in precision agriculture application. Appl. Soft Comput. **11**(4), 3643–3657 (2011)
39. Jayashree, S., Nikhil P., Papageorgiou E.I., Papageorgiou K.: Application of fuzzy cognitive maps in precision agriculture: a case study of coconut yield prediction in India. Neural Comput. Appl. (2015). doi:10.1007/s00521-015-1864-5
40. Halbrendt, J., Steven, A., Gray, B., Crow, S., Radovich, T., Kimura, A.H., Tamang, B.B.: Differences in farmer and expert beliefs and the perceived impacts of conservation agriculture. Glob. Environ. Change **28**, 50–62 (2014)
41. Christen, B., Kjeldsen, C., Dalgaard, T., Martin-Ortega, J.: Can fuzzy cognitive mapping help in agricultural policy design and communication? Land Use Policy **45**, 64–75 (2015)
42. Zimmermann, H.J.: Advanced Review: Fuzzy set theory. Wiley, New York (2010). doi:10.1002/wics.82
43. Ross, T.: Fuzzy Logic in Engineering Applications. McGraw-Hill, New York (1995)
44. Yager, R., Filev, D.: Essentials of Fuzzy Modeling and Control. Wiley, New York (1994)
45. Tagarakis, A., Koundouras, S., Papageorgiou, E.I., Dikopoulou, Z., Fountas, S., Gemtos, T.A.: A fuzzy inference system to model grape quality in vineyards. Precis. Agric. Int. J. Adv. Precis. Agric. **15**(5), 555–578 (2014)
46. Vitoriano, B., Montero, J., Ruan, D.: Decision Aid Models for Disaster Management and Emergencies, p. 325. Springer Science & Business Media (2013)
47. Arabacioglu, B.C.: Using fuzzy inference system for architectural space analysis. Appl. Soft Comput. **10**(3), 926–937 (2010)
48. Papageorgiou, E.I.: Learning algorithms for fuzzy cognitive maps—A review study. IEEE Trans. Syst. Man Cybern. Part C Appl. Rev. **42**(2), 150–163 (2012)
49. Dai, Z.W., Ollat, N., Gomès, E., Decroocq, S., Tandonnet, J.P., Bordenave, L., Pieri, P., Hil-bert, G., Kappel, C., van Leeuwen, C., Vivin, P., Delrot, S.: Ecophysiological, genetic, and molecular causes of variation in grape berry weight and composition. Am. J. Enol. Viticulture **62**, 413–425 (2011)
50. Ribéreau-Gayon, P., Dubourdieu, D., Donèche, B., Lonvaud, A.: Handbook of Enology, Microbiology of Wine and Vinification. Wiley, West Sussex (2006)
51. Ruffner, H.P.: Metabolism of tartaric and malic acids in Vitis: a review. Part B Vitis **21**, 346–358 (1982)
52. van Leeuwen, C., Tregoat, O., Choné, X., Bois, B., Pernet, D., Gaudillère, J.-P.: Vine water status is a key factor in grape ripening and vintage quality for red Bordeaux wine. How can it be assessed for vineyard management purposes? Journal International des Sciences de la Vigne et du Vin **43**, 121–134 (2009)
53. Bramley, R.G.V., Trought, M.C.T., Praat, J.P.: Vineyard variability in Marlborough, New Zeland: characterizing variation. Aust. J. Grape Wine Res. **17**, 72–78 (2011)
54. Koundouras, S., Marinos, V., Gkoulioti, A., Kotseridis, Y., van Leeuwen, C.: Influence of vineyard location and vine water status on fruit maturation of nonirrigated cv. Agiorgitiko

(*Vitis vinifera* L.). Effects on wine phenolic and aroma components. J. Agric. Food Chem. **54**, 5077–5086 (2006)
55. Kennedy, J.A., Saucier, C., Glories, Y.: Grape and wine phenolics: history and perspective. Am. J. Enol. Viticulture **57**, 239–248 (2006)
56. Mazza, G., Francis, F.J.: Anthocyanins in grapes and grape products. Crit. Rev. Food Sci. Nutr. **35**, 341–371 (1995)
57. Sannakki, S.S., Rajpurohit, V.S., Arunkumar, R.: A survey on applications of fuzzy logic in agriculture. J. Comput. Appl. (JCA) **4**(1) (2011)
58. Salleh, M., Nawi, N., Ghazali, R.: Uncertainty analysis using fuzzy sets for decision support system. Efficient Decision Support Systems—Practice and Challenges in Multidisciplinary Domains. InTech, pp. 273–290 (2011)

Part IV
Applications of Fuzzy Sets

Solving a Multiobjective Truck and Trailer Routing Problem with Fuzzy Constraints

Isis Torres, Alejandro Rosete, Carlos Cruz and José L. Verdegay

Abstract The Truck and Trailer Routing Problem uses trucks pulling trailers as a distinctive feature of the Vehicle Routing Problem. Recently, this problem has been treated considering the capacity constraints as fuzzy. This situation means that the decision maker admits the violation of these constraints according to a value of tolerance. This relaxation can generate a set of solutions with very low costs but its non-fulfillment grade of the capacity constraints can be high and vice versa. This fuzzy variant is generalized in this work from a multiobjective approach by incorporating an objective to minimize the violation of constraints. We present and discuss the computational experiments carried out to solve the multiobjective Truck and Trailer Routing Problem with fuzzy constraint using benchmark instances with sizes ranging from 50 to 199 customers.

Keywords Truck and trailer routing problem · Multiobjective · Fuzzy sets · Optimization

I. Torres (✉) · A. Rosete
Facultad de Ingeniería Informática, Instituto Superior Politécnico José Antonio Echeverría, 11500 La Habana, Cuba
e-mail: itorres@ceis.cujae.edu.cu

A. Rosete
e-mail: rosete@ceis.cujae.edu.cu

C. Cruz · J.L. Verdegay
Department of Computer Science and Artificial Intelligence, University of Granada, 18071 Granada, Spain
e-mail: carloscruz@decsai.ugr.es

J.L. Verdegay
e-mail: verdegay@decsai.ugr.es

C. Kahraman et al. (eds.), *Fuzzy Logic in Its 50th Year*,
Studies in Fuzziness and Soft Computing 341,
DOI 10.1007/978-3-319-31093-0_11

1 Introduction

The Truck and Trailer Routing Problem (TTRP) is an extension of the well-known Vehicle Routing Problem (VRP). In the TTRP a heterogeneous fleet composed of trucks and trailers is used to serve a set of customers geographically dispersed from a central depot. Some customers with limited maneuvering space or accessible through narrow roads must be served only by truck (TC), while others customers can be served either by truck or by truck with trailer (VC). A solution of the TTRP is generally composed of different types of routes (see Fig. 1). In particular, we can distinguish three types of routes: Pure Vehicle Route (PVR) is the tour traveled by a complete vehicle and contains only vehicle customer. Pure Truck Route (PTR) is traveled by a truck alone and are visited both customer type and Complete Vehicle Route (CVR) consisting of a main tour traveled by a complete vehicle, and at least one sub-tour traveled by the truck alone. A sub-tour has to start and end at the same parking area (root) belonging to the main tour. In each sub-tour the trailer is uncoupled from the truck, and, when the sub-tour has been completed, the trailer is coupled again to the pure truck. On a sub-tour, can be visited both types of customers. A route is defined as a vehicle route (R_{cv}) if the assigned vehicle is a complete vehicle; otherwise the route is referred to as a truck route (R_{pt}) since it is serviced by a truck. Also, each route is limited by capacity of vehicle used. In general, the goal of the TTRP is to find a set of least cost vehicle routes that start

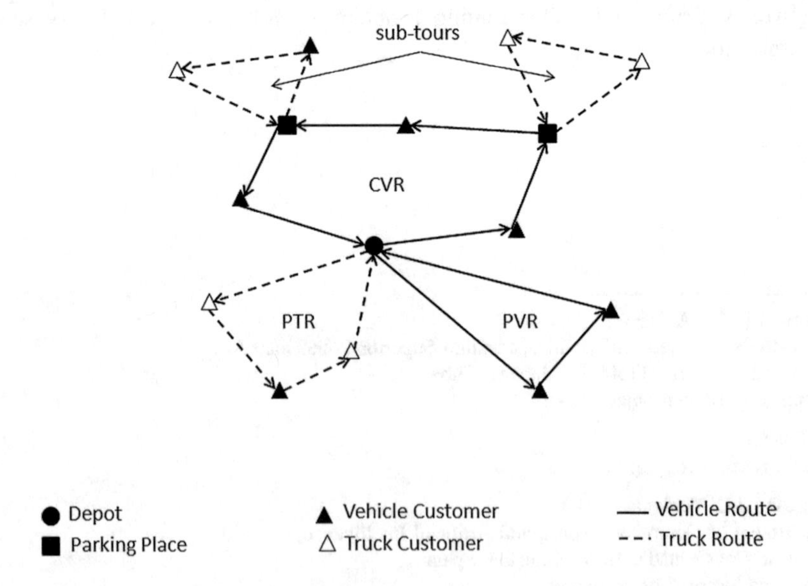

Fig. 1 Types of routes in the TTRP (reprinted from Torres et al. [23])

and end at the central depot such that each customer is serviced exactly once; the total demand of any vehicle route does not exceed the total capacity of the allocated vehicles used in that route; and the number of required trucks and trailers is not greater than available vehicles in the fleet.

The term "Truck and Trailer Routing Problem" was coined in 2002 by Chao [4]. Nevertheless, problems with characteristics similar to the TTRP but using other denominations were modeled in some previous works. For example, Semet and Taillard [20] in 1993 presented an approach to solve a transportation problem using a heterogeneous fleet (trucks and trailers) of one of the major chain stores in Switzerland. Later, Semet in 1995 extended the VRP taking into account the partial accessibility constraint introduced in their previous work [19]. In 1996 Gerdessen [13] conducted a study on the vehicle routing problem with trailer and presented two real world applications. Also, the TTRP has been extended in several ways. Other problem is proposed by Lin et al. in [16] where the limited-fleet constraint is dropped, this problem is known as the relaxed TTRP (RTTRP). Derigs et al. [9] tackled the TTRP with Time Windows (TTRPTW) and also introduced the so-called TTRP without load transfer. Other related problems are tackled in [11, 17, 26].

An exact algorithm that can optimally solve a large-size TTRP problem is unlikely. The solution approaches published in the literature about this topic can be divided into three groups: exact approaches [11], approximated approaches (including heuristic and metaheuristics) [4, 9, 15, 18, 27], or a combination of these approaches (the so-called matheuristics) [28]. An interesting review about models and solutions for the TTRP appear in [22].

In most papers, the data of the problem are assumed exact; even when the available information can present a high grade of vagueness or uncertainty. Generally, these procedures simplify the problem forcing these vague values to be exact, so obliging us to formulate and to solve a problem with a (precise) nature different from the (vague) original one. This simplification of the problem causes that the nature of the model is changed and can produce serious errors for obtaining the solution [3].

An approach based on Fuzzy Lineal Programming (FLP) was studied by the authors [21, 23] in recent years. These proposals model and solve the TTRP when the constraints have a fuzzy nature, and obtain a fuzzy solution, i.e., a set of solutions that has the same nature that the original problem. Although, these solutions have low costs, its non-fulfillment grade of the constraints of the problem can be high and vice versa. Even when the TTRP with fuzzy constraints is treated as single-objective problem (where the objective is to minimize total distances the routes); this fuzzy variant can be generalized from a multiobjective approach by incorporating an objective to minimize the violation of constraints. In consequence, this chapter presents a multiobjective approach for the TTRP when the constraints of vehicle capacity are considered fuzzy.

2 Truck and Trailer Routing Problem with Fuzzy Constraints

A real-world context of the TTRP is very complex and the information is not always available with sufficient precision and completeness as desired for adequate planning and management. For example, it is usual that when we ask for the customer demand, the expressions are "about 50 units" or "no more than 65 units" etc. And the same can happen with the rest of the parameters that define the problem such as the travel times ("around 2–2:30 h"), vehicle capacity ("no less than 3000 kg", "about 100 kg extra load"), etc. The application of the Soft Computing techniques is a recognized way to describe mathematically all these si-tuations. Thus, a TTRP under fuzzy environments was proposed in [23, 21], in which a general parametric approach [25] we used for solving the problem.

This parametric approach is composed of two phases: the first one transforms the fuzzy problem into several parametric problems in which a parameter α-cut (which belongs to the interval [0, 1]) represents the decision maker's satisfaction level. Then, in the second phase each of these α-problems is solved by means of classical optimization techniques. The obtained results, to the different α-values, generate a set of solutions and then all of these particular α-solutions are integrated by the Representation Theorem for fuzzy sets. In this way the solution to a fuzzy problem has the same nature as the problem at hand, as stated by Delgado et al. [8].

An example of application of this approach is the case in which a decision maker assumes that he can tolerate violations in the accomplishment of the constraints; i.e., he permits the constraints to be satisfied "as well as possible". Then, each constraint is represented as follows:

$$a_i x \leq_f b_i \quad i \in M = \{1, \ldots, m\} \tag{1}$$

The symbol $\leq_f$ indicates imprecision in the constraints, being measured these imprecision by means of membership functions $\mu_i: \mathbb{R} \rightarrow [0, 1]$, $i = 1, \ldots, m$. Each membership function will give the membership (satisfaction) degree with which any $x \in \mathbb{R}$ accomplishes the corresponding fuzzy constraint on which it is defined. This degree is equal to one when the constraint is perfectly accomplished (no violation), and decreases to zero according to greater violations. Finally for not admissible violations the accomplishment degree will equal zero in all the cases. These membership functions can be formulated as follows:

$$\mu_i(x) = \begin{cases} 1 & \text{if } a_i x \leq b_i \\ f_i(a_i x) & \text{if } b_i \leq a_i x \leq b_i + \tau_i \\ 0 & \text{if } b_i + \tau_i \leq a_i x \end{cases} \tag{2}$$

These functions express that the decision maker is tolerating violations in each constraint up to a value of $b_i + \tau_i$ (τ is referred to as a violation tolerance level). The fulfillment of the constraints can be verified using the following auxiliary model:

$$a_i x \leq b_i + \tau_i(1 - \alpha) \quad (3)$$

3 TTRP with Fuzzy Constraints as Multiobjective Optimization Problem

This parametric approach described above requires that the decision maker establish the values of a parameter α-cut. One of the limitations of the fuzzy variant of the problem is the number of solutions to be obtained, which depends directly on the total α-values. Also, regarding the quality of the solutions, the tendency is to obtain solutions with low costs, but its non-fulfillment grade of the constraints of the problem can be high and vice versa. This fuzzy variant can be generalized from a multiobjective approach by incorporating an objective to minimize the violation of constraints. In this manner it seeks a compromise between the cost of the solution and the non-fulfillment grade of the capacity constraints. With this approach the decision maker no longer provides the level of satisfaction (α) because this parameter is part of the solution sought. The decision maker obtains a set of solutions which cannot be improved in any dimension and correspond to the non-dominated solutions composing the Pareto Front. In this section a multi-objective model for TTRP with fuzzy constraints is presented.

The model discussed in this paper is an adaptation of the standard TTRP model proposed in [4]. This model can be formally defined on an undirected graph $G = (V, A, C)$, where $V = \{v_0, v_1, \ldots, v_n\}$ is a set of vertex representing the customers, except the vertex v_0 that be corresponds to central depot. Each vertex $v_i \in V \backslash \{0\}$ is associated with a non-negative demand q_i and a customer type t_i $\{0,1\}$, where $t_i = 1$ indicates that customer i is served by a truck only (TC) while $t_i = 0$ means customer i can be serviced by a complete vehicle (VC). The access constraints create a partition of $V = V_c \cup V_t \cup 0$ into two subsets, where $V_c \cap V_t =$. $V_c = \{v_1, v_2, \ldots, v_p\}$ is the subset of vertex representing the vehicle customers and $V_t = \{v_{p+1}, v_{p+2}, \ldots, v_n\}$ is the subset of vertex representing the truck customers. $A = \{(v_i, v_j): v_i, v_j \in V, i \neq j\}$ is the set of possible travel edges between customers. And, $C = \{c_{ij}\}$ is a matrix of non-negative cost. Each edge (v_i, v_j) is associated with a cost c_{ij} that represents the travel distance between vertex v_i and vertex v_j. A heterogeneous fleet composed of m_c trucks and m_r trailers ($m_c \geq m_r$) serves a set of customers. All trucks and trailers have identical capacity Q_c and Q_r respectively. In general there are m_r complete vehicles (truck pulling a trailer) and $m_c - m_r$ are trucks without trailers. A complete vehicle has a capacity $Q = Q_c + Q_r$. A route is composed of a partition of V: $R_1, R_2, \ldots, R_{mc}$ and a permutation of σ_k specifying the order of the customers on route. Each route originating from and terminating at a central depot: $R_k = \{v_0, v_1, v_2, \ldots, v_{n+1}\}$, where $v_0 = v_{n+1}$ denotes the depot. In the TTRP, a set of routes consisting of m_r pure and complete vehicle routes and $m_c - m_r$ pure truck routes can be constructed so that the total distance traveled by the fleet over all three types of routes is minimized and all constraints are satisfied.

The proposed formulation of the TTRP with fuzzy constraints uses the following indices, parameters and variables:

- n—the number of customers.
- i, j—customer index which represent the localization of the customer.
- q_i—the demand of the customer i.
- c_{ij}—the cost of arc (i, j).
- k—vehicle index that represents the vehicle.
- m_c—the number of trucks.
- m_r—the number of trailers.
- Q_c—the capacity of a truck.
- Q_c—the capacity of a trailer.
- x_{ij}^{kl}—a binary variable equal to 1 if and only if the vehicle k with (l=0) or without trailer (l = 1) is used from i to j, and 0 otherwise.

Mathematically, the multiobjective TTRP with fuzzy constraints considered here is defined in the following way:

$$\begin{aligned} &\min(f_1, f_2, f_3) \\ &f_1 = \sum_{k=1}^{m} C(R_k) \\ &f_2 = 1 - \alpha_{ccvr} \\ &f_3 = 1 - \alpha_{ccpr} \end{aligned} \tag{4}$$

where α_{ccvr} and α_{ccpr} are degrees of relaxation of each capacity constraint. $C(R_k)$ denotes the cost in distance for the route performed with the vehicle k and may be formulated by:

$$C(R_k) = \sum_{i=0}^{n} \sum_{j=0}^{n} \sum_{k=1}^{m} \sum_{l=0}^{1} c_{ij} x_{ij}^{kl} \quad k = [1, \ldots, m_c] \tag{5}$$

The set of constraints for the TTRP with fuzzy constraints formulation is then stated as follows:

$$\sum_{i=0}^{n} \sum_{k=1}^{m_c} x_{ij}^{k0} = 1 \quad j = [1, \ldots, n] \tag{6}$$

$$\sum_{i=0}^{n} x_{ij}^{k1} \geq 1 \quad j = [1, \ldots, n], \quad k = [1, \ldots, m_c] \tag{7}$$

$$\sum_{i=0}^{n} \sum_{k=1}^{m_c} x_{ji}^{k0} = 1 \quad j = [1, \ldots, n] \tag{8}$$

$$\sum_{i=0}^{n} x_{ji}^{k1} \geq 1 \quad j = [1, \ldots, n], \quad k = [1, \ldots, m_c] \tag{9}$$

$$\sum_{i=1}^{n} \sum_{l=0}^{1} x_{0i}^{kl} \leq 1 \quad k = [1, \ldots, m_c] \tag{10}$$

$$\sum_{i=1}^{n} \sum_{l=0}^{1} x_{i0}^{kl} \leq 1 \quad k = [1, \ldots, m_c] \tag{11}$$

$$\sum_{i=0}^{n} \sum_{l=0}^{1} x_{ij}^{kl} - \sum_{i=0}^{n} \sum_{l=0}^{1} x_{ji}^{kl} = 0 \quad j = [1, \ldots, n], \quad k = [1, \ldots, m_c] \tag{12}$$

$$\sum_{i=0}^{n} \sum_{j=1}^{n} q_j x_{ij}^{k1} + \sum_{i=0}^{n} \sum_{j=1}^{n} q_j x_{ij}^{k0} \leq Q_c + Q_r \quad k = [1, \ldots, m_c] \tag{13}$$

$$\sum_{i=0}^{n} \sum_{j=1}^{n} q_j x_{ij}^{k1} \leq Q_c \quad k = [1, \ldots, m_c] \tag{14}$$

$$\sum_{i=1}^{n} \sum_{k=1}^{m_c} x_{0i}^{k0} \leq m_r \tag{15}$$

$$\sum_{i=1}^{n} \sum_{k=1}^{m_c} x_{0i}^{k0} + \sum_{i=1}^{n} \sum_{k=1}^{m_c} x_{0i}^{k1} \leq m_c \tag{16}$$

$$x_{ij}^{kl} \in \{0, 1\} \tag{17}$$

$$i \in [0, \ldots, n] j \in [1, \ldots, n] k \in [1, \ldots, m_c] l \in \{0, 1\} \tag{18}$$

The objective functions of the problem (4) minimize the total cost (not include any further cost components) (f_1) and the nonfulfillment grade of capacity constraints for vehicle route (f_2) and pure route (f_3). Constraints (6-9) guarantee that only one vehicle enters and leaves from one node or that each customer is served exactly once. Constraints (10) and (11) ensure that each vehicle leaves the depot and returns to it, thereby limiting vehicle use to one trip. Constraint (12) establishes the conditions to maintain continuity of the route. Constraint (13) force the maximum demand load of each pure or complete vehicle route to be less than or equal to the complete vehicle capacity. Constraint (14) force the maximum demand load of each truck route to be less than or equal to the truck capacity. The constraints (15) and (16) ensure that both the number of vehicle and pure routes are not greater than the number of trailers and trucks respectively. Finally (17) and (18) establishes the conditions of the variables.

Note that constraints (13) and (14) are considered as fuzzy. Therefore, they can be replaced by the following constraints:

$$\sum_{i=0}^{n}\sum_{j=1}^{n} d_j x_{ij}^{k0} + \sum_{i=0}^{n}\sum_{j=1}^{n} d_j x_{ij}^{k1} \leq (Q_c + Q)_r + \tau_{ccvr}(1 - \alpha_{ccvr}) \quad k = [1, \ldots, m_c] \tag{19}$$

$$\sum_{i=0}^{n}\sum_{j=1}^{n} d_j x_{ij}^{k1} \leq Q_c + \tau_{ccpr}(1 - \alpha_{ccpr}) \qquad k = [1, \ldots, m_c] \tag{20}$$

$$\alpha_{ccvr} \in [0, 1]$$

$$\alpha_{ccpr} \in [0, 1]$$

where τ_{ccvr} and τ_{ccpr} are tolerance levels of each capacity constraint. Also, in the constraint (19) the tolerance level τ_{ccvr} incorporates the tolerance level of constraint (20), because this constraint verifies the truck and trailer capacity.

4 Experiments and Computational Results

In this section, we present and discuss the computational experiments carried out to solve the multiobjective TTRP with fuzzy constraints. In order to present them, this section is organized as follows:

1. In Sect. 4.1, we describe the test instances that are used in these experiments.
2. In Sect. 4.2, we explain the solution strategy applied for to solve TTRP with fuzzy constraints for each α-value.
3. In Sect. 4.3, we introduce a brief description of the algorithms used for comparison and we show the configuration of the algorithms (determining all the parameters used).
4. In Sect. 4.4, we explain the quality indicators used to evaluate the quality of the output produced by multiobjective algorithms.
5. In Sect. 4.5, we present the configurations of the general approach in these experiments.
6. In Sect. 4.6, we analyze the solutions obtained and the results of the non-parametric statistical test.

Table 1 Chao instances considered for the experimental study

Problem	VC	TC	m_c	q_c	m_r	q_r
1	38	12	5	100	3	100
2	25	25				
3	13	37				
4	57	18	9	100	5	100
5	38	37				
6	19	56				
7	75	25	8	150	4	100
8	50	50				
9	25	75				
10	113	37	12	150	6	100
11	75	75				
12	38	112				
13	150	49	17	150	9	100
14	100	99				
15	50	149				
16	90	30	7	150	4	100
17	60	60				
18	30	90				
19	75	25	10	150	5	100
20	50	50				
21	25	75				

4.1 Test Instances

In order to test our model, we used 21 TTRP benchmark problems reported by Chao [4]. These problems are converted from seven basic VRP test problems given by Christofides et al. [6]. The authors use the following procedure: for each customer i, the distance between i and its nearest neighbor customer is calculated and denoted by A_i. The generation procedure creates three TTRP instances by defining in the first problem 25 % of the customers with the smallest Ai values as TC. For the second and third problem the percentage of the nodes as TC is 50 and 75 % respectively. Table 1 shows the characteristics of the problems.

4.2 Implementation

The TTRP is a combinatorial optimization problem and is computationally more difficult to solve compared with the VRP. An exact algorithm that can optimally solve a large-size TTRP problem is unlikely. Therefore, the main alternative to solving this problem is using approximate algorithms. Our implementation of TTRP is based in the heuristic model proposed by Lin et al. [15]. The proposal was coded in Java and compiled using Eclipse 4.2.1.

4.2.1 Solution Representation and Initial Solution

A solution is represented by a string of numbers consisting of a permutation of n customers and N_{dummy} zeros, followed by the service vehicle types of individual VCs and α-values of the constraints of capacity (α_{ccpr} and α_{ccvr}). The N_{dummy} zeros (artificial depot) are used to separate routes or terminate a sub-tour. The service vehicle type of a VC determines the type of the vehicle used to service the VC. If the VC is serviced by a complete vehicle, its service vehicle type is set to be 0. Otherwise, it is serviced by a truck alone is set to be 1. The α-values indicate the degree of relaxation in each capacity constraint.

The initial solution is randomly generated. It is comprised of a randomly ordered sequence of the customers and the dummy zeros, and randomly set service vehicle types of VCs and α-values of the constraints of capacity.

4.2.2 Neighborhood Structure

The approach uses a random neighborhood structure that includes insertion, swap, and change of service vehicle type. Also, we include an operator that inverts a subset of elements of the solution and other to modify the α-values. We define the set N(X) to be the set of solutions neighboring a solution X. In each iteration, the next solution Y is generated from N(X) either by any of the operators mentioned above as follows.

- Insertion: Position-based operator where an element at one position is removed and put at another position. Both positions are randomly selected.
- Swap: Order-based operator where arbitrarily is selected two elements and are swapped.
- Invert: Order-based operator where two elements are randomly selected and the sequence of elements between these two elements are inverted.
- Change of service vehicle type: This operator is performed by randomly selecting a VC and changing its service vehicle type from 1 to 0 or vice versa.
- Change of α-value: This operator is performed by randomly selecting a α-value of the constraints of capacity and generating a new value.

4.2.3 Evaluate Objective Functions

A process of decoding is used for evaluate the first objective function (f_1). In this process has been determined the customers on each route and the service vehicle type of each VC. Once this is done, it is easy to calculate the objective value (total distance traveled) as the sum of the distances of all routes. A penalty strategy was applied, which involves adding a penalty cost to the objective function for each extra vehicle requiring the solution. The remaining objective functions (f_2 and f_3) are calculated as the difference between the maximum degree (no violation) and the obtained α-value in each capacity constraints in the solution: $\Delta = 1 - \alpha_i$.

4.3 Algorithms Considered for Comparison

In these experiments we compare four multiobjective algorithms: Multiobjective Stochastic Hill Climbing (MSHC) [10], Multiobjective Hill Climbing with Restart (MHCR) [10], Multiple Objective Tabu Search (MOTS) [1] and Multi-Case Multi-Objective Simulated Annealing (MC-MOSA) [14]. The codification of the algorithms was using the BiCIAM library[1] [12], which implements a unified model of metaheuristics algorithm. A brief description of these algorithms follows and the parameters of the analyzed algorithms are shown.

- Multiobjective Stochastic Hill Climbing (MSHC): This algorithm follows the original Stochastic Hill Climbing same principle. To be able to face problems multiobjective, the algorithm incorporates a list with the non-dominated solutions found in the search process [10].
- Multiobjective Hill Climbing with Restart (MHCR): This method works similar to the previous one. The main difference consists on the substitution of the current solution when the candidate solution is not accepted. In this case it is verified if all the possible solution neighbors have already been generated, then the current solution is substituted by other obtained randomly [10].
  - Neighborhood size: 2
- MOTS algorithm redefines selection and updating stages of classical Tabu Search (TS) to work with more than one objective [1].
  - Tabu list: visited solutions
  - Tabu list size: 20 solutions
- Multi-Case MultiObjective Simulated Annealing (MC-MOSA): It is an adaptation of a well-known MultiObjective Simulated Annealing (U-MOSA) proposed by Ulungu et al. [24]. This variant uses some basics of U-MOSA and substitutes others, especially in the acceptance decision criterion [14].
  - Maximum allowable number of reductions in temperature ($N_{non\text{-}improving}$): 100.0
  - Coefficient controlling the cooling scheme (α): 0.9

[1]BiCIAM library can be downloaded in http://modo.ugr.es/algorithmportfolio/index.html.

– Final temperature (T_f): 0.0
– Initial temperature (T_0): 20.0

The parameters of the MHCR, MC-MOSA and MOTS algorithms were selected based the opinion of the author.

4.4 Quality Indicators

Taking into account that there are conflicting goals in the objective function, we conducted an analysis of the results from the multiobjective point of view. For this analysis, we calculate the Pareto front (**PF**) [5] generated by each algorithm for each test problem. There are several well-known measures in the literature to assess the quality of the output produced by multiobjective procedures [7]. The following four measures are considered in this paper.

- Number of True Non-dominated Solutions (N): N measures the total number of non-dominated vectors that belong to $\mathbf{PF_{true}}$. We define this metric as shown in Eq. (21):

$$N \triangleq |PF_{known}| - \sum_{i=1}^{|PF_{known}|} e_i \tag{21}$$

- Error Ratio (ER): This metric reports the proportion of vectors in $\mathbf{PF_{known}}$ that are not members of $\mathbf{PF_{true}}$. Mathematically, this metric is represented in Eq. (22):

$$ER \triangleq \frac{\sum_{i=1}^{|PF_{known}|} e_i}{|PF_{known}|} \tag{22}$$

- Generational Distance (GD): This measure is a value representing how far $\mathbf{PF_{known}}$ is from $\mathbf{PF_{true}}$. It is mathematically defined in Eq. (23):

$$GD \triangleq \frac{\left(\sum_{i=1}^{n} d_i^p\right)^{1/p}}{|PF_{known}|} \tag{23}$$

- Spacing (S): This metric numerically describes the spread of the vectors in $\mathbf{PF_{known}}$. Equation (24) define this metric:

$$S \triangleq \sqrt{\frac{1}{n-1} \sum_{i+1}^{n} (\bar{d} - d_i)^2} \tag{24}$$

The first three metrics require use the true Pareto front ($\mathbf{PF_{true}}$), but in many scenarios this front is unknown. For this reason we generate an unified Pareto Front ($\mathbf{PF_{unified}}$), which is the set of all non-dominated solutions obtained by all metaheuristics in all executions.

4.5 Setting Experimental

All computational tests were conducted on an Intel Core Duo running at 2.66 GHz with 2 GB of RAM processor and operative system Windows 7. Results reported in the following subsection were obtained with 30 independent runs with 100 000 fitness evaluations for each problem. The tolerance levels (τ) for each capacity constraint was calculated as 10 % of the capacity of the used vehicle. Lastly, the penalty cost associated with the number of extra trucks and trailers used is equal to 125. This penalty value reported the best results in [15].

4.6 Result

Tables 2 and 3 show the results obtained by the analyzed algorithms. For each test problem are presented the number of solution members of $\mathbf{PF_{known}}$ in the $\mathbf{PF_{unified}}$ (N), the best and average value for the metrics: Error Ratio (ER), Generational Distance (GD) and Spacing (S). We also include the number of solutions in $\mathbf{PF_{unified}}$ and the number of solutions in $\mathbf{PF_{known}}$ for each algorithm. The numbers set in bold face give the best value achieved for each problem among all algorithms. Letters i, j, k, and l used in Table 3 represent the best algorithm in each case. i: MSHC, j: MHCR, k: MC-MOSA, l: MOTS.

We can present the following conclusions based on an analysis of the results presented in Table 2:

1. MOTS algorithm generates the biggest number of solutions in more problems than the rest of the algorithms. However, MHCR algorithm has the best average of generated solutions (142.33).
2. At least an algorithm obtains more solutions in its $\mathbf{PF_{known}}$ than the total of solutions of the $\mathbf{PF_{unified}}$ (this occurs in 71.43 % of the problems). According with ER values this means that some solutions of these algorithms are dominated by the solutions of the remaining algorithms.
3. In 28.57 % of the problems, the total of generated solutions with MHCR algorithm are members of $\mathbf{PF_{unified}}$ (ER = 0). The remaining cases (with MHCR) averaged 86.27 % of solutions found. Also, for this algorithm the results of S metric suggest that the solutions are not spaced evenly.

Table 2 Summarizes of the results for 21 instances (metric N)

ID	$PF_{unified}$	PF_{MSHC}	N_{MSHC}	PF_{MHCR}	N_{MHCR}	$PF_{MC\text{-}MOSA}$	$N_{MC\text{-}MOSA}$	PF_{MOTS}	N_{MOTS}
1	663	403	141	478	**465**	453	54	98	3
2	765	285	74	826[a]	**489**	367	200	100	2
3	391	117	25	404[a]	**293**	86	73	97	0
4	101	194[a]	5	122[a]	**82**	118[a]	12	85	2
5	339	333	107	120	**119**	289	113	98	0
6	258	148	112	182	**142**	158	4	119	0
7	123	37	5	141[a]	**104**	35	14	107	0
8	88	82	1	82	**73**	41	14	117[a]	0
9	73	57	3	61	**59**	26	11	96[a]	0
10	45	31	12	31	**31**	29	2	127[a]	0
11	45	42	7	36	**19**	19	**19**	115[a]	0
12	56	33	12	48	**36**	10	8	82[a]	0
13	25	25	6	19	**19**	18	0	88[a]	0
14	54	49	1	44	**44**	14	8	103[a]	1
15	28	37[a]	2	26	**25**	27	1	95[a]	0
16	44	23	**17**	21	14	42	13	110[a]	0
17	103	49	30	73	**70**	73	3	93	0
18	67	24	7	58	**58**	32	2	109[a]	0
19	111	146[a]	1	103	**130**	73	7	104	0
20	83	82	4	62	**60**	60	19	113[a]	0
21	112	89	49	52	**52**	50	10	78	1
Avg.	170.19	109	29.57	142.33	**112.23**	96.19	27.95	101.62	0.43

[a]Cases where the PF_{known} overcomes the number of solutions of the $PF_{unified}$

4. In 76.19 % of the analyzed cases, the MOTS algorithm did not obtain any solution members of $\mathbf{PF_{unified}}$. Also, it presents the values of higher GD in 19 test problems.
5. Regarding the S metric the average in the 21 problems for MSHC and MC-MOSA algorithms is quite similar.

This result is confirmed by Figs. 2, 3, 4 and 5; where you can observe the behavior of the algorithms in the four metrics.

In order to assess whether significant differences exist among the results, we adopt statistical analysis and, in particular, nonparametric tests. We decided to apply the statistical tests to the results obtained for the measures N, ER, GD and S. We employ Friedman's test. It is a nonparametric tests for multiple comparisons, that allow checking if the results obtained by the algorithms present any significant differences. Then, if there are differences, we use Holm's method, Rom's method, Finner's method and Li procedure to compare the best ranking algorithm with the remaining algorithms. The level of confidence is 0.05 in all cases.

Table 3 Summarizes of the results for 21 instances (metric ER, GD and S)

ID	$PF_{unified}$	ER_{Best}	ER_{Avg}	GD_{Best}	GD_{Avg}	S_{Best}	S_{Avg}
1	663	0.03[j]	0.63	0.05[i]	1.26	33.94[k]	89.50
2	765	0.41[j]	0.65	0.03[j]	1.59	21.72[k]	92.55
3	391	0.15[k]	0.55	0.21[j]	4.28	27.01[i]	121.81
4	101	0.33[j]	0.80	0.23[j]	2.66	54.51[i]	196.66
5	339	0.01[j]	0.58	0.01[j]	5.10	67.99[k]	185.14
6	258	0.22[j]	0.61	0.37[j]	2.64	59.29[k]	190.97
7	123	0.26[j]	0.68	0.46[j]	9.54	66.27[k]	266.85
8	88	0.11[j]	0.69	0.13[j]	4.03	44.24[i]	228.83
9	73	0.03[j]	0.54	0.28[j]	3.50	36.33[k]	265.87
10	45	0.00[j]	0.64	0.00[j]	5.77	98.90[i]	218.15
11	45	0.00[k]	0.58	0.00[k]	36.82	107.19[k]	201.41
12	56	0.20[k]	0.52	0.20[k]	5.09	124.47[i]	429.62
13	25	0.00[j]	0.69	0.00[j]	32.56	145.65[k]	404.48
14	54	0.00[j]	0.60	0.00[j]	3.64	120.10[k]	556.44
15	28	0.04[j]	0.74	2.54[i]	5.34	80.03[i]	476.22
16	44	0.26[i]	0.57	6.39[i]	15.06	75.98[i]	571.10
17	103	0.04[j]	0.60	0.62[j]	6.30	128.92[k]	506.06
18	67	0.00[j]	0.66	0.00[j]	12.58	124.70[i]	378.47
19	111	0.00[j]	0.72	0.00[j]	4.90	40.53[i]	260.90
20	83	0.03[j]	0.67	0.33[j]	3.90	58.11[i]	280.16
21	112	0.00[j]	0.56	0.00[j]	3.30	80.19[i]	268.89

i MSHC
j MHCR
k MC-MOSA
l MOTS

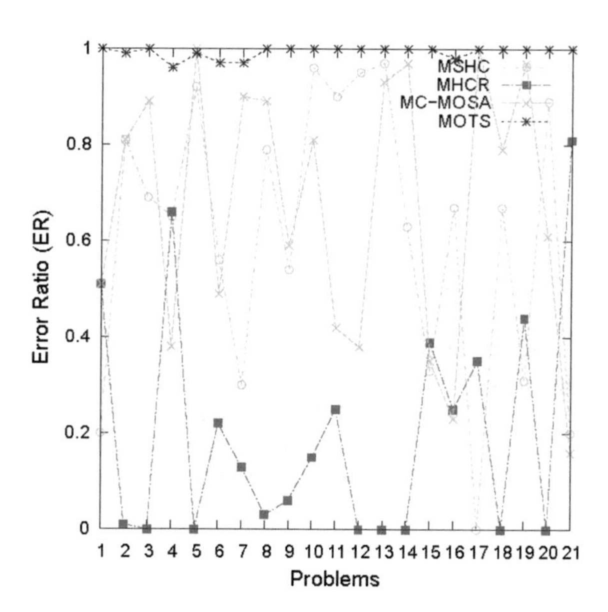

Fig. 2 Performance of the metric error ratio (ER) in the 21 problems

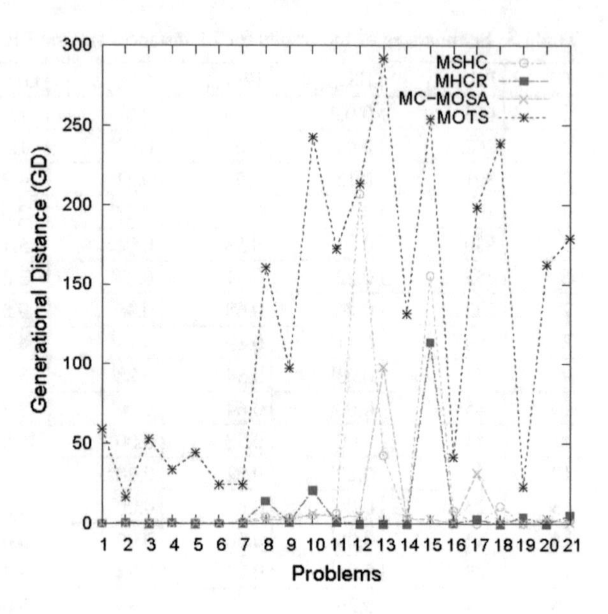

Fig. 3 Performance of the metric generational distance (GD) in the 21 problems

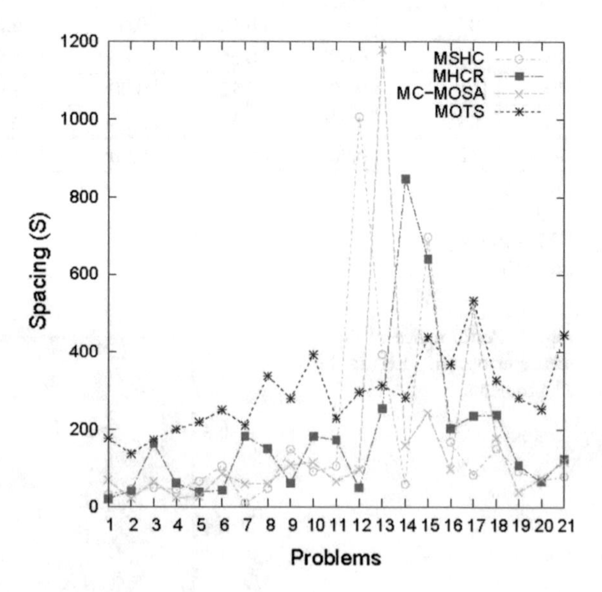

Fig. 4 Performance of the metric spacing (S) in the 21 problems

Table 4 shows the results of Friedman test for the four measures. The MHCR algorithm has achieved the highest rankings for the metrics N, ER and GD, while that MC-MOSA and MSHC are better in the metric S. In all the cases the p-value computed through the statistic of the test considered strongly suggest the existence of significant differences among the algorithms considered.

Fig. 5 Performance of the metric number of true non-dominated solutions (N) in the 21 problems

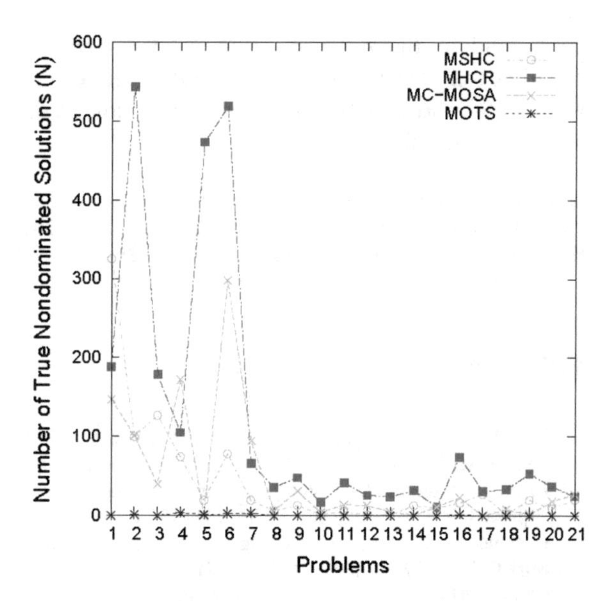

Table 4 Average ranking of the multiobjective algorithm

Algorithm	N	ER	GD	S	Global
MSHC	2.5	2.5238	2.4524	**1.5714**	2.2619
MHCR	**1.0714**	**1.1905**	**1.2619**	3.619	**1.7619**
MC-MOSA	2.4762	2.3095	2.3333	**1.5714**	2.1845
MOTS	3.9524	3.9762	3.9824	3.2381	3.7917

According to used procedures (Rom, Holm, Finner and Li) in the metrics N, ER and GD the Friedman test shows a significant improvement of MHCR algorithm over MSHC, MC-MOSA and MOTS for all the post hoc procedures considered. For S metric, the test does not find any significant difference between the algorithms MC-MOSA and MSHC, but there are significant difference with MHCR and MOTS. Also, the last column in Table 4 shows an overall ranking (we considered the results of all metrics) that indicates the improvement of MHCR over the remaining algorithms.

5 Conclusions

In this chapter we focus on the TTRP with fuzzy constraints from a multi-objective viewpoint. In this manner it seeks a compromise between the cost of the solution and the nonfulfillment grade of the capacity constraints. With this approach the decision maker no longer provides the level of satisfaction (α) but this parameter is part of the solution is sought. The decision maker obtains a set of solutions that correspond to the non-dominated solutions composing the Pareto Front. The experimental study shows of multiobjective algorithms MSCH, MHCR, MC-MOSA and MOTS. The metrics used indicate that the best performance is the algorithm MHCR.

Acknowledgments This work was supported in part by the projects P11-TIC-8001 from the Andalusian Government (including FEDER funds), TIN2014-55024-P from Spanish Ministry of Economy and Competitiveness, and PYR 2014-9 from the GENIL, University of Granada.

References

1. Baykasoglu, A., Owen, S., Gindy, N.: A taboo search based approach to find the Pareto optimal set in multiple objective optimization. J. Eng. Optim. **31**(6), 731–748 (1999)
2. Brito, J., Moreno, J.A., Verdegay, J.L.: Transport route planning models based on fuzzy approach. Iran. J. Fuzzy Syst. **9**(1), 141–158 (2012)
3. Cadenas, J., Verdegay, J.L.: Using fuzzy numbers in linear programming. IEEE Trans. Syst. Man Cybern. **27**(6), 1017–1022 (1997)
4. Chao, I.-M.: A tabu search method for the truck and trailer routing problem. Comput. Oper. Res. **29**(1), 33–51 (2002)
5. Chen, Y., Zou, X., Xie, W.: Convergence of multi-objective evolutionary algorithms to a uniformly distributed representation of the pareto front. Inf. Sci. **181**(16), 3336–3355 (2011)
6. Christofides, N., Mingozzi, A., Thot, P.: The vehicle routing problem. In: Christofides, N., Mingozzi, A., Thot, P. (eds.) Combinatorial Optimization, pp. 315–338. Wiley, Chichester (1979)
7. Coello, C.A., Van Veldhuizen, D.A., Lamont, G.B.: Evolutionary Algorithms for Solving Multi-objective Problems, 1st edn. Kluwer Academic Publishers, New York (2002)
8. Delgado, M., Verdegay, J.L., Vila, M.A.: A general model for fuzzy linear programming. Fuzzy Sets Syst. **29**(1), 21–29 (1989)
9. Derigs, U., Pullmann, M., Vogel, U.: Truck and trailer routing-problems, heuristics and computational experience. Comput. Oper. Res. **40**(2), 536–546 (2013)
10. Díaz, R. Estudio de la capacidad del algoritmo escalador de colinas estocástico para enfrentar problemas multiobjetivo. Master's thesis, Department Artificial Intelligence and Infrastructure of Informatic Systems, Instituto Superior Politécnico José Antonio Echeverría (2001)
11. Drexl, M.: Branch and price and heuristic column generation for the generalized truck and trailer routing problem. J. Quant. Methods Econ. Bus. Adm. **12**(1), 5–38 (2011)
12. Fajardo, J., Rosete, A.: Algoritmo multigenerador de soluciones para la competencia y colaboración de generadores metaheuristicos. Revista Internacional de Investigación de Operaciones (RIIO) **1**(1), 57–63 (2011)
13. Gerdessen, J.C.: Vehicle routing problem with trailers. Eur. J. Oper. Res. **93**(1), 135–147 (1996)
14. Haidine, A., Lehnert, R.: Multi-case multi-objective simulated annealing (mc mosa): new approach to adapt simulated annealing to multi-objective optimization. Int. J. Inf. Technol. **4** (3), 197–205 (2008)
15. Lin, S.-W., Yu, V.F., Chou, S.-Y.: Solving the truck and trailer routing problem based on a simulated annealing heuristic. Comput. Oper. Res. **36**(5), 1638–1692 (2009)
16. Lin, S.-W., Yu, V.F., Chou, S.-Y.: A note on a the truck and trailer routing problem. Expert Syst. Appl. **37**(1), 899–903 (2010)
17. Lin, S.-W., Yu, V.F., Chou, S.-Y.: A simulated annealing heuristic for the truck and trailer routing problem with time windows. Expert Syst. Appl. **38**(12), 15244–15252 (2011)

18. Scheuerer, S.: A tabu search heuristic for the truck and trailer routing problem. Comput. Oper. Res. **33**(4), 894–909 (2006)
19. Semet, F.: A two-phase algorithm for the partial accessibility constrained vehicle routing problem. Ann. Oper. Res. **61**(1), 45–65 (1995)
20. Semet, F., Taillard, E.: Solving real-life vehicle routing problems efficiently using tabu search. Ann. Oper. Res. **41**(4), 469–488 (1993)
21. Torres, I., Rosete, A., Cruz, C., Verdegay, J.L.: Fuzzy constraints in the truck and trailer routing problem. In: Proceedings of Fourth International Workshop on Knowledge Discovery, Knowledge Management and Decision Support, pp. 71–78 (2013)
22. Torres, I., Verdegay, J.L., Cruz, C., Rosete, A.: Models and solutions for truck and trailer routing problems: an overview. Int. J. Appl. Metaheuristic Comput. **4**(2), 31–43 (2013)
23. Torres, I., Cruz, C., Verdegay, J.L.: Solving the truck and trailer routing problem with fuzzy constraints. Int. J. Comput. Intell. Syst. **8**(4), 713–724 (2015)
24. Ulungu, E.L., Teghem, J., Fortemps, P.H., Tuyttens, D.: MOSA method: a tool for solving multiobjective combinatorial optimization problems. J. Multicriteria Decis. Anal. **8**(4), 221–236 (1999)
25. Verdegay, J.L.: Fuzzy mathematical programming. In: Gupta, M.M., Sanchez, E. (eds.) Fuzzy Information and Decision Processes, pp. 231–237 (1982)
26. Villegas, J.G., Prins, C., Prodhon, C., Medaglia, A.L., Velasco, N.: GRASP/VND and multi-start evolutionary local search for the single truck and trailer routing problem with satellite depots. Eng. Appl. Artif. Intell. **23**(5), 780–794 (2010)
27. Villegas, J.G., Prins, C., Prodhon, C., Medaglia, A.L., Velasco, N.: A GRASP with evolutionary path relinking for the truck and trailer routing problem. Comput. Oper. Res. **38** (9), 1319–1334 (2011)
28. Villegas, J.G., Prins, C., Prodhon, C., Medaglia, A.L., Velasco, N.: A matheuristic for the truck and trailer routing problem. Eur. J. Oper. Res. **230**(2), 231–244 (2013)

Health Service Network Design Under Epistemic Uncertainty

Mohammad Mousazadeh, S. Ali Torabi and Mir Saman Pishvaee

Abstract If a health system wants to achieve its strategic goal known as "reducing health inequalities", making health services available and accessible to all people is an essential prerequisite. Health service network design (HSND) is known as one of the most critical strategic decisions that affects performance of health systems to the great extent. Important decisions such as location of health service providers (i.e. clinics, hospitals, etc.), allocation of patient zones to health service providers and optimal designing of patients flow via the network are some of the main strategic and tactical decisions that should be made when configuring a health service network. On the other hand, coping with uncertainty in data is an inseparable part of strategic and tactical problems. More specifically, the complex structure of health service networks alongside the volatile environment surrounding the health systems would impose a higher degree of uncertainty to the decision makers and health network designers. Among different methods to cope with uncertainty, possibilistic programming approaches are well-applied methods that can handle epistemic uncertainty in parameters.

Keywords Health care · Service network design · Possibilistic programming · Fuzzy sets

1 Introduction to Healthcare and Healthcare Systems

The most-frequently quoted definition of health is the one that has been defined by World Health Organization (WHO) in 1946 as "state of complete physical, mental and social well-being and not merely the absence of disease or infirmity".

M. Mousazadeh · S. Ali Torabi (✉)
School of Industrial Engineering, College of Engineering,
University of Tehran, Tehran, Iran
e-mail: satorabi@ut.ac.ir

M.S. Pishvaee
School of Industrial Engineering, Iran University of Science and Technology,
Tehran, Iran

C. Kahraman et al. (eds.), *Fuzzy Logic in Its 50th Year*,
Studies in Fuzziness and Soft Computing 341,
DOI 10.1007/978-3-319-31093-0_12

Regarding this definition, all the efforts and activities that are directly and/or indirectly related to maintaining, improving or recovering health status of a society would be accounted as a part of health services. This organization also defines the health system as "all the activities whose primary purpose is to promote, restore or maintain health" [1].

According to Porter and Teisberg [2], the healthcare delivery value chain consists of 6 phases starting with monitoring/preventing, and followed by diagnosing, preparing, treating, rehabbing/recovering and ends to monitoring/managing. These phases could be crossed with two types of activities i.e. primary activities and supportive activities. The primary activities are those that directly address care delivery to patients including screening, organizing tests and imaging, pre-intervention preparations, performing procedures, inpatient recovery and monitoring compliance of patient with therapy. On the other hand, the supportive activities have a supportive rule for all the primary activities throughout the whole health value chain. These activities include knowledge development, informing activities, measuring activities and patient accessing activities. The visual presentation of the adopted health value chain is depicted in Fig. 1.

Healthcare sector has been developed notably in the last 50 years and healthcare spending keep outpacing economic growth across the world. Let us begin by reviewing some facts about historical trends in global economy and health sector economy. The collected data from World Bank and WHO indicates that (1) the gross world product (GWP) has been increased from 52.2 Trillion Dollars in the year 2000 to 77.2 Trillion Dollars in 2011, (2) the total expenditure on health as

		Monitornig/ Preventing	Diagnosing	Preparing	Intervening	Recovering/ Rehabilitating	Monitoring/ managing
Primary activities		- Medical history - Screening - ID risk factors - Prevention program	- Medical history - Specifying and organizing tests - Interpreting data - Determining treatment plan	- Choosing the team - Pre-intervention preparation	- Ordering and administrating drug therapy - Performing procedures	- Inpatient recovery - Therapy fine tuning - Developing a discharge plan	- Monitoring and managing the patients condition - Monitoring compliance with therapy - Monitoring lifestyle modifications
supportive activities	Knowledge Development	Results measurement & tracking, staff/physician training, technology development, process improvement					
	Informing	Patient education, patient counseling, pre-intervention education, patient compliance counseling					
	Measuring	Tests, imaging, patient records management					
	Accessing	office visits, lab visits, hospital sites of care, patient transport, visiting nurses, remote consultation					

Fig. 1 The healthcare delivery value chain adopted from Porter and Teisberg [2]

percentage of gross domestic product (GDP) has been increased from 8.2 % in 2000 to 9.1 % in 2011 globally, and (3) the total expenditure on health as percentage of gross domestic product (GDP) has been increased from 9.8 % in 2000 to 11.9 % in 2011 in high-income countries, revealing that both the percentage rate and a rate of increase in high-income countries is higher than global values. Now, putting these facts together, one can infer that the total health economy has significantly been increased from 4.28 Trillion Dollars in 2000 to 7.02 Trillion Dollars in 2011 which is around 64 % increase.

The reasons for growth in the healthcare industry are numerous. Decreasing fertility rates and increasing life expectancy could be concerned as major factors that increase the total spending on healthcare. As an instance, average worldwide life expectancy was 46 years in 1965, increased to 61 years by 1980, rose to 67 years by 1998 and according to final statistics was around 70 years by 2012 [3]. It is obvious that the aged population not only will put strong demand upon health systems but also consume more intensive health services. On the other hand, although technological advancements have improved health sector, they have increased healthcare costs as well.

This drastic change could be regarded as an opportunity for some of the healthcare stakeholders and as a threat for other stakeholders. In detail, the growing market of health has persuaded the private sectors and independent investors to play a more active role in this section than some years ago. However, if a value of close and mutual cooperation between private sector and government is not appreciated, the government will be faced an astronomical expenses in providing health services to people. In addition, the governments are in charge of cutting unnecessary costs in this sector while maintaining or improving the quality of health services. Also, they face different challenges such as improper configuration of the health system, increased complexity of processes, the need for efficient utilization of resources, increased pressure to improve the quality of services and the need to control the workload of healthcare personnel, most of them will be mounted if a systematic approach could be adopted [4].

Operations research (OR) science can enormously help the health sector in solving these complex but solvable problems. To do so, according to Brailsford and Vissers [5], the application of OR in healthcare has been improved not only in the number of OR applications but also in the scopes of topics covered. As an instance, problems ranging from strategic level decisions such as location of different health service providers (e.g. [6–9]) to operational decisions such as patients/nurse/operating room scheduling (e.g. [10–12]) are among many that are mathematically modeled and solved. Figure 2 shows a more classified review of OR applications in the healthcare sector proposed by Rais and Viana [13].

For more information about applications of OR in healthcare, the interested readers are referred to [3, 5, 13–15].

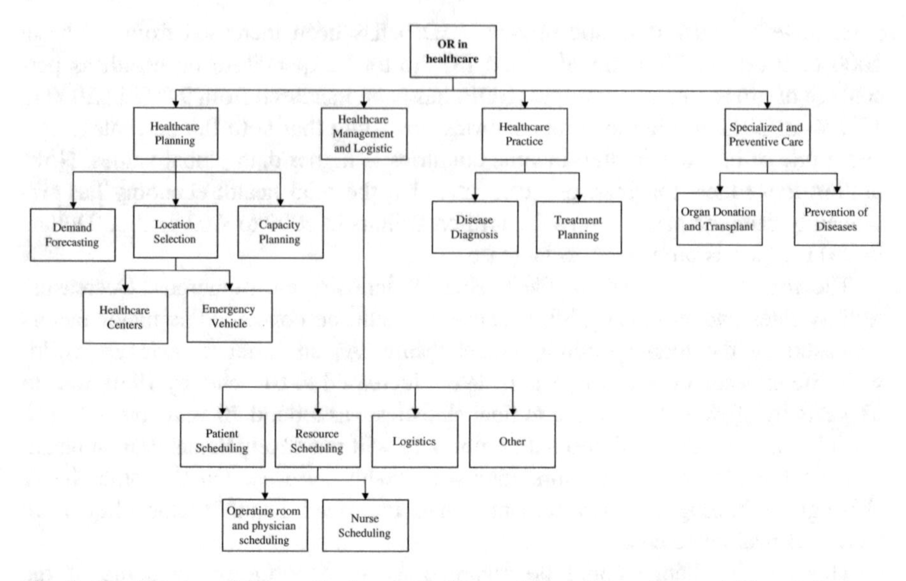

Fig. 2 Classification of OR applications in healthcare (adopted from Rais and Viana [13])

2 Health Service Network Design

The importance and drivers of health service network design (HSND) problem is briefly described in this section, followed by a typical structure of a health service network.

2.1 *Importance and Drivers*

As mentioned before, reaching a health system its ultimate goal known as "reducing health inequalities", necessitates making health services available and accessible to all people. An excellent HSND is the first and main step for achieving this goal. In fact, HSND can be regarded as one of the most critical strategic decisions that its poor design can completely ruin the functionality of the health system while its good design can greatly support equality in access to health services, quality of provided services to people as well as pursuing cost reduction policies of governments.

According to WHO [16], although health service networks are much more extensive than 1990s, large group of people are not covered or are partially covered yet. For example, in some countries, war and civil strife have partly or completely destroyed healthcare infrastructure, in others, although the health service is available, unregulated commercialization leads to providing those services that are not necessarily needed. In addition, supply gaps are still an unpleasant truth in many countries; making extension of health service networks a main concern, as was the case 30 years ago. These facts clearly reflect the immediate but constant need for optimal health services network design, which can be well responded through applying OR tools and techniques.

2.2 Definitions and Scope

The application of network design practices in the health care sector not only relates to the product/non-service problem such as pharmaceuticals [17], blood collection [18], organ transplant [19], etc., but also is related to the medical and health service network design [9].

The basic context of health service network design problem is similar to the problem of supply chain/logistics network design in many assumptions, parameters and decision variables. In fact, logistics network design addresses the locations, numbers and capacities of needed facilities in the logistics network alongside aggregate material flow between the facilities [20]. Similarly, the health service network design addresses location, number and capacity of different health service centers (including primary health centers, clinics, hospitals, etc.), allocation of patient zones to primary health centers, designing referral system throughout the different layers/levels of the network and finally designing the best flow pattern of patients via the network.

Most of the network design problems regardless of their concerned sector or industry, have been mathematically modeled based on the facility location theory. As an instance, in the field of logistics network design, the basic works start with classic facility location models (e.g. [21]) which are then followed by more complex models taking into account new set of assumptions such as multiple products (e.g. [22]), multiple capacities for network facilities (e.g. [23]), direct and indirect shipments (e.g. [24]) and green aspects (e.g. [25]). Interestingly, a similar trend could be observed in the field of health service network design.

2.3 Typical Structure of Health Service Networks

Marmot et al. [26] declared that healthcare systems have better health outcomes when built on primary health centers (PHCs). In fact, if the entrance point of patients to a health system changes from hospitals and specialists service providers' levels into PHCs, it would have positive effects on total functionality of the health system and also will improve performance of health network. Taking this fact into consideration, many countries have redesigned their current health networks so that people first visit their family physicians/general practitioners and then, in the case of need, would be referred to the upper level of health network. Conducted researches (e.g. [16, 27, 28]) show that success of a health network reconfiguration

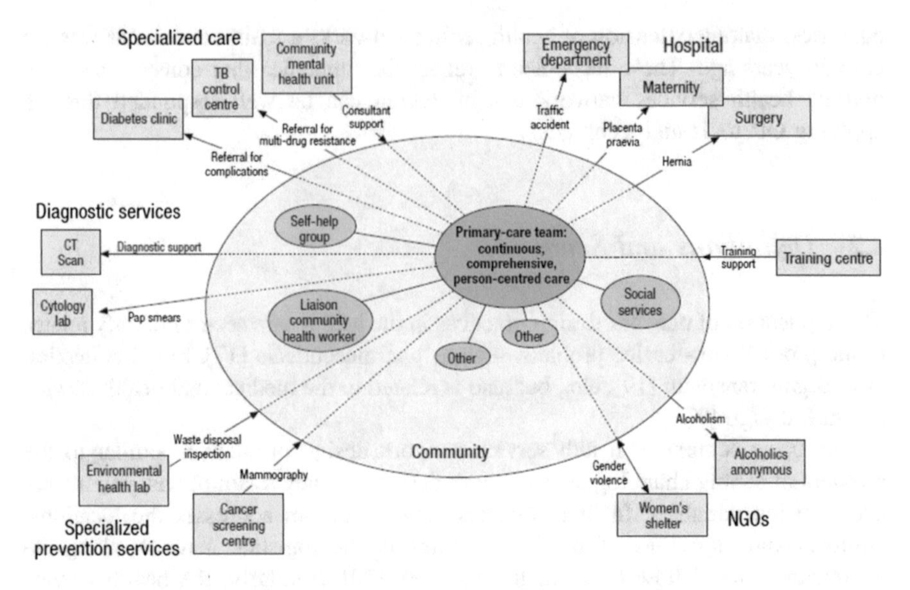

Fig. 3 Typical structure of health service network adopted from WHO [16]

based on the PHCs' main role, highly depends on three main arrangements i.e. (i) bringing care closer to people, (ii) giving primary-care providers the responsibility for the health of the assigned population and (iii) strengthening primary-care providers' role as coordinators of the inputs of other levels of care. Taking these main arrangements into account, WHO [16] by inspiring from the work of Criel et al. [29], proposed a typical structure of health service network in which PHCs are defined as entering point of patients to health systems in which PHCs are viewed as coordination hubs in the network. The concerned health service network is depicted in Fig. 3.

3 Literature Review

In this section, a comprehensive review of related papers to the HSND problem is provided. To do so, the methodology of selecting the relevant papers is first described which is followed by a detailed description of the selected papers.

3.1 Selection Methodology

If one wants to translate the three aforementioned arrangements into the location science keywords, the "approximation of service" could be viewed in the form of location problem, the "clear responsibility" would be addressed via location-allocation problem and "coordination role", could be translated as a hierarchical facility location problem. Thus, in order to review the most relevant papers to the concerned topic of this study, those papers addressing the facility location/network design and healthcare/medical were first extracted. Then, those papers dealing with qualitative and simulation aspect of the concerned problem have been eliminated and finally the papers in which a mathematical programming model is proposed, are selected for final review. These papers are briefly reviewed in a chronological order as follows.

3.2 Detailed Review of Related Papers

Among the earlier works on healthcare facility network design, Dökmeci [30] presents a quantitative model to determine the number, size and locations of regional health facilities including the medical center, intermediate and local hospitals, and health centers with respect to the minimization of the total transportation and facility costs. Also, a heuristic model is developed to solve the proposed model which is then followed by a numerical example in order to illustrate the solution procedure.

Schweikhart and Smith-Daniels [31] present a nonlinear integer model for determining the number, location and service offerings of facilities with respect to maximization of market share subject to a budgetary constraint. In detail, their model addresses a two-level hierarchical referral delivery network in which, each opened facility offers primary care services and may offer one or more of the specialty care services as well. Moreover, specialized healthcare services are provided by physicians in several fields such as allergy, urology, orthopedics, otolaryngology, and dermatology. In order to solve the proposed integrated model, they proposed an interchange heuristic.

Galvao et al. [32] present a 3-level hierarchical location model for maternal and perinatal health care facilities in Rio de Janeiro. In addition, flow of mothers from demand points to facilities and from maternity home to neonatal clinic are two other decision variables that are taken regarding minimization of total travelled distances. Also, some relaxations and heuristics are developed to solve the model using a real data extracted from a case study.

Marianov and Serra [33] propose an extension of their earlier model for the location of congested facilities Marianov and Serra [34], the former model is intended for the design of health and emergency medical services (EMS), banking or distributed ticket-selling services. Their model addresses a set covering problem, aiming to locate the least number of facilities and assigns the minimum number of servers to the established centers in order to minimize queuing effects. They also developed and tested a novel heuristic algorithm which had a good performance on a 55-node network.

Verter and Lapierre [35] focus on the Maximal Covering Location Problem (MCLP) for preventive health carefacilities to maximize participation of inhabitants to prevention programs. The main decisions in the proposed model are twofold: (i) location of new facilities and (ii) allocation of population centers to the new established centers. Performance of the proposed methods is then assessed in a real case of locating public health centers in Fulton County, Georgia and mammography screening centers in Montreal, Quebec.

Galvão et al. [36] propose an uncapacitated, three-level hierarchical model for location of perinatal facilities in the municipality of Rio de Janeiro. The model incorporates capacity constraints and service referrals in which a fraction of demand served at a lower-level perinatal facility may be referred directly to a higher-level facility, while single-assignment is not obliged and split assignments can take place. Finally, the authors propose a Lagrangian heuristic to solve the proposed model.

Yasenovskiy and Hodgson [37] address the problem of designing optimal hierarchical facility systems, in which the location of new facilities and allocation of zones to the facilities are decision variables of the proposed model. Unlike the classic p-median model in which it is always assumed that people will travel to the closest facility to be served, the proposed model defines the facility size, distance, and neighborhood accessibility as the main criteria for travelling preferences of people. They verified their model for locating healthcare facilities in Suhum District, Ghana including 150 nodes.

Griffin et al. [38] propose an optimization model for determining the number and the location of new Community Health Centers (CHCs) in a health network. Another decision variable in the proposed model that differentiates this model from other works is that what services each CHC should offer at which capacity level. These decisions must be made under the budget and capacity constraints with the aim of maximizing the demand coverage of the population.

Ndiaye and Alfares [39] formulate a binary integer programming model to determine the optimal number and locations of primary health centers for serving seasonally moving populations. Objective function of the proposed model consists of two contradictory terms, (i) opening costs and (ii) total transportation costs throughout the network. They also applied their model to the populations of 17 nomadic groups in the Oman and UAE which are located in southwestern Asia, next to the Persian Gulf.

Smith et al. [40] propose a number of hierarchical location-allocation models for the planning of community health schemes by non-governmental organizations in rural areas of developing countries. The concerned decisions are taken with respect to objectives pertaining to both efficiency and equity of provision. They also present modelling of the location of a maximal number of self-sustainable primary healthcare workers in a rural region of India as an additional case study.

Zhang et al. [41] incorporate congestion in preventive healthcare facility network design via a nonlinear programming model. Taking this fact into account that the level of participation of a given population to preventive healthcare programs is a critical factor to the effectiveness and efficiency of preventive health network, they propose a methodology for designing a network of preventive healthcare facilities aiming at maximizing the population participation. The number and location of facilities to be established in the configuration of a healthcare facility network are the main decision variables in the proposed model. In addition, the proposed model covers three main characteristics of the concerned problem i.e. (i) elastic demand, (ii) congestion, and (iii) user-choice environment. Since the proposed model is highly nonlinear, they provide four heuristic solutions for this problem which are then compared in terms of accuracy and required computational time. Finally, they assessed the performance of the model in solving a real case study that involves the breast cancer screening center network in Montreal.

Syam and Côté [8] develop a location-allocation model for the non-for-profit organizations that provides specialized health care services such as traumatic brain injury (TBI) treatment. The model incorporates two primary criteria i.e. (i) a cost minimization objective that includes establishment costs, variable costs of treatment (variable labor cost, and patient travel and patient family lodging costs) and penalty cost for failing to serve patients, and (ii) a minimum service proportion requirement. In addition, the model is solved using simulated annealing approach for one of the Department of Veterans Affairs' integrated service networks.

Mahar et al. [42] propose a mathematical model to determine how many and which hospitals should be established to deliver a specialized service such as magnetic resonance imaging (MRI), transplants, or neonatal intensive care. Moreover, both financial considerations and patient service levels are taken into account in the model. They used commercial solvers in order to solve the proposed Mixed Integer Non-Linear Programming model (MINLP) under two different problem sets i.e. (i) hospital network in eastern Pennsylvania, and (ii) hospital network in Indiana.

Fo and da Silva Mota [43] propose four mathematical models based on the most important location models, i.e. P-median model, set and maximal covering models and P-center model. An experiment was conducted in a real situation, in a Brazilian city, in order to show differences between the optimal solutions achieved from the proposed models as well as the current sites configuration with these optimal configurations.

Mestre et al. [44] propose a hierarchical multiservice model to make decisions on the location of district hospitals (DHs)/central hospitals (CHs), ascendant/descendent flow of population in the network and define service capacities in DHs and CHs, with respect to maximization of patients' geographical access to a hospital network. They have applied the proposed model in a real life problem in the South region of the Portuguese NHS, in which three scenarios are generated to show the usefulness and the behavior of the proposed model.

Syam and Côté [6] develop and apply a location-allocation model for specialized health careservices, i.e. the treatment and rehabilitation necessary for strokes or traumatic brain injuries. The model minimizes the total cost of health system and its patients and also includes many additions to a classic healthcare location-allocation model including the multiple acuity levels, multiple service level mandates by acuity, and facility utilization targets by acuity. The applicability of the model and the impact of several factors, i.e. lost admission cost, service level mandate, target utilization percent, and overloading penalty cost on total cost and service level is investigated using data derived from one of the Department of Veterans Affairs specialized healthcare services.

Zhang et al. [45] propose two alternative models for the problem of preventive care facility network design. In the first model entitled as "probabilistic-choice model", a client is free to go to one of the established facilities under this assumption that he/she may go to the nearest facility with the highest probability but may visit the farthest facility with the least probability. On the other hand, in the second proposed model entitled as "optimal-choice model", it is assumed that each client will patronize only the nearest facility. In addition, they impose an upper bound on the mean waiting time and lower bound on workload requirement at each open facility in order to ensure the quality of provided care. The aim of both proposed models is to maximize the total participation of community to a preventive care programs while the number of open facilities as well as the location and the number of servers of each open facility are the main determinants of the facility network configuration.

Benneyan et al. [46] develop a single and multi-period location-allocation integer programming models for specialty care facilities that minimize total procedure, travel, non-coverage, and start-up costs subject to access constraints. Selecting facilities among the existing centers as provider of sleep apnea services, allocation of patients' demand to these facilities and the number of additional sleep beds to add are main decision variables in the models. These models have been utilized to aid the Veterans Health Administration (VHA) to find a tradeoff surface between costs, coverage, service location, and capacity. An application of the proposed models to planning short and long-term sleep apnea care across the VHA New England integrated network resulted in around 10–15 % improvements in each performance measure.

Shariff et al. [7] formulate a location-allocation model for healthcare facility planning in Malaysia. The proposed model is formulated as Capacitated Maximal Covering Location Problem (CMCLP) and as the main contribution; the authors develop an effective genetic algorithm to solve a large network consisting of 809 nodes.

Song et al. [47] address the problem of facility network design for long-term care services. Optimal location of long-term care facilities as well as the type of new facility are the main determinants of the proposed integer programming model, with the objective of minimizing the total construction cost. For the sake of simplicity, the closest assignment rule is imposed to reflect the preference of patients in selecting long term care facility. As a solution approach, the authors developed a branch and bound (B&B) algorithm for exact solution and a genetic algorithm to solve large-scale problems.

And more recently, Shishebori and Babadi [9] address the problem of robust and reliable medical services network design in which uncertainty in input parameters, system disruptions, and investment budget constraints are taken into account in the proposed Mixed Integer Linear Programming (MILP) model. They applied the developed model for medical network design in one of the deprived provinces in Iran.

3.3 Literature Analysis

From reviewing 21 papers, the following conclusions could be drawn:

- A wide range of facilities/services are addressed in the literature including preventive healthcare facilities, primary healthcare facilities, community health centers, regional health facilities, maternal and perinatal healthcare facilities, district and central hospitals, specialty healthcare facilities, emergency medical services, specialized healthcare services and long-term care services. However, most of the papers have only focused on a single type of facility/service rather than addressing various facilities belonging to the different levels of the health system.
- Most of the papers simply assumed that the exact (i.e. crisp) input data are available, and among the very limited papers considering the data uncertainty, stochastic programming and robust programming approaches have been applied to handle uncertainty. Interestingly, none of the papers used fuzzy programming approaches to handle (epistemic) uncertainty in data where enough historical data are not available.
- In order to solve the proposed models, usual commercial solvers i.e. CPLEX in GAMS, LINGO, etc. are mostly applied, however, heuristic methods, genetic algorithm (GA), and simulated annealing (SA) approach are used in 5, 2 and 1 papers, respectively.
- Interestingly, the proposed models in 15 papers have been applied for a real case study. It clearly shows applicability of the proposed models in real life situations.

4 Typical HSND Mathematical Models

In this section, inspired from the typical health service network proposed by WHO [16], a basic mixed-integer linear programming (MILP) model is proposed and developed gradually by adding new logical assumptions into the problem.

The concerned problem is a multi-level health service network designincluding patient zones, primary health centers, clinics and hospitals. The entering point of the patients to the health network is via PHCs, which are responsible for delivering primary health services to the assigned population. After visiting PHCs, three alternative scenarios are likely to happen for patients including: (1) they exit the health network because no further treatment is needed, (2) they will be referred to the clinics for receiving other non-surgical treatments or for further diagnostic purposes and (3) they will be referred to the hospitals for receiving surgical interventions. Several candidate locations are assumed to be available for establishing PHCs, clinics and hospitals. Hence, the main decisions to be made by the proposed model are the optimal selection of locations for opening different health service provider centers. In addition, allocation of patient zones to PHCs and referral pattern from PHCs to clinics and hospitals must be determined optimally by

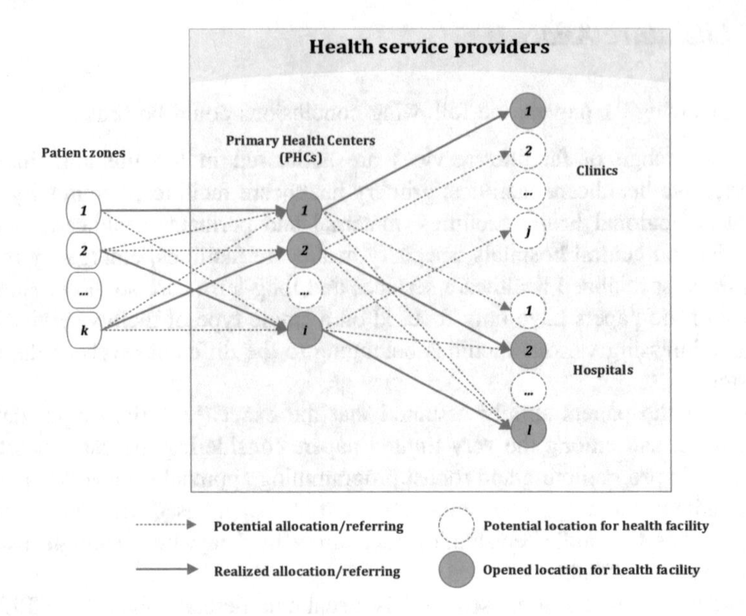

Fig. 4 The structure of the described health service network

the proposed model. Moreover, capacity of each health center should be determined regarding the amount of patients' flow throughout the whole network. All these decisions are made with respect to minimization of total costs while the complete fulfillment of demands is guaranteed via a set of hard constraints.

Figure 4 simply illustrates the structure of the discussed problem. It is worth noting that in real situations, patients will directly refer to clinics or even hospitals without visiting PHCs when facing immediate problem such as severe dental problems, episodes of acute illnesses or receiving severe injuries such as road accidents, different strokes, etc. However, for the sake of simplicity and for educational purposes, these rather exceptional flows are not considered in the model to avoid unnecessary complexity.

In abstract, the main assumptions of the proposed problem can be described as follows:

- There is not any pre-established facility.
- All the demands for all three types of health centers i.e. PHCs, clinics and hospitals must be fulfilled.
- At most one new center from each type of facilities can be located at each candidate location.
- All centers are capacitated and each center can be opened in one of the available capacity levels.
- The maximum demand of each patient zone for PHCs, clinics and hospitals is equal to or less than the capacity of the largest PHC, clinic and hospital, respectively.

- Patients will go to the health service providers, so the network will be designed based on customer-to-service system.
- Each patient zone is free to be served by more than once PHC and each PHC is free to refer patients to more than one clinic and/or one hospital.

The indices, parameters and variables used to mathematically formulate the basic health service network design problem are described below.

Indices

i index of candidate location for primary health centers (PHCs)
j index of candidate locations for clinics
l index of candidate locations for hospitals
n, n' indices of capacity levels
k index of patients zones

Parameters

f_i^n fixed cost of opening a PHC at candidate location i with capacity level n
g_j^n fixed cost of opening a clinic at candidate location j with capacity level n
h_l^n fixed cost of opening a hospital at candidate location k with capacity level n
d_k number of visits to PHC in patient zone k
α_k fraction of people in patient zone k that are referred to a clinic by PHCs
β_k fraction of people in patient zone k that are referred to a hospital by PHCs
cax_i^n capacity of opened PHC with capacity level n at candidate location i
cay_j^n capacity of opened clinic with capacity level n at candidate location j
caz_l^n capacity of opened hospital with capacity level n at candidate location l

Variables

W_i^n $\begin{cases} 1 & \text{if a PHC with capacity level } n \text{ is opened at candidate location } i \\ 0 & \text{otherwise} \end{cases}$

Y_j^n $\begin{cases} 1 & \text{if a clinic with capacity level } n \text{ is opened at candidate location } j \\ 0 & \text{otherwise} \end{cases}$

Z_l^n $\begin{cases} 1 & \text{if a hospital with capacity level } n \text{ is opened at candidate location } j \\ 0 & \text{otherwise} \end{cases}$

x_{ki} flow of patients from patient zone k to PHC i
u_{ij} flow of patients from PHC i to RHC j
q_{il} flow of patients from RHC j to DHC k in order to get special service category m

Using the abovementioned notations, the MILP model of the concerned problem is as follows:

$$\min \sum_{i,n} f_i^n W_i^n + \sum_{j,n} o_j^n Y_j^n + \sum_{l,n} h_l^n Z_l^n \quad (1)$$

s.t:

$$\sum_k x_{k,i} \leq \sum_n caw_i^n W_i^n \quad \forall i \tag{2}$$

$$\sum_i u_{i,j} \leq \sum_n cay_j^n Y_j^n \quad \forall j \tag{3}$$

$$\sum_i q_{i,l} \leq \sum_n caz_l^n Z_l^n \quad \forall l \tag{4}$$

$$\sum_i x_{k,i} \geq d_k \quad \forall k \tag{5}$$

$$\sum_j u_{i,j} \geq \sum_k \alpha_k x_{k,i} \quad \forall i \tag{6}$$

$$\sum_l q_{i,l} \geq \sum_k \beta_k x_{k,i} \quad \forall i \tag{7}$$

$$\sum_n W_i^n \leq 1 \quad \forall i \tag{8}$$

$$\sum_n Y_j^n \leq 1 \quad \forall j \tag{9}$$

$$\sum_n Z_l^n \leq 1 \quad \forall l \tag{10}$$

$$W_i^n, Y_j^n, Z_l^n \in \{0, 1\} \quad \forall i, j, l, n \tag{11}$$

$$x_{k,i}, u_{i,j}, q_{i,l} \geq 0 \quad \forall i, j, l, k \tag{12}$$

The objective function (1) tries to minimize the total opening costs of PHCs, clinics and hospitals. Equations (2)–(4) ensures that the total incoming flow to a given health center will not exceed the related capacity. Equations (5)–(7) guarantee that all the demands must be responded by PHCs, clinics and hospitals, respectively. Equations (9)–(11) indicate that each health center could be opened in one of the available capacity levels. Finally, Eqs. (11) and (12) enforce the binary and non-negativity restrictions on the corresponding decision variables.

The proposed model tends to set up a health network as centralized as possible, with the least number of established centers, under this condition that all the incoming demands must be responded by different levels of the network i.e. PHCs, clinics and hospitals. As a result, accessibility of the network which is among the main criteria of well-designed health networks is completely neglected in the model. Hence, in order to satisfy cost and accessibility criteria simultaneously, some new parameters are defined as follows:

tx_{ki} travel cost/time between customer zone k and PHC i
tu_{ij} travel cost/time between PHC i and clinic j
tq_{il} travel cost/time between PHC i and hospital l.

Accordingly, the former objective function is replaced by the objective function (13)

$$\min \sum_{i,n} f_i^n W_i^n + \sum_{j,n} o_j^n Y_j^n + \sum_{l,n} h_l^n Z_l^n + \sum_{k,i} tx_{k,i} x_{k,i} + \sum_{i,j} tu_{i,j} u_{i,j} + \sum_{i,l} tq_{i,l} q_{i,l} \tag{13}$$

Building long-term relationships between patients and care providers can create more positive interaction and better communication, which also can establish a feeling of empathy, understanding and trust. This relationship can be realized if the individual has an access to the same team of health-care providers over time [48, 49]. However, in the proposed model, each patient zone can be assigned partially to different PHCs and each PHC can deal with different clinics or hospitals. Hence, if one wants to assign all inhabitants in a given patient zone to exactly one PHC and also refer patients from a given PHC to only one clinic from opened clinics and just one hospital from opened hospitals, new set of decision variables must be added to the basic model as follows:

X_{ki} $\begin{cases} 1 & \text{if customer zone } k \text{ is assigned to PHC } i \\ 0 & \text{otherwise} \end{cases}$

U_{ij} $\begin{cases} 1 & \text{if PHC } i \text{ will be refer patients to clinic } j \\ 0 & \text{otherwise} \end{cases}$

Q_{il} $\begin{cases} 1 & \text{if PHC } i \text{ will be refer patients to hospital } l \\ 0 & \text{otherwise} \end{cases}$

Moreover, while keeping the objective function (13), new complementary constraints i.e. Eqs. (14)–(19) must be added to the basic set of constraints as follows:

$$x_{k,i} \leq M.X_{k,i} \quad \forall k, i \tag{14}$$

$$u_{i,j} \leq M.U_{i,j} \quad \forall i, j \tag{15}$$

$$q_{i,l} \leq M.Q_{i,l} \quad \forall i, l \tag{16}$$

$$\sum_i X_{k,i} = 1 \quad \forall k \tag{17}$$

$$\sum_j U_{i,j} \leq 1 \quad \forall i \tag{18}$$

$$\sum_{l} Q_{i,l} \leq 1 \quad \forall i \tag{19}$$

Equations (14) ensure that only a patients' flow from patient zone k to PHC i could occur if and only if this patient zone is assinged to that PHC. Similarly, Eqs. (15) and (16) guarantee there will be a patients' flow from PHC i to clinic j and from PHC i to hospital l if the concerned PHC is assigned to the corresponding clinic and hospital. In addition, Eq. (17) assures that each patient zone is assigned to exactly one PHC, while Eqs. (18) and (19) assure that each PHC will refer patients only to a single clinic and to a single hospital, respectively.

5 Accounting for Data Uncertainty in HSND

Because of the dynamic nature of the concerned health service network designproblem over the strategic horizon, all of the parameters i.e. demand parameters for all levels of the health network (PHCs, clinics and hospitals), health centers opening costs, transportation costs/times and capacities of health centers are subject to unforeseen fluctuations in the long-term and thus are tainted with a high degree of uncertainty.

According to Mula et al. [50] and Mousazadeh et al. [51], uncertaintyin data are twofold: (1) "randomness" that stem from the random nature of parameters and stochastic programming methods can be applied to handle this type of uncertainty and (2) "epistemic uncertainty" that deals with ill-known and imprecise parameters and arises from lack of knowledge about the exact value of uncertain parameters. Possibilistic programming approaches are the most applied approaches to cope with this sort of uncertainty. Although the uncertain parameters of the concerned problem involve deal both sorts of the uncertainties, because this research mainly concerns about the fuzzy programming approach, different versions of the proposed model are formulated as fuzzy mathematical models.

5.1 *Fuzzy Programming Approach*

Fuzzy mathematical programming can be categorized into two main classes [50, 52] i.e. "flexible programming" and "possibilistic programming". Flexible programming deals with flexibility in the objectives' goals and/or elasticity in soft constraints. On the other hand, possibilistic programming is used when the parameters are tainted with epistemic uncertaintyand there is lack of knowledge about exact values of input parameters due to unavailability or insufficiency of objective data. In this approach, one sets a minimum confidence level (α) as safety margin for satisfaction of each possibilistic chance constraint.

Possibility (*Pos*), necessity (*Nec*) and credibility (*Cr*) measures are the most applied fuzzy measures in possibilistic chance constrained programming models. If a decision maker (DM) adopts optimistic/pessimistic attitude towards level of occurrence of possibilistic parameters, using possibility/necessity measure is mainly suggested. Moreover, if a DMs' attitude fluctuates between optimistic and pessimistic extremes, a credibility measure (*Cr*) can be well matched for this situation.

Given a trapezoidal fuzzy variable $\tilde{\xi} = (r_1, r_2, r_3, r_4)$, $r_1 < r_2 < r_3 < r_4$ with the following membership function:

$$\mu_{(x)} = \begin{cases} \dfrac{x - r_1}{r_2 - r_1} & if \quad r_1 \leq x \leq r_2 \\ 1 & if \quad r_2 \leq x \leq r_3 \\ \dfrac{r_4 - x}{r_4 - r_3} & if \quad r_3 \leq x \leq r_4 \\ 0 & if \quad otherwise \end{cases} \tag{20}$$

As stated in Inuiguchi and Ramík [52] and Liu and Iwamura [53], for all confidence levels equal or greater than 0.5 ($\alpha \geq 0.5$), the crisp counterpart of the possibility and necessity measure are as follows:

$$Pos\left\{x \leq \tilde{\xi}\right\} \geq \alpha \Leftrightarrow \frac{r_4 - x}{r_4 - r_3} \geq \alpha \Leftrightarrow x \leq \alpha r_3 + (1 - \alpha) r_4 \tag{21}$$

$$Pos\left\{x \geq \tilde{\xi}\right\} \geq \alpha \Leftrightarrow \frac{x - r_1}{r_2 - r_1} \geq \alpha \Leftrightarrow x \geq (1 - \alpha) r_1 + \alpha r_2 \tag{22}$$

$$Nec\left\{x \leq \tilde{\xi}\right\} \geq \alpha \Leftrightarrow \frac{r_2 - x}{r_2 - r_1} \geq \alpha \Leftrightarrow x \leq \alpha r_1 + (1 - \alpha) r_2 \tag{23}$$

$$Nec\left\{x \geq \tilde{\xi}\right\} \geq \alpha \Leftrightarrow \frac{x - r_3}{r_4 - r_3} \geq \alpha \Leftrightarrow x \geq (1 - \alpha) r_3 + \alpha r_4 \tag{24}$$

In addition, the expected value of fuzzy variable $\tilde{\xi}$ with the condition that $r_1 \geq 0$, would be as follows:

$$E^{Pos}[\xi] = \int_0^{+\infty} Pos\{\xi \geq x\} dx - \int_{-\infty}^{0} Pos\{\xi \leq x\} dx = \frac{r_3 + r_4}{2} \tag{25}$$

$$E^{Nec}[\xi] = \int_0^{+\infty} Nec\{\xi \geq x\} dx - \int_{-\infty}^{0} Nec\{\xi \leq x\} dx = \frac{r_1 + r_2}{2} \tag{26}$$

Thereafter, Liu and Liu [54] introduced the credibility measure in order to support compromise attitude of a DM over both extremes, such that:

$$Cr\{A\} = \frac{1}{2}(Pos\{A\} + Nec\{A\}) \tag{27}$$

Similarly, for all confidence levels equal or greater than 0.5 ($\alpha \geq 0.5$), the crisp counterparts of the credibility measure are as follows:

$$Cr\left\{x \leq \tilde{\xi}\right\} \geq \alpha \Leftrightarrow \frac{2r_2 - r_1 - x}{2(r_2 - r_1)} \geq \alpha \Leftrightarrow x \leq (2\alpha - 1)r_1 + (2 - 2\alpha)r_2 \tag{28}$$

$$Cr\left\{x \geq \tilde{\xi}\right\} \geq \alpha \Leftrightarrow \frac{x - 2r_3 + r_4}{2(r_4 - r_3)} \geq \alpha \Leftrightarrow x \geq (2 - 2\alpha)r_3 + (2\alpha - 1)r_4 \tag{29}$$

and for $r_1 \geq 0$, the expected value of the fuzzy variable $\tilde{\xi}$ using credibility measure is as follows:

$$E^{Cr}[\xi] = \int_0^{+\infty} Cr\{\xi \geq x\}dx - \int_{-\infty}^{0} Cr\{\xi \leq x\}dx = \frac{r_1 + r_2 + r_3 + r_4}{4} \tag{30}$$

5.2 *Fuzzy Counterparts of the Concerned HSND Problem*

The compact form of the concerned HSND problem can be articulated as below:

$$\begin{aligned} & \min\ z = F.Y + C.X \\ & s.t: \\ & X \leq N.Y \\ & A.X \geq D \\ & T.Y \leq 1 \\ & X \leq M.Y \\ & P.Y \leq 1 \\ & Y \in \{0,1\}, X \geq 0 \end{aligned} \tag{31}$$

in which the vectors F and C and matrices N and D address fixed opening costs, travel costs, capacity of centers and demand parameters, respectively. Moreover, matrices A, T and P are coefficient matrices of the constraints while M is a large number. Finally, variable Y refers to location decisions and variable X addresses patients' flow throughout the whole network.

Now assume that parameters F, C, N and D are tainted with epistemic uncertainty. Accordingly, the basic possibilistic chance constraint programming (BPCCP) model is as follows:

$$\begin{aligned}
&\min E[z] = E[\tilde{F}].Y + E[\tilde{C}].X \\
&s.t: \\
&Ch\{X \leq \tilde{N}.Y\} \geq \alpha \\
&Ch\{A.X \geq \tilde{D}\} \geq \beta \\
&T.Y \leq 1 \\
&X \leq M.Y \\
&P.Y \leq 1 \\
&Y \in \{0,1\}, X \geq 0
\end{aligned} \tag{32}$$

Taking into account the possibility measure as a chance constraint transformation approach, the crisp counterpart of the BPCCP model can be stated as follows:

$$\begin{aligned}
&\min E[z] = \left(\frac{F_{(3)} + F_{(4)}}{2}\right).Y + \left(\frac{C_{(3)} + C_{(4)}}{2}\right).X \\
&s.t: \\
&x \leq [\alpha N_{(3)} + (1-\alpha)N_{(4)}].Y \\
&A.X \geq (1-\alpha)D_{(1)} + \alpha D_{(2)} \\
&T.Y \leq 1 \\
&X \leq M.Y \\
&P.Y \leq 1 \\
&Y \in \{0,1\}, X \geq 0
\end{aligned} \tag{33}$$

Also, the crisp counterpart of the BPCCP model under the necessity measure is as follows:

$$\begin{aligned}
&\min E[z] = \left(\frac{F_{(1)} + F_{(2)}}{2}\right).Y + \left(\frac{C_{(1)} + C_{(2)}}{2}\right).X \\
&s.t: \\
&x \leq [\alpha N_{(1)} + (1-\alpha)N_{(2)}].Y \\
&A.X \geq (1-\alpha)D_{(3)} + \alpha D_{(4)} \\
&T.Y \leq 1 \\
&X \leq M.Y \\
&P.Y \leq 1 \\
&Y \in \{0,1\}, X \geq 0
\end{aligned} \tag{34}$$

And finally, the crisp counterpart of the BPCCP model under the credibility measure would be as follows:

$$\min E[z] = \left(\frac{F_{(1)} + F_{(2)} + F_{(3)} + F_{(4)}}{4}\right).Y + \left(\frac{C_{(1)} + C_{(2)} + C_{(3)} + C_{(4)}}{4}\right).X$$

$$\begin{aligned}
&s.t: \\
&x \leq \left[(2\alpha - 1)N_{(1)} + (2 - 2\alpha)N_{(2)}\right].Y \\
&A.X \geq (2 - 2\alpha)D_{(3)} + (2\alpha - 1)D_{(4)} \\
&T.Y \leq 1 \\
&X \leq M.Y \\
&P.Y \leq 1 \\
&Y \in \{0, 1\}, X \geq 0
\end{aligned} \tag{35}$$

6 Case Studies

As was mentioned in the "literature review" section earlier, there is not any research focusing on mathematical modeling of HSND problem under epistemic uncertaintyof input data. However, the work by Shishebori and Babadi [9] can be regarded as the closest article to the concerned problem of this paper, in which a mixed integer programming model (MIP) for a reliable medical service network designunder uncertainty is proposed. Nevertheless, the main difference is that they used robust optimization approach to protect the network against uncertainty and also just focused on medical service rather than the whole health service including preventive and non-medical centers. Thus, in this part, a proposed case study provided in Shishebori and Babadi [9] is briefly reviewed to show a practical application of HSND mathematical modeling under uncertain environment.

The main goal of the proposed model in Shishebori and Babadi [9] is to improve accessibility of urban residents to MS centers in Chaharmahal-Bakhtiari, one of the 31 provinces in Islamic Republic of Iran. Since, the healthcare system conditions have a significant effect on promotion of province development indicators and improvement of public health welfare, the stated aim of the proposed model is to provide the most robust and reliable medical network design/plan to promote the medical system of the province regarding to the limited allocated budget.

In order to optimally design the medical service network in this study, four main decision variables are taken into account i.e. (1) the optimum locations of new MS centers, (2) the transfer links that should be constructed/improved in the existing network, (3) the amount of demands of nodes that should be transferred by links, and (4) the fraction of every demand that should be supplied by new and exciting medical service (MS) centers (i.e., clinics). These decisions must be made with respect to minimization of total costs, including fixed facility opening costs, links' construction/improvement costs and the transfer costs between nodes.

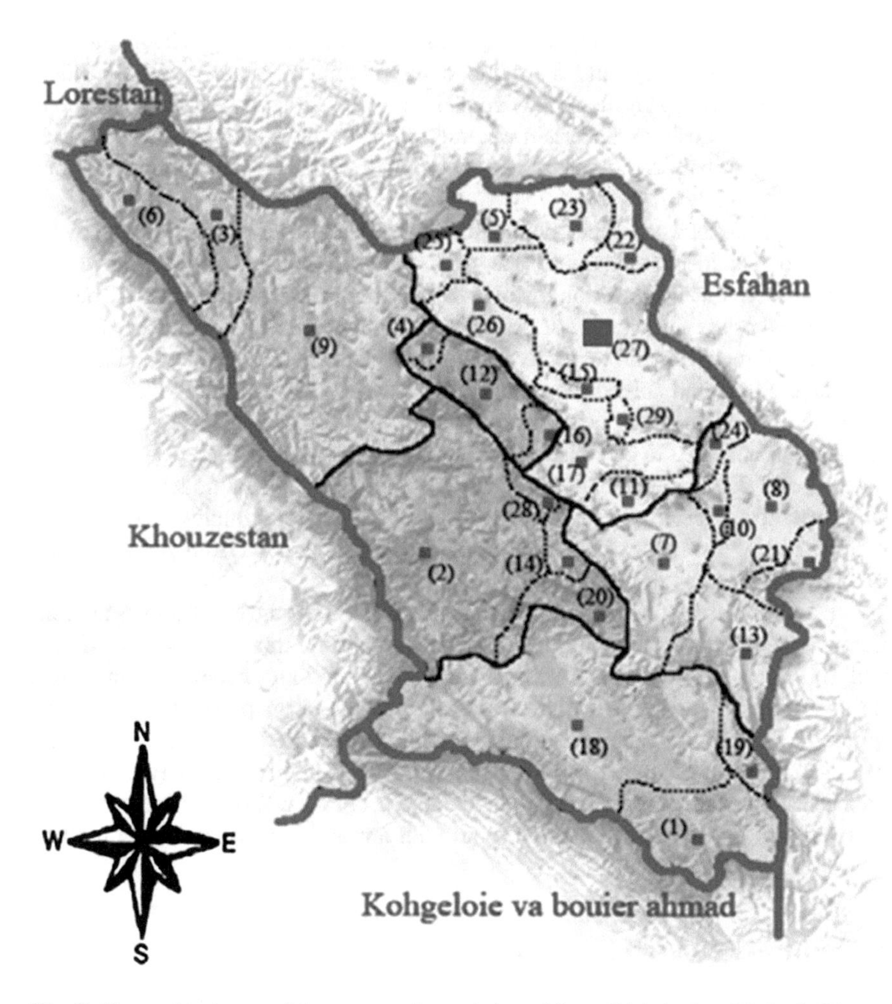

Fig. 5 Geographical map of the concerned area (adopted from Shishebori and Babadi [9])

Figure 5 clearly depicts the geographical map of Chaharmahal-Bakhtiari province, which consists of 29 nodes (towns) with a total population of 895,263. Each town is a client node whose demand equals its patients. They stated that there is two active MS centers which are located in the nodes 12 (Farsan) and 27 (Shahrekord), the latter is known as a reliable MS center. Since other 27 towns (nodes) do not have any healthcare service centers, they can be considered as potential candidate locations to open new MS centers. However, because of some reasons ranging from bad weather conditions, delay in drug supply and lack of specialists to the occurrence of natural disasters such as floods and earthquakes, all of them are unreliable. Moreover, there are 29 existing and 22 potential roads which can be classified into four categories in terms of quality: high, good, medium and low and constructing/improving costs of these links highly depends on the quality status of the roads.

In addition, since the changing conditions (e.g. climate change, fluctuations in stock market prices, etc.) will lead to uncertainty in input parameters of the medical service network designproblem, they consider the demands for medical service and the transfer unit costs as uncertain parameters. They then defined four scenarios with different probability of occurrence including excellent, good, medium and bad situations.

They also defined an upper bound for investment in medical service network construction/expansion activities, due to the budget constraints of Ministry of Health and Medical Education (MoHME) and Ministry of Road and Transfer (MoRT), who are in charge of investment in MS centers and road network construction/improvement, respectively. They also set the predetermined number of new centers as 3 and solved the model using CPLEX solver in GAMS.

In the optimal solution, new MS centers were established at nodes 2, 8 and 16 and also 6 roads must to be constructed between nodes 15 and 27, nodes 15 and 17, nodes 2 and 20, nodes 7 and 18, nodes 10 and 11, and also nodes 18 and 19. Moreover, it has been reported that the quality of 8 "good quality", 9 "medium quality" and 8 "low quality" existing roads should be improved from good, medium and low quality to high quality, respectively.

Noteworthy, they finally declared that, if all the existing MS centers, all candidate locations for establishing new MS centers and all transfer links are considered 100 % reliable and also neglect uncertaintyof the input parameters and solve the deterministic model with the mean value of uncertain parameters, then, the optimal value of the objective function will be increased only 3.025 % (from 401,453.231 to 413,598.511 Monetary Unit) which is highly impressive. This fact emphasizes that taking real practical factors into accounts can substantially improve the efficiency of the achieved solution but with a modest growth at the total costs.

7 Future Research Directions

Health service network design (HSND) as one of the critical strategic decisions can affects the performance of health systems to the great extent. The main determinants of HSND problem are location of different healthcare facilities, allocation of regions to new established or existing facilities, flow of patients throughout the network. In this research, after explaining health, health system and HSND problem briefly, a comprehensive review of the related literature to the HSND problem is provided, which is followed by a typical mathematical model for the concerned problem. Finally, fuzzy programming approach is described in brief and different fuzzy measures i.e. possibility, necessity and credibility measures are applied to the compact form of the proposed mathematical programming model.

With respect to the current state-of-the-art literature in HSND problem, there are various avenues for further research i.e. (i) addressing various healthcare facilities/services via a single HSND problem, (ii) considering input parameters as uncertain data and using a mixture of uncertaintyprogramming approaches like the fuzzy programming, stochastic programming and robust optimization approaches in order to handle different types of uncertainties in data, (iii) developing heuristic, metaheuristic and exact algorithms to solve the HSND problem in large-scale problems.

References

1. WHO: The World Health Report 2000: Health Systems: Improving Performance. WHO (2000)
2. Porter, M.E., Teisberg, E.O.: Redefining Health Care: Creating Value-Based Competition on Results. Harvard Business Press (2006)
3. Brandeau, M.L., Sainfort, F., Pierskalla, W.P.: Operations Research and Health Care: A Handbook of Methods and Applications, vol. 70. Springer Science & Business Media (2004)
4. De Vries, G., Bertrand, J., Vissers, J.: Design requirements for health care production control systems. Prod. Plann. Control **10**(6), 559–569 (1999)
5. Brailsford, S., Vissers, J.: OR in healthcare: a European perspective. Eur. J. Oper. Res. **212**(2), 223–234 (2011)
6. Syam, S.S., Côté, M.J.: A comprehensive location-allocation method for specialized healthcare services. Oper. Res. Health Care (2012)
7. Shariff, S.R., Moin, N.H., Omar, M.: Location allocation modeling for healthcare facility planning in Malaysia. Comput. Ind. Eng. **62**(4), 1000–1010 (2012)
8. Syam, S.S., Côté, M.J.: A location–allocation model for service providers with application to not-for-profit health care organizations. Omega **38**(3), 157–166 (2010)
9. Shishebori, D., Babadi, A.Y.: Robust and reliable medical services network design under uncertain environment and system disruptions. Transp. Res. Part E Logistics Transp. Rev. **77**, 268–288 (2015)
10. Lamiri, M., et al.: A stochastic model for operating room planning with elective and emergency demand for surgery. Eur. J. Oper. Res. **185**(3), 1026–1037 (2008)
11. Chien, C.-F., Tseng, F.-P., Chen, C.-H.: An evolutionary approach to rehabilitation patient scheduling: a case study. Eur. J. Oper. Res. **189**(3), 1234–1253 (2008)
12. Burke, E.K., et al.: A scatter search methodology for the nurse rostering problem. J. Oper. Res. Soc. **61**(11), 1667–1679 (2010)
13. Rais, A., Viana, A.: Operations research in healthcare: a survey. Int. Trans. Oper. Res. **18**(1), 1–31 (2011)
14. Hulshof, P.J., et al.: Taxonomic classification of planning decisions in health care: a structured review of the state of the art in OR/MS. Health syst. **1**(2), 129–175 (2012)
15. Xing, Y., et al.: Operations research (OR) in service industries: a comprehensive review. Syst. Res. Behav. Sci. **30**(3), 300–353 (2013)
16. WHO: World Health Report 2008. Primary Health Care, Now More than Ever. WHO, Geneva (2008)
17. Mousazadeh, M., Torabi, S., Zahiri, B.: A robust possibilistic programming approach for pharmaceutical supply chain network design. Comput. Chem. Eng. (2015)
18. Zahiri, B., et al.: Blood collection management: a robust possibilistic programming approach. Appl. Math. Model. (2015)
19. Zahiri, B., Tavakkoli-Moghaddam, R., Pishvaee, M.S.: A robust possibilistic programming approach to multi-period location–allocation of organ transplant centers under uncertainty. Comput. Ind. Eng. **74**, 139–148 (2014)
20. Pishvaee, M.S., Farahani, R.Z., Dullaert, W.: A memetic algorithm for bi-objective integrated forward/reverse logistics network design. Comput. Oper. Res. **37**(6), 1100–1112 (2010)

21. Melkote, S., Daskin, M.S.: An integrated model of facility location and transportation network design. Transp. Res. Part A Policy Pract. **35**(6), 515–538 (2001)
22. Miranda, P.A., Garrido, R.A.: Incorporating inventory control decisions into a strategic distribution network design model with stochastic demand. Transp. Res. Part E Logistics Transp. Rev. **40**(3), 183–207 (2004)
23. Amiri, A.: Designing a distribution network in a supply chain system: formulation and efficient solution procedure. Eur. J. Oper. Res. **171**(2), 567–576 (2006)
24. Lin, L., Gen, M., Wang, X.: Integrated multistage logistics network design by using hybrid evolutionary algorithm. Comput. Ind. Eng. **56**(3), 854–873 (2009)
25. Pishvaee, M., Torabi, S., Razmi, J.: Credibility-based fuzzy mathematical programming model for green logistics design under uncertainty. Comput. Ind. Eng. **62**, 624–632 (2012)
26. Marmot, M., et al.: Closing the gap in a generation: health equity through action on the social determinants of health. Lancet **372**(9650), 1661–1669 (2008)
27. Saltman, R., Rico, A., Boerma, W.: Primary Care in the Driver's Seat: Organizational Reform in European Primary Care. McGraw-Hill International (2006)
28. Nutting, P.A.: Population-based family practice: the next challenge of primary care. J. Fam. Pract. **24**(1), 83 (1987)
29. Criel, B., De Brouwere, V., Dugas, S.: Integration of vertical programmes in multi-function health services. ITG Press Antwerp, Belgium (1997)
30. Dökmeci, V.F.: A quantitative model to plan regional health facility systems. Manage. Sci. **24** (4), 411–419 (1977)
31. Schweikhart, S.B., Smith-Daniels, V.L.: Location and service mix decisions for a managed health care network. Socio-Econ. Plan. Sci. **27**(4), 289–302 (1993)
32. Galvao, R.D., Espejo, L.G.A., Boffey, B.: A hierarchical model for the location of perinatal facilities in the municipality of Rio de Janeiro. Eur. J. Oper. Res. **138**(3), 495–517 (2002)
33. Marianov, V., Serra, D.: Location–allocation of multiple-server service centers with constrained queues or waiting times. Ann. Oper. Res. **111**(1–4), 35–50 (2002)
34. Marianov, V., Serra, D.: Hierarchical location–allocation models for congested systems. Eur. J. Oper. Res. **135**(1), 195–208 (2001)
35. Verter, V., Lapierre, S.D.: Location of preventive health care facilities. Ann. Oper. Res. **110** (1–4), 123–132 (2002)
36. Galvão, R.D., et al.: Load balancing and capacity constraints in a hierarchical location model. Eur. J. Oper. Res. **172**(2), 631–646 (2006)
37. Yasenovskiy, V., Hodgson, J.: Hierarchical location-allocation with spatial choice interaction modeling. Ann. Assoc. Am. Geogr. **97**(3), 496–511 (2007)
38. Griffin, P.M., Scherrer, C.R., Swann, J.L.: Optimization of community health center locations and service offerings with statistical need estimation. IIE Trans. **40**(9), 880–892 (2008)
39. Ndiaye, M., Alfares, H.: Modeling health care facility location for moving population groups. Comput. Oper. Res. **35**(7), 2154–2161 (2008)
40. Smith, H.K., et al.: Planning sustainable community health schemes in rural areas of developing countries. Eur. J. Oper. Res. **193**(3), 768–777 (2009)
41. Zhang, Y., Berman, O., Verter, V.: Incorporating congestion in preventive healthcare facility network design. Eur. J. Oper. Res. **198**(3), 922–935 (2009)
42. Mahar, S., Bretthauer, K.M., Salzarulo, P.A.: Locating specialized service capacity in a multi-hospital network. Eur. J. Oper. Res. **212**(3), 596–605 (2011)
43. Fo, A.R.A.V., da Silva Mota, I.: Optimization models in the location of healthcare facilities: a real case in Brazil. J. Appl. Oper. Res. **4**(1), 37–50 (2012)
44. Mestre, A.M., Oliveira, M.D., Barbosa-Póvoa, A.: Organizing hospitals into networks: a hierarchical and multiservice model to define location, supply and referrals in planned hospital systems. OR Spectrum **34**(2), 319–348 (2012)
45. Zhang, Y., Berman, O., Verter, V.: The impact of client choice on preventive healthcare facility network design. OR Spectrum **34**(2), 349–370 (2012)
46. Benneyan, J.C., et al.: Specialty care single and multi-period location–allocation models within the Veterans Health Administration. Socio-Econ. Plann. Sci. **46**(2), 136–148 (2012)

47. Song, B.D., Ko, Y.D., Hwang, H.: The design of capacitated facility networks for long term care service. Comput. Industr. Eng. (2015)
48. Naithani, S., Gulliford, M., Morgan, M.: Patients' perceptions and experiences of 'continuity of care' in diabetes. Health Expect. **9**(2), 118–129 (2006)
49. Beach, M.C., et al.: Are physicians' attitudes of respect accurately perceived by patients and associated with more positive communication behaviors? Patient Educ. Couns. **62**(3), 347–354 (2006)
50. Mula, J., Poler, R., Garcia, J.: MRP with flexible constraints: a fuzzy mathematical programming approach. Fuzzy Sets Syst. **157**(1), 74–97 (2006)
51. Mousazadeh, M., Torabi, S.A., Pishvaee M.S.: Green and reverse logistics management under fuzziness. In: Supply Chain Management Under Fuzziness, pp. 607–637. Springer (2014)
52. Inuiguchi, M., Ramík, J.: Possibilistic linear programming: a brief review of fuzzy mathematical programming and a comparison with stochastic programming in portfolio selection problem. Fuzzy Sets Syst. **111**(1), 3–28 (2000)
53. Liu, B., Iwamura, K.: Chance constrained programming with fuzzy parameters. Fuzzy Sets Syst. **94**(2), 227–237 (1998)
54. Liu, B., Liu, Y.-K.: Expected value of fuzzy variable and fuzzy expected value models. Fuzzy Syst. IEEE Trans. **10**(4), 445–450 (2002)

Robotics and Control Systems

M.H. Fazel Zarandi and H. Mosadegh

Abstract Robots are of those intelligent systems created to do a wide range of activities with the aim of human aid and productivity improvement. Besides, many different fields of studies such as engineering, healthcare, computer science, mathematics and management are involved in order to increase the efficiency and effectiveness of robots. Generally speaking, robotics and control systems is a branch of engineering science that deals with all aspects of robot's design, operation and control. More precisely, the concept of control in this paper is knowing the techniques required for programming robot's activities such as its physical movements, rotations, decisions and planning. In addition to mathematical modeling optimization and scheduling, there are a lot of control theory based approaches dealing with physical movement control of the robot at every moment of time. Due to the uncertainties, fuzzy set theory, applicable for all control techniques, is extensively used for robots. The role of fuzzy modeling becomes more evident when one can include human expertise and knowledge via fuzzy rules in the control system. Without loss of generality, this paper presents fuzzy control techniques as well as fuzzy mathematical scheduling model for an m-machine robotic cell with one manipulator robot. Furthermore, it proposes an integrated fuzzy robotic control system, in which the fuzzy optimization model is solved at every predetermined period of time such as beginning of shifts or days, etc. Then, based on the solutions obtained, input parameters and unpredictable disturbances, the autonomous fuzzy control is executed continuously. These two modules transfer information and feedback to each other via an intermediate collaborative module. The explanations are supported via an example.

Keywords Robotic · Control · Fuzzy · Optimization · Rule-based

M.H. Fazel Zarandi (✉) · H. Mosadegh
Department of Industrial Engineering and Management Systems,
Amirkabir University of Technology, 424 Hafez Avenue, 1591634311 Tehran, Iran
e-mail: zarandi@aut.ac.ir

C. Kahraman et al. (eds.), *Fuzzy Logic in Its 50th Year*,
Studies in Fuzziness and Soft Computing 341,
DOI 10.1007/978-3-319-31093-0_13

1 Introduction

Human knowledge is increasingly growing and the details of surrounding environment are discovered fast. Robots are of those human-made instruments with the goal of aiding in a lot of life and working activities. Todays, many types of robotsare designed and built in different areas like healthcare and surgery, manufacturing, human-aid, agriculture, urban search and rescue, research and other engineering applications [1–7]. One of the most important aspects that robots have in common is that they are all programmable and can be designed and controlled according to their capabilities and limitations. Without loss of generality, this paper will focus on manufacturing manipulator robots, namely robotic cell.

Robots are intelligent agents that are guided by computer programs to do some activities automatically. As well as electrical circuitry, they are equipped with mechanical mechanism and facilities. Robotics is a branch of engineering science that deals with all aspects of any kind of robot including its design, operation and control. The physical details and kinematics of a robot may be found in [8]. Additionally, all aspects of robot intelligence as well as techniques and features are described and studied by [9] and [10]. The expansion of robot studies is growing fast so that for the manipulator of robot, also known as manipulator robot, is becoming a branch of technology and business [11, 12].

One of the most important areas in the field of robotics is investigating the approaches for controlling a robot. Autonomous or semi-autonomous control of a robot has been studied by many researches and engineers, but there are a lot of problems remained unsolved. As the concept of control can be interpreted as the decision made for the robot action, there are many books and papers dealing with robotic control systems. In fact, the control could be categorized into static and dynamic control. In the static approach, an optimization problem is defined and solved at the beginning of predetermined periods e.g., shifts, days etc. Many researchers have focused on optimization and scheduling of robot activities such as [13–22]. However, to the best of authors' knowledge, some aspects of real problems like an *m*-machine robotic cell have not been considered yet.

On the basis of mechanical and electrical engineering methods, most of dynamic activities of the robot can be predicted and studied. Therefore, by the means of control theory [23], robot's actions are managed and controlled at each unit of time. However, due to the uncertainties, there might be some uncontrollable parameters by which disturbances may occur. In this regard, fuzzy set theory is the most superior technique that is employed to deal with this issue [24–29]. In addition to coping with uncertainties, fuzzy rules are applicable for those aspects of the robot which cannot be modeled mathematically. Technically, some problems are not handled with scientific methods, but human expertise and knowledge will help in many practical situations. By the use of fuzzy logic, human experience can be modelled as a knowledge base for controlling issues [30, 31]. This approach is the basis of the most of conventional fuzzy control techniques.

The rest of the paper is organized as follows. Section 2 describes the dynamic and static modelling of a robot on the basis of control theory, mathematical modeling and expert system approaches. The concept of control theory, fuzzy control theory and one supporting example of robotic system is provided in Sect. 3. Section 4 proposes a robotic control system including combination of fuzzy optimization models as well as fuzzy controllers. Finally, Sect. 5 concludes the paper.

2 Dynamics and Modeling

2.1 State Space Modeling

Modeling is a systematic process that use some information about the real entity which is modeled. This information come from many sources of knowledge and data. In some cases, where data is not available but there is enough knowledge to describe its features and behavior, the modeling approach is called state space modeling. Before description of the modeling, a definition of the system is required which is provided as follows.

2.1.1 System

Understanding of a system is an intuitive concept, but some different definitions are provided. As a simple statement, a system is a combination of interacting items that work together to achieve a goal, not reachable by the absence of at least one of those item. Regarding many system analysis approaches, engineers prefer to measure quantitative aspects of a system and control assessable outputs as much as possible. More precisely, three concepts each system should be clearly identified: boundary of the system, inputs and outputs. As another definition of the system, it could be stated as an aggregation of components which transfer the inputs to the outputs.

Figure 1 depicts a system with its model. As the boundary of the model becomes wider the complexity of the model increases. However, more details of the system can be considered and hence, the accuracy of the output results will improve. Normally, the boundary of the model is established with assumptions. As a result, the more simplifying assumptions the more unreliable output results.

2.1.2 Dynamics of Robot Motions

Using kinematics and energy relations of physical body of the robot, its dynamics could be formulated, typically as differential equations systems. As a system point of view, the robot has two parameters, i.e., inputs and outputs as well as one

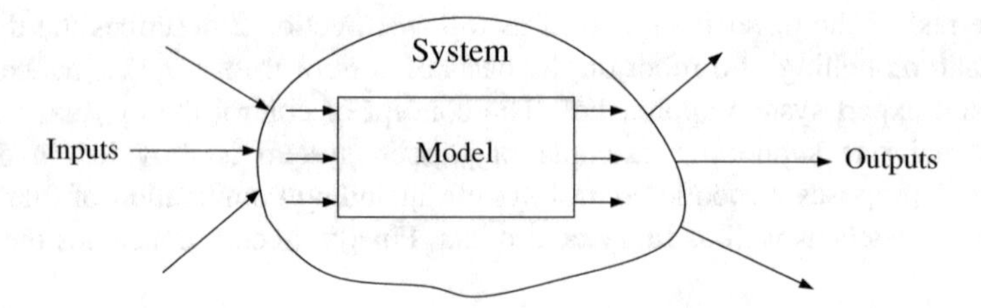

Fig. 1 A system with boundary, inputs and outputs. The model of the system has much narrower boundary than the real system. Inputs of the model may differ from the real inputs of the system. Normally, the boundary and inputs of the model are defined with establishing assumptions about the systems. Consequently, the outputs of the model may differ from the outputs of the real system. Each model would be valid for its relative system as much as its outputs approach to the system's outputs

variable, namely state. Because they are all measurable, it is convenient to use some characters for their formulations. For this purpose, let u be the vector of inputs, x be the state of the system and y be the vector of outputs. Equations (1) and (2) represent the state space modeling of the robot.

$$\dot{x} = Ax + Bu \quad (1)$$

$$y = Cx + Du \quad (2)$$

where A, B, C and D are coefficient matrices of the model. In fact, the state of the robot is one or more quantitative variables which continuously changes when the robot moves. As a simple example, illustrated in Fig. 2, one can assume a robot with two links for which one is fixed and the other one is oscillating like a pendulum. For simplification, the weight of handing arm is assumed to be negligible, but it carries a manufacturing part with mass m.

The state variable of the robot is its angle between the oscillating link and the vertical line, i.e., θ. After some calculations, the differential equations of the state space modeling of the robot can be written as Eq. (3).

$$ml^2\ddot{\theta} + mgl \sin \theta = 0 \quad (3)$$

Dynamics of a robot with only one oscillating link can be stated as Eq. (3). However, by simultaneously moving both links, the state and differential equations of the robot will significantly change and therefore its dynamic formulations become more complicated. The modeling issue goes worst when the mass of each link is included. Figure 3 represents a robot for which both links are able to move in different directions of a plane. In this case, the robot is assumed to be free of load. m_1 and m_2 are germs of the main and the handling links of the robot, respectively.

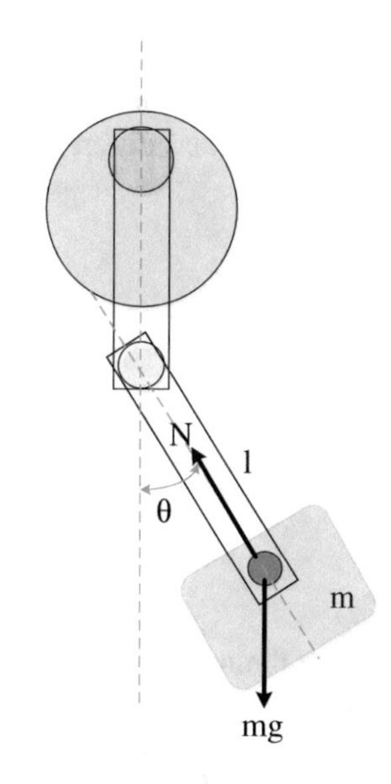

Fig. 2 A robot with two arms, say links. The main link is fixed and the other link with manipulator can oscillate. Both links are connected with joints. The robot movements is assumed to occur in a two-dimensional space

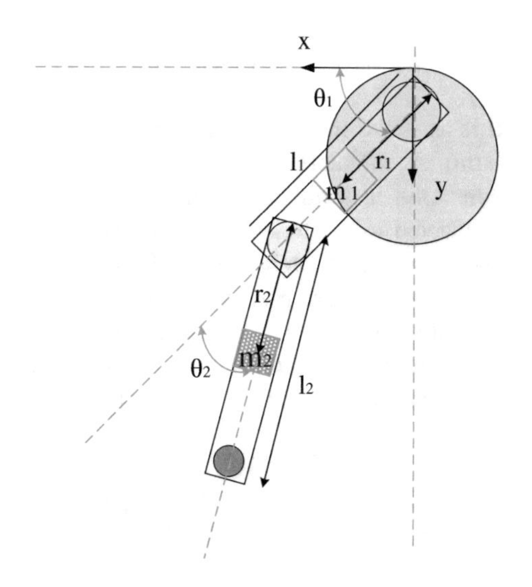

Fig. 3 A two-link planar robot. Both links can oscillate independently in different directions

Normally, Lagrange's equations are applied for developing differential equations of robots' movements. For this purpose, assume each link with mass of m_i and inertia tensor moment of z, I_{z_i}. Furthermore, let τ_1 be the applied torque for the acceleration of the joints and τ_2 be the applied torque for the Coriolis and centrifugal forces. After some calculations, following results are obtained. For convenience, some terms, written in the first three equations, i.e., Eqs. (4)–(6), are represented as parameters α, β and δ. The last two equations, i.e., Eqs. (7) and (8), are the resulted differential state space models for the two-link planar robot.

$$\alpha = I_{z_1} + I_{z_2} + m_1 r_1^2 + m_2 \left(l_1^2 + r_2^2\right) \tag{4}$$

$$\beta = m_2 l_1 r_2 \tag{5}$$

$$\delta = I_{z_2} + m_2 r_2^2 \tag{6}$$

$$(\alpha + 2\beta c_2)\ddot{\theta}_1 + (\delta + \beta c_2)\ddot{\theta}_2 - \beta s_2 \dot{\theta}_2 \dot{\theta}_1 - \beta s_2 \left(\dot{\theta}_1 + \dot{\theta}_2\right)\dot{\theta}_2 = \tau_1 \tag{7}$$

$$(\delta + \beta c_2)\ddot{\theta}_1 + \delta\ddot{\theta}_2 + \beta s_2 \dot{\theta}_1^2 = \tau_2 \tag{8}$$

By the use of above differential equations, behavior of the robot can be studied based on the changing parameters. However, in real cases, there are many other aspects which should be considered in modeling and investigations. The next section presents dynamics of a general manipulator robot.

2.1.3 Dynamics of a General Manipulator Robot

It is assumed that a general robot has a tree structure containing a fixed base link (arm), N movable links that are connected with N joints. The robot links are numbered in sequence from the base to the hand. In other word, the robot link connected to the base link is numbered as 1, and its immediate next link is numbered as 2, etc. Each robot link is considered as the parent of its immediate following link. More precisely, the ith arm is the parent of the $(i + 1)$th arm.

Unlike aforementioned planar cases, a 3-dimensional state space is considered here. The state variables of the robot are its angular and linear values of velocity, acceleration and force. Dynamics of the general robot is written as Eqs. (9)–(11) [32].

v_i	Velocity of link i
w_i	The motion freedom matrix of joint i
$\dot{k}_i$	Velocity of joint i
a_i	Acceleration of link i
$\ddot{k}_i$	Acceleration of joint i
I_i	Inertia of link i
F_i	Sum of all forces acting on link i

$$v_i = v_{i-1} + w_i \dot{k}_i \tag{9}$$

$$a_i = a_{i-1} + \dot{w}_i \dot{k}_i + w_i \ddot{k}_i \tag{10}$$

$$F_i = I_i a_i + v_i \times I_i v_i \tag{11}$$

The state space modeling of the general robot is represented as differential equations for each moving arm *i*. For detail information of robot dynamics one can refer to [33]. In the next sections, it is described that how these equations are used to achieve control of the robot at each unit of time.

2.2 Mathematical Modeling

Another approach of controlling the robot is developing mathematical optimization models. In this procedure the nature of modeling completely differs from the one that considers dynamics of the robot. In fact, the mathematical model is a static model that is solved in some points of times, e.g. at the beginning of shifts. Figure 4 shows a robotic cell for loading/unloading manufacturing parts on/from *m* machines working in the cell. In what follows, a mathematical optimization model for an *m*-machine robotic cell scheduling problem with sequence-dependent setup times is developed. The details of modeling can be found in [20].

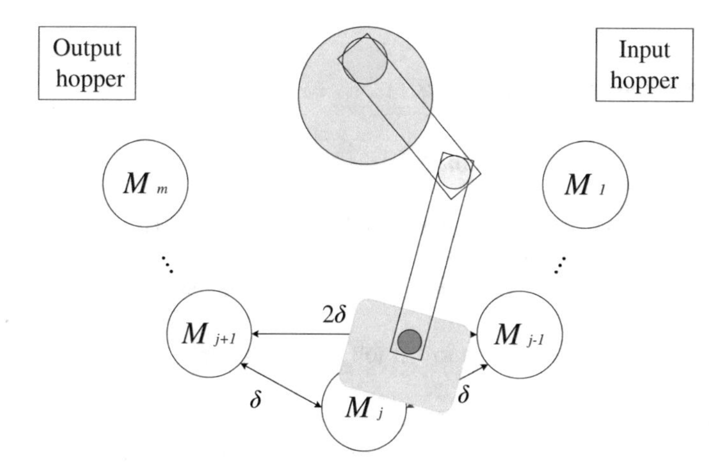

Fig. 4 An *m*-machine robotic cell. Manufacturing parts, say jobs, are entered into the cell through the Input hopper. All of the manufacturing parts are processed from machine M_1 to machine M_m sequentially

2.2.1 Nomenclature

MPS the minimal part set,
T set of part types to be manufactured,
M set of machines including the Input and Output hoppers with $m = 0, 1, \ldots, |M|$,
d_t demand of the *t*th part type with $t = 1, 2, \ldots, |T|$,
r_t minimum ratio of the *t*th part type with $t = 1, 2, \ldots, |T|$,
J set of jobs, i.e., the Minimal Part Set (**MPS**), where $|J| = |\mathbf{MPS}| = r_1 + r_2 + \cdots + r_{|T|}$,
P_j^m processing time of the *j*th job on the *m*th machine, where $j = 1, 2, \ldots, |J|$ and $m = 0, 1, \ldots, |M|$,
K set of steps with $k = 1, 2, \ldots, |J| \times |M|$,
$\varepsilon_{ij}^{m(l)}$ loading time of part *j* on machine *m*, scheduled after part *i*, where $i = 1, 2, \ldots, |J|$, $j = 1, 2, \ldots, |J|$ and $m = 0, 1, \ldots, |M|$,
$\varepsilon_j^{m(u)}$ unloading time of part *j* from machine *m*, where $j = 1, 2, \ldots, |J|$, $m = 0, 1, \ldots, |M|$,
δ traveling time of robot between two consecutive machines,
Ψ a very large positive number.

It should be mentioned that symbol $|\,.\,|$ might have two meanings. If it contains a set like *J* or *M* it will return the cardinality of the set. However, containing a mathematical term stands for the absolute value of that term.

2.3 *Mathematical Modeling*

2.3.1 Parameters and Variables

Parameters were previously introduced in the nomenclature. The decision variables of the mathematical model are presented as follows.

x_j^{km} binary variable that equals to 1 if job *j* at the *k*th step is loaded on machine *m* or 0 otherwise,
y_{ij} binary variable that equals to 1 if job *j* is immediately scheduled after job *i* or 0 otherwise,
l_j^{km} loading time of the *j*th job on the *m*th machine at step *k*,
ct cycle time, the time between loading a job on a machine and loading that job again on the machine after on cycle completed,
$l_{\max}$ the maximum loading time before the last loading

There are some auxiliary variables used for linearizing the model, introduced as follows. The supportive description about these variables are mentioned later.

r_k number of machines between departure and destination of the robot at the kth step when robot moves forward that means the departure index is fewer than the destination index

s_k number of machines between departure and destination of the robot at the kth step when robot moves backward in which departure index is larger than the destination index

The constraint to be formulated for the mathematical model are categorized as follows.

2.3.2 Initialization

In the cyclic modeling, the starter job should be appeared at the end of the cycle again, or at the beginning of the next cycle. Without loss of generality and in order to prevent enlarging the model as much as possible, it is recommended here to fix some variables, namely initializers. There is no difference in fixing variables, but it should be noticed that the initializers should be traced for some constraints explained later. As a general formulation, *initialization* is done through Eqs. (12) and (13).

$$x_1^{11} = 1 \tag{12}$$

$$l_1^{11} = 0 \tag{13}$$

Equation (12) states that job 1 id the starter of the cycle that is loaded on machine 1 at the first step. The following sub-sections will declare that l_1^{11} could be fixed at any value, however it has been assigned equal to zero in Eq. (13).

2.3.3 Part Sequence Constraints

The easiest group of constraints are the sequential restrictions in which each job should be scheduled after and before exactly one job in the **MPS**.

$$\sum_{i=1,\, i\neq j}^{|J|} y_{ij} = 1 \quad j = 1, 2, \ldots, |J| \tag{14}$$

$$\sum_{j=1, j \neq i}^{|J|} y_{ij} = 1 \quad i = 1, 2, \ldots, |J| \tag{15}$$

$$y_{ij} + y_{ji} \leq 1 \quad i = 1, 2, \ldots, |J|, \quad j = 1, 2, \ldots, |J| \tag{16}$$

Equation (14) ensures that only one job precedes job j. Similarly, in Eq. (15) there is exactly one job that follows job i. Inequality (16) indicates that if job i precedes job j, it cannot appear after it in the cycle and vice versa.

2.3.4 Robot Sequence Constraints

It is obvious that at each step exactly one part should be loaded on a machine. This restriction is considered in Eq. (17).

$$\sum_{j=1}^{|J|} \sum_{m=1}^{|M|} x_j^{km} = 1 \quad k = 1, 2, \ldots, |J| \times |M| \tag{17}$$

Each part like j is loaded on each machine like m in only one step. Then the following equation should be considered.

$$\sum_{k=1}^{|J| \times |M|} x_j^{km} = 1 \quad j = 1, 2, \ldots, |J|, m = 1, 2, \ldots, |M| \tag{18}$$

Since there are no buffers between machines, jobs are blocked on machines if their processing ends earlier than robot arriving. Moreover, the cell environment consists of a flow shop with m machines in the cell and the objective is to minimize the cycle time, i.e., the $RF_m|block|ct$ problem [20]. As a result, each part should pass through all machines according to the sequence that all jobs are entered into the cell. In order to take this consideration to account, the loading step of each part on each machine should be greater than the one on the previous machine, i.e., Eq. (19).

$$\sum_{k=1}^{|J| \times |M|} k x_j^{k,m-1} \leq \sum_{k=1}^{|J| \times |M|} k x_j^{k,m} \quad j = 1, 2, \ldots, |J|, \; m = 2, 3, \ldots, |M| \tag{19}$$

2.3.5 Relation Constraints

As the model simultaneously formulates the sequence of parts and the sequence of robot movements, it is necessary to make connection between corresponding variables by means of some constraints, namely *relation constraints*. For this purpose, it is ensured that when part j follows part i, it cannot visit a machine before part i. This consideration is taken into account by means of inequality (20).

$$\sum_{k=1}^{|J|\times|M|} kx_i^{k,m-1} \leq \sum_{k=1}^{|J|\times|M|} kx_j^{k,m} + (1 - y_{ij})\Psi$$
$$i = 1, 2, \ldots, |J|, j = 2, \ldots, |J|, i \neq j, m = 2, 3, \ldots, |M| \quad (20)$$

Since job 1 is the starter of the cycle, in constraint (20), index j starts from 2, where the step number, k, of the first job is less than that of other jobs.

If job i precedes job j in the sequence, then job j cannot be loaded on machine m before unloading job i from that machine. In other words, loading of part j on the mth machine should be after loading of part i on machine $(m+1)$. This limitation is modeled as inequality (21).

$$\sum_{k=1}^{|J|\times|M|} kx_i^{k,m+1} \leq \sum_{k=1}^{|J|\times|M|} kx_j^{k,m} + (1 - y_{ij})\Psi$$
$$i = 1, 2, \ldots, |J|, j = 2, \ldots, |J|, i \neq j, m = 1, 2, \ldots, |M| - 1 \quad (21)$$

2.3.6 Completion Time Constraints

Referring to the classical scheduling, each loading time can be represented as a completion time either. In this regard, the loading time of each job on each machine starts from completion time of that job on the previous machine. More precisely, two situations are possible. First, the processing has completed and the job is blocked on the machine until the robot arrives, that is leaving another machine. Then the loading time should be greater than the loading time of machine which has been just left as well as traveling time of robot from that machine to the machine on which the job has been blocked.

In the second case, the robot should wait for the job to be processed and after that start its loading on the next machine. Therefore, the loading time of job j on machine m should be greater than its loading time on the previous machine plus its processing time on that machine. For simplification, temporarily assume that ζ_k is the traveling time of robot from the machine that has been just loaded at step $(k-1)$ to the machine which is going to be loaded at step k. As a result, constraint (22) formulates the two aforementioned cases.

$$l_j^{km} = \left(\max\left\{ l_{j'}^{k-1,m'} + \zeta_k, l_j^{k',m-1} + P_j^{m-1} \right\} + \varepsilon_j^{m-1,2} + \delta + \sum_{i=1}^{|J|} \varepsilon_{ij}^{m1} y_{ij} \right) x_j^{km}$$
$$j = 1, 2, \ldots, |J|, m = 2, 3, \ldots, |M|, m' = 1, 2, \ldots, |M|,$$
$$k = 2, 3, \cdots, |J| \times |M|, k' = 1, 2, \ldots, |J| \times |M| \quad (22)$$

Assume that robot has loaded part j' on machine m' at the kth step and at step k it loads part j on machine m, i.e., $x_{j'}^{k-1,m\prime} = x_j^{km} = 1$. It is possible that $j = j'$ and $m = m'$ as well. But, it is assumed that part j has been loaded on machine $(m-1)$ at step k' so that $k \neq k'$. As a result, the term $\max\left\{l_{j'}^{k-1,m'} + \zeta_k\delta, l_j^{k',m-1} + P_j^{m-1}\right\}$ gives the time that robot reaches part j, ready for pickup, on machine $(m-1)$. In the rest of actions, robot picks up the part with time of $\varepsilon_j^{m-1,2}$, carries it toward machine m with time of δ and loads the part with time of $\sum_{i=1}^{|J|} \varepsilon_{ij}^{m1} y_{ij}$. The last term includes the concept of sequence dependent setup times. In other words, when $y_{ij} = 1$ the loading time of job j on machine m is affected by its preceding job, i, which is presented as ε_{ij}^{m1}.

Now, ζ_k should be computed via the aforementioned parameters and variables. According to additive-travel metric [20] ζ_k is obtained as follows.

$$\zeta_k = \left| \sum_{j=1}^{|J|} \sum_{m=1}^{|M|} m\, x_j^{k-1,m} - \sum_{j=1}^{|J|} \sum_{m=1}^{|M|} (m-1)\, x_j^{km} \right| \delta \quad k = 2, 3, \ldots, |J| \times |M| \tag{23}$$

Substituting Eq. (23) in Eq. (22) results the following nonlinear equation.

$$\begin{aligned} l_j^{km} = {} & \left(\max\left\{ l_{j'}^{k-1,m'} + \left| \sum_{j=1}^{|J|} \sum_{m=1}^{|M|} m x_j^{k-1,m} - \sum_{j=1}^{|J|} \sum_{m=1}^{|M|} (m-1) x_j^{km} \right| \delta, l_j^{k',m-1} + P_j^{m-1} \right\} \right. \\ & \left. + \varepsilon_j^{m-1,2} + \delta + \sum_{i=1, i \neq j}^{|J|} \varepsilon_{ij}^{m1} y_{ij} \right) x_j^{km} \\ & j = 1, 2, \ldots, |J|, m = 2, 3, \ldots, |M|, m' = 1, 2, \ldots, |M|, \\ & k = 2, 3, \ldots, |J| \times |M|, k' = 1, 2, \ldots, |J| \times |M| \end{aligned} \tag{24}$$

2.3.7 Linearization of Completion Time Constraints

Equation (24) is nonlinear because of including a *max* operator, an absolute term and multiplication of decision variables. To linearize the absolute term, Eq. (25) is inserted into the model. It is obvious that two variables r_k and s_k are linearly dependent that implicitly means $r_k s_k = 0$.

$$\left| \sum_{j=1}^{|J|} \sum_{m=1}^{|M|} m\, x_j^{k-1,m} - \sum_{j=1}^{|J|} \sum_{m=1}^{|M|} (m-1)\, x_j^{km} \right| = r_k - s_k \quad k = 2, 3, \ldots, |J| \times |M| \tag{25}$$

In order to remove the *max* operator, Eq. (24) is replaced with the two following inequalities, while the absolute term is substituted by term $r_k - s_k$.

$$l_j^{km} \geq \left(l_{j'}^{k-1,m'} + (r_k - s_k)\delta + \varepsilon_j^{m-1,2} + \delta + \sum_{i=1,\, i\neq j}^{|J|} \varepsilon_{ij}^{m1} y_{ij} \right) x_j^{km}$$
$$j = 1, 2, \ldots, |J|, m = 2, 3, \ldots, |M|,$$
$$m' = 1, 2, \ldots, |M|, k = 2, 3, \ldots, |J| \times |M| \quad (26)$$

$$l_j^{km} \geq \left(l_j^{k',m-1} + P_j^{m-1} + \varepsilon_j^{m-1,2} + \delta + \sum_{i=1,\, i\neq j}^{|J|} \varepsilon_{ij}^{m1} y_{ij} \right) x_j^{km}$$
$$j = 1, 2, \ldots, |J|, m = 2, 3, \ldots, |M|,$$
$$= 1, 2, \ldots, |M|, k'1, 2, \ldots, |J| \times |M| \quad (27)$$

Inequalities (26) and (27) are still nonlinear because of variables' multiplication. Dealing with this issue, an approach should be applied which activates the inequalities as one of the aforementioned situations happens. The linearized inequalities are presented as inequalities (28) and (29).

$$l_j^{km} \geq l_{j'}^{k-1,m'} + (r_k - s_k)\delta + \varepsilon_j^{m-1,2} + \delta + \sum_{i=1,\, i\neq j}^{|J|} \varepsilon_{ij}^{m1} y_{ij} + \left(x_j^{km} + x_{j'}^{k-1,m'} - 2 \right)\Psi$$
$$j = 1, 2, \ldots, |J|, m = 2, 3, \cdots, |M|,$$
$$m' = 1, 2, \cdots, |M|, k = 2, 3, \ldots, |J| \times |M| \quad (28)$$

$$l_j^{km} \geq l_j^{k',m-1} + P_j^{m-1} + \varepsilon_j^{m-1,2} + \delta + \sum_{i=1,\, i\neq j}^{|J|} \varepsilon_{ij}^{m1} y_{ij} + \left(x_j^{km} + x_j^{k',m-1} - 2 \right)\Psi$$
$$j = 1, 2, \ldots, |J|, m = 2, 3, \cdots, |M|, k = 1, 2, \ldots, |M|,$$
$$k' = 1, 2, \ldots, |J| \times |M| \quad (29)$$

2.3.8 Cycle Time Constraints

In the final set of constraints, the cycle time is derived via constraints (30) and (31). A cycle is the time between loading job 1 on machine 1 at the first step and loading that job again on machine 1 at the $(|J| \times |M| + 1)th$ step so that all parts have been processed.

$$l_{\max} \geq l_j^{|J|\times|M|,\,|M|} + \left(x_j^{|J|\times|M|,\,|M|} - 1 \right)\Psi \quad j = 1, 2, \ldots, |J| \quad (30)$$

$$ct = l_{\max} + |M|\delta + \varepsilon_1^{02} + \sum_{i=2}^{|J|} \varepsilon_{i1}^{11} y_{i1} - l_1^{11} \quad (31)$$

2.3.9 Variables' Domain Constraints

Constraints (32)–(36) determine domain of the variables.

$$x_j^{km} \in \{0,1\} \quad j = 1,2,\ldots,|J|, m = 1,2,\ldots,|M|, k = 1,2,\ldots,|J| \times |M| \tag{32}$$

$$y_{ij} \in \{0,1\} \quad i = 1,2,\ldots,|J|, j = 1,2,\ldots,|J| \tag{33}$$

$$l_j^{km} \geq 0 \quad j = 1,2,\ldots,|J|, m = 1,2,\ldots,|M|, k = 1,2,\ldots,|J| \times |M| \tag{34}$$

$$r_k, s_k \geq 0 \quad k = 1,2,\ldots,|J| \times |M| \tag{35}$$

$$l_{\max},\ ct \geq 0 \tag{36}$$

Besides all formulation, the value of Ψ also has an incredible effect on the time consumption of finding solution. It, however, should be determined as small as possible.

2.3.10 Objective Function

In many traditional scheduling problems, minimizing maximum completion time, $C_{\max}$, is a common objective function. $C_{\max}$ is not considered in this study, but it can be easily formulated as $\max\left\{l_j^{km} + P_j^m\right\}$, where $j = 1, 2, \ldots, |J|, m = 1, 2, \ldots, |M|$, $k = 1, 2, \ldots, |J| \times |M|$.

In fact, there is a difference between minimizing maximum completion time, $C_{\max}$, maximum loading time, $l_{\max}$ and cycle time, ct. The last objective sometimes is referred to maximizing the steady state throughput rate of the cell that has a different nature in comparison with the other criteria, i.e., $l_{\max}$ and $C_{\max}$. As a conclusion, since the cycle is repeated for a long time, this paper considers minimizing ct as the objective function, illustrated in Eq. (37), which incredibly improves the efficiency of m-machine robotic cell. The problem is now formulated as a Mixed-Integer Programming (MIP) model as follows.

$$\textit{minimize} \quad ct \tag{37}$$

Subject to: constraints (12)–(21) and (28)–(37).

The above mathematical model can be solved using optimization commercial solvers such as CPLEX.

2.4 Expert Systems Modeling

Despite mathematical modeling, some system developers believe in human experience and expertise. The ability of a human expert in problem solving is modeled via a computer program, i.e., the expert system (ES). Most expert systems contain three main modules. The principle module of an ES, obtained from a real expert, is its knowledge, which is stored in the knowledge base. Another main characteristic of an ES is the method of reasoning that it has, which is performed by the inference engine. In order to keep all the information required for problem solving, a working memory is designed for the ES as its third module [34].

Modeling of the human knowledge is an important issue in all expert systems. However, rule-base approach is one of the most common techniques, by which one can represent the knowledge of the system. In fact, a rule-base is a procedural knowledge.

Basically, a rule consists of two main parts, namely IF part and THEN part. However, there are some other expressions used for these two parts. For example, the IF part is also known as the Antecedent, Conditions, Premises and for the THEN part one may use Consequent, Conclusions and Actions. There are also some operators that concatenate or relate some IF expressions together, i.e., AND, OR and ELSE. An example of a rule-base knowledge representation is provided as follows.

IF *the robot is idle* AND *the processing of all jobs have not finished* THEN *turn the robot to standby mode*.

In the knowledge base, there are a huge pool of IF-THEN rules representing the human knowledge, but utilizing these rules to solve a problem is another issue of an ES. More precisely, there are some techniques for reasoning that search through the knowledge base to solve the problem. In this regard, forward chaining and backward chaining are of the most popular methods which construct the main structure of ES' inference engine.

Rule-base knowledge representation are applicable for every expert system kind modeling such as fuzzy control systems. In what follows, main framework of fuzzy controllers are described and as discussed, rule-base approach play a significant role in modelling the fuzzy parameters and rules.

3 Fuzzy Control Techniques

Using state space modeling and differential equations of the system, it can be analyzed and studied by which the control is applicable. The input, output and dynamics of a linear system can be represented as Fig. 5.

Since the movements of a manipulator robot is highly nonlinear, without loss of generality, in following formulations the state space equations are rewritten as Eqs. (38) and (39).

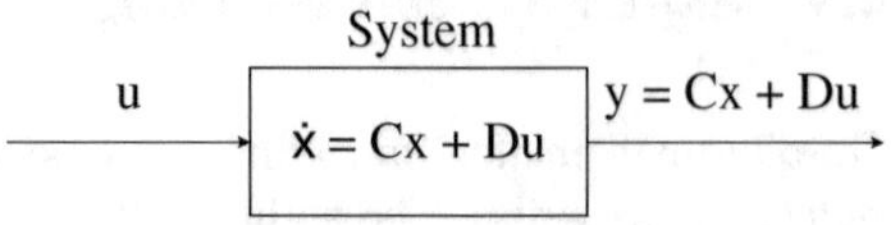

Fig. 5 The inputs, dynamics and outputs of a system. The state space modeling of the system is linear

System

u

$\dot{x} = Cx + Du$

$y = Cx + Du$

$$\dot{x} = f(x, u) \tag{38}$$

$$y = g(x, u) \tag{39}$$

In what follows the concept of control theory and applications of state space modeling is described.

3.1 Control Theory

The concept of control theory is concerned with how one can feed right input to the system to achieve desired output. In this regard, a reference signal is defined as a measure of state control of the system. Figure 6 represents a system with a control module that justifies the input variables.

In Fig. 6, the control module can adjust the input values at any time, but there is not any feedback from the system to check how far it is from the desired state. This type of control is named as open-loop control system. However, the feedback of system is necessary to evaluate error of the control system comparing system's state with the reference signal. Unlike open-loops, closed-loop control systems utilize feedback of the system at each moment of action. The input is a function of reference signal and state of the system. In general, when feedback is determined, the error is computed by subtracting the current state from the desired state. This error is used for further analysis and investigations. Figure 7 illustrates a closed-loop control system. Recently, use of Proportional Integral Derivative (PID) controller as a closed-loop feedback mechanism controller is extensively used in industrial control systems.

Apparently, the robot with aforementioned dynamic formulations is controllable using control theory. For this purpose, it should be identified which parameters of the system can be changed by the controller, e.g., the external forces and tensions.

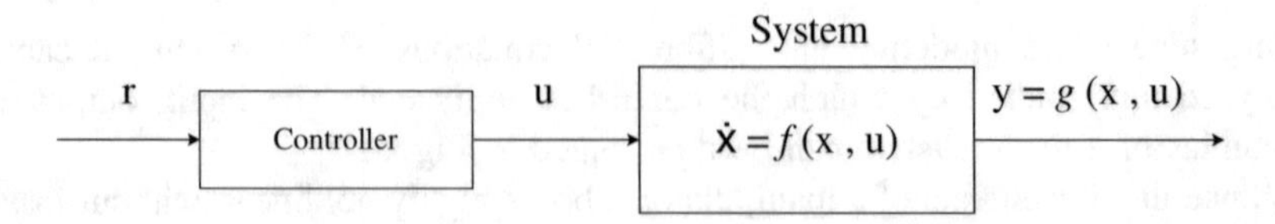

Fig. 6 A system with controller. Since there is not any feedback to the system, it is called an open-loop system

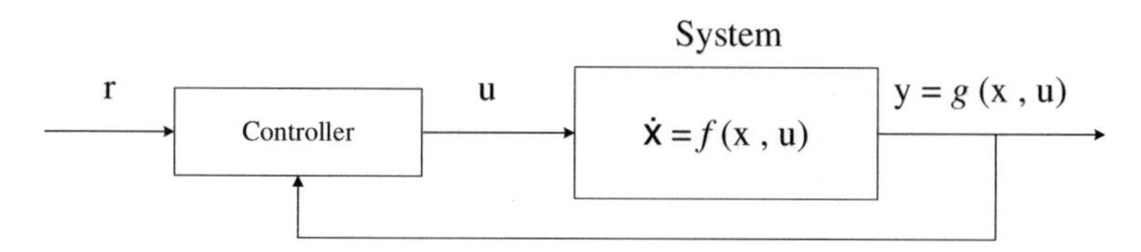

Fig. 7 A closed-loop control system. The feedback of the system is evaluated and analyzed for improving the control actions

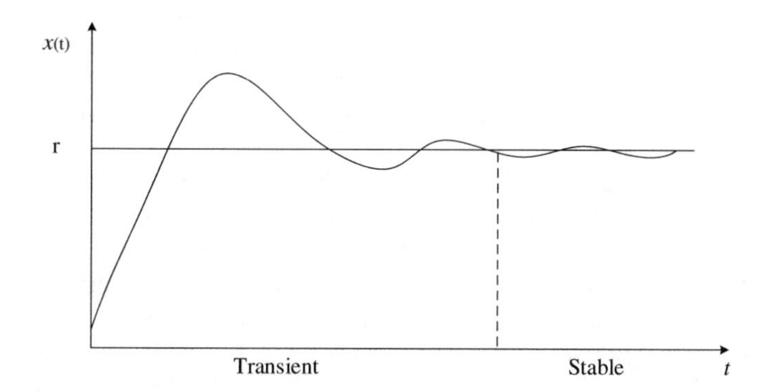

Fig. 8 Trajectory of a system during its transient and stable periods

The aim of the controller is to guide the system to reach its desirable state and maintaining in that state for some amounts of time. Figure 8 provides the trajectory of a system under transient and stable periods of time, where x(t) is the state of the system and r stands for the reference signal or the ideal point. As a predefinition, the control is the decisions one can take during some periods of time to bring the system toward its ideal situation as close as possible.

In general, there are other external parameters, some of which are not controllable, but influence on the output results, e.g., the inertia effect, maintenance program and unknown disturbances. The issue of these non-controllable factors become worst when uncertainty phenomena presents. Most of imprecise input parameters are dealt with fuzzy set theory. Hence, the concept of fuzzy control has been developed in order to include non-predictable factors in the state space modeling and achieve desirable results [31].

3.2 Fuzzy Controller

Uncertainties can be grouped into structured and unstructured parameters. Unlike unstructured, the structured parameters are of those that are identified but because

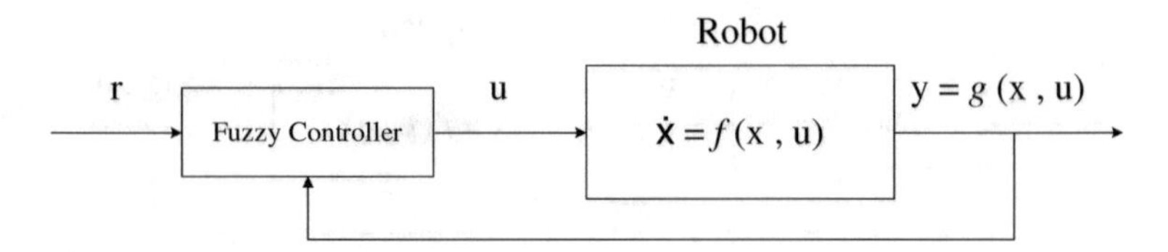

Fig. 9 A system with fuzzy controller

of their random nature no mathematical model is available to involve them in dynamics of the system. For example, unknown loads, unpredictable failures and processing times may cause an unwanted disturbance. Fortunately, fuzzy set theory as a model-free approach is applicable to deal with these types of uncertainties. However, the other unstructured uncertainties remained unsolved for further investigations and studies.

Fuzzy control is such a technique by which not only many unknown variables can be captured and investigated, but also it has the ability of including the experts' knowledge and experience in robotic planning and control. When the mathematical models and dynamic state space equations cannot handle complicated systems, the human knowledge and expertise will deal with most of troubles and situations. Fuzzy logic is the approach of including heuristics and human knowledge in such a servomechanism system. Figure 9 is a block diagram of a fuzzy control system in its general form.

As the knowledge of experts are stated linguistically, they are converted into a rule-based framework with If-Then structure. The rule-based knowledge of the system is the basis of fuzzy controller, which should be converted to a set of fuzzy rules. There are many different types of fuzzy rules of which Mamdani is the first with the linguistic structure, which is widely used in many fuzzy control designs. For example, one control rule can be stated as follows.

IF "the *temperature* is *very low* AND the *heater* is *of* " THEN "turn on the heater".

In the above statement *temperature* and *heater* are fuzzy variables and *very low* and *off* are fuzzy linguistic values. The IF-part sentences are called antecedent and the THEN-part sentence is called the consequent. Some rules may have more than one statement as antecedent which are connected with AND/OR operators. All rules and reasoning of a fuzzy controller belong to its inference engine.

In general, a fuzzy controller consists of three main modules, namely fuzzifier, inference engine and defuzzifier. The role of fuzzifier and defuzzifier is to transform crisp data to fuzzy variable(s) and vice versa, respectively. Inference engine, like many expert systems with the use of knowledge-based through reasoning procedure takes the control action.

Including most uncontrollable external factors, say disturbances, is the other important advantage of fuzzy controllers. Disturbances may not be modelled with mathematical programming tools, but the knowledge-based can be designed in such a way that many rules support the system in disturbance situations. Figure 10

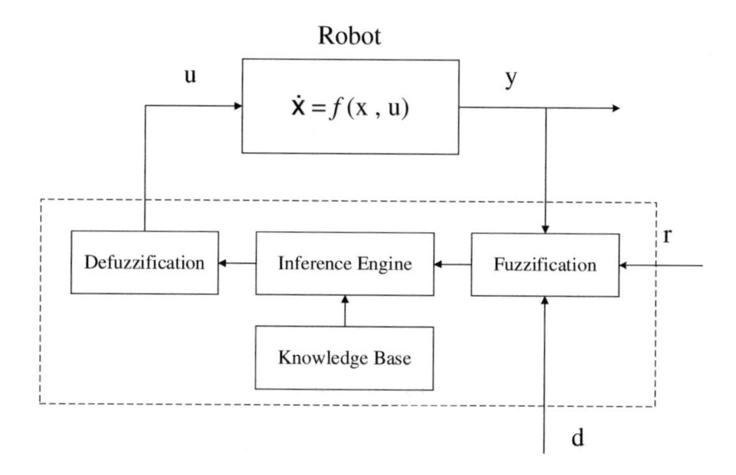

Fig. 10 A fuzzy control system with fuzzifier, inference engine, knowledge base and defuzzifier

illustrates a robot control system with detail modules of its fuzzy controller, where parameter d stands for the disturbance.

3.2.1 Example

Here, for more understanding of the fuzzy controllers, an example is provided. Consider the following knowledge base of a robot.

IF *processing time of part A is very long* THEN *remove it from the machine.*
IF *an unfinished part is chosen to be removed* THEN *put it part in the rework station.*
IF *all machines are processing* THEN *go to the standby mode.*
IF *a failure happens* AND *it is not preventing* THEN *finish just high priority jobs.*

There are also more priority scheduling rules that can be used in the knowledge base of the controller in order to manage the sequence of all jobs in the workshop. However, it is important to transform crisp data to fuzzy variables and vice versa, which is performed via fuzzifier and defuzzifier modules. As an example, assume that processing time of a job has a fuzzy linear membership function as follows.

$$\mu_P(t) = \begin{cases} 1 & t \geq b \\ \frac{t-a}{b-a} & a < t < b \\ 0 & t \leq a \end{cases}$$

Figure 11 illustrates the membership function of the example. A fuzzifier can be programmed as follows.

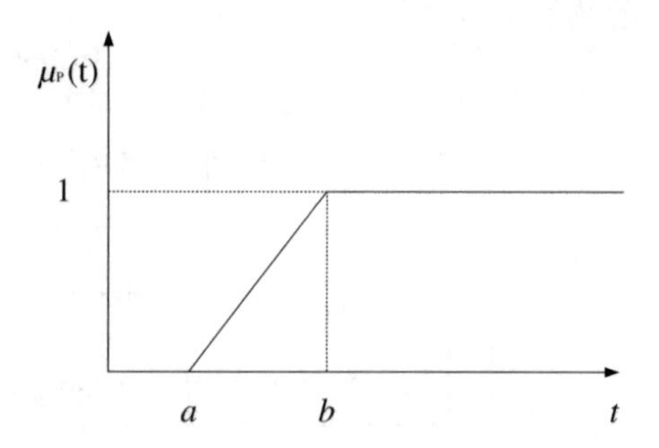

Fig. 11 The membership function of the illustrative example

IF $0 \leq \mu_P(t) < 0.33$ THEN *the processing time is short*,
IF $0.33 \leq \mu_P(t) < 0.67$ THEN *the processing time is normal*,
IF $0.67 \leq \mu_P(t) \leq 1$ THEN *the processing time is long*.

When the processing of a job starts, every unit of time, every second for example, its processing time is checked and it is reported to the controller. According to the knowledge base and the inference engine, the controller takes a control action in that time. For the mentioned example, assume that the processing of a particular job took more than usual. Then, based on the aforementioned rule, the robot can use preemption and interrupt processing of that job.

There are much more examples of fuzzification, that might be converted to the other fuzzy variables through reasoning process as the outputs. For example, consider x_i as the position of the ith link of the robot, where $i = 1, 2, \ldots, I$. Assume that there are R rules in the knowledge base as follows.

If $x_1 \in A_1^r$ And $x_2 \in A_2^r$ … And $x_I \in A_I^r$ Then $y^r \in C^r$ for $r = 1, 2, \ldots, R$.

The membership function of each consequent C^r depends on the inference engine and the knowledge base. As an example, employing a min-max rule can conclude another fuzzy output which is required to be defuzzified. One of the most common defuzzification approach is the center of gravity defuzzifier. However, if the output is not a fuzzy variable, the defuzzifier may be programmed simpler than other procedures such as follows.

IF *a job have to be removed from a machine*, THEN *remove the part from the MPS list.*

3.3 Fuzzy Mathematical Optimization and Control

Along with control techniques, the fuzzy logic has been employed to embed uncertainty in mathematical models as well. The application of fuzzy theory in these models, especially in optimization models, provide valid and valid solutions. The general form of a fuzzy mathematical model is provided as follows.

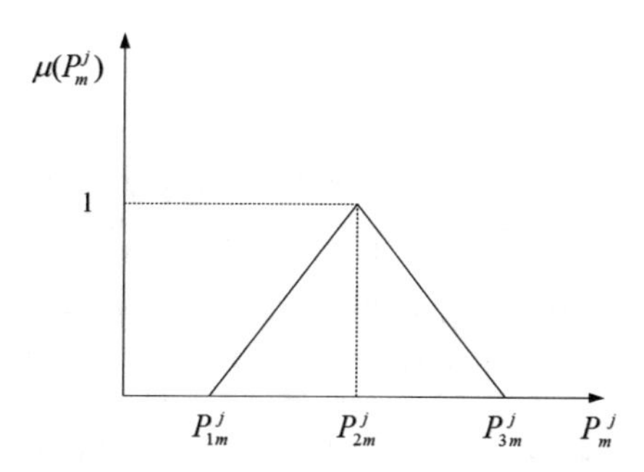

Fig. 12 Membership function of the processing times

$$\begin{aligned} &\min \tilde{c}x \\ &st: \\ &\quad \tilde{A}x \le \tilde{b} \\ &\quad x \in X \end{aligned}$$

where the symbol $\sim$ represents the fuzzy parameters. In the above model, it is not necessary to consider all parameters as fuzzy variables, but some of them will be helpful. In the robotic cell scheduling problem, processing times as well as robot activity times such as loading/unloading may be fuzzified, i.e., $\tilde{P}_j^m$ and $\tilde{\varepsilon}_{ij}^{m(l)}/\tilde{\varepsilon}_j^{m(u)}$. To do so, depending on the expert knowledge the membership function of each parameter is identified. Figures 12 and 13 show triangular membership functions for the processing and the loading/unloading times.

Using the mathematical formulation of triangular membership functions, they could be included in the optimization model. As a general formulation, the membership functions could be represented as follows.

$$\mu_m^j\left(P_m^j\right) = f_m^j\left(P_m^j\right) \quad j = 1, 2, \ldots, |J| \quad m = 1, 2, \ldots |M| \tag{40}$$

$$\mu_{ij}^m\left(\varepsilon_{ij}^{m(l)}\right) = g_{ij}^m\left(\varepsilon_{ij}^{m(l)}\right) \quad i = 1, 2, \ldots, |J| \quad j = 1, 2, \ldots, |J| \quad m = 1, 2, \ldots |M| \tag{41}$$

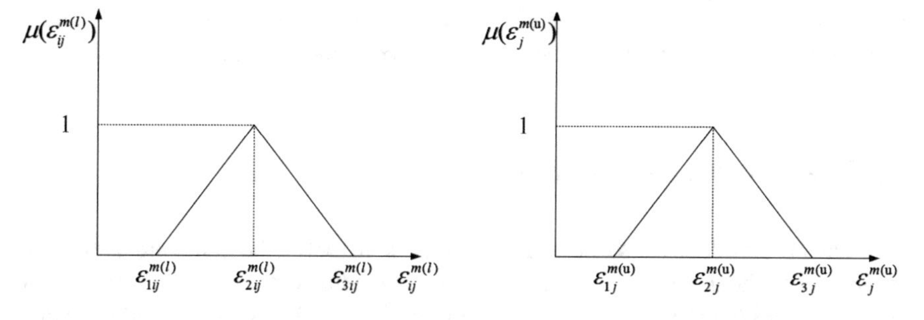

Fig. 13 Membership functions of loading and unloading times

$$\mu_j^m\left(\varepsilon_j^{m(u)}\right) = h_j^m\left(\varepsilon_j^{m(u)}\right) \quad j = 1,2,\ldots,|J| \quad m = 1,2,\ldots|M| \tag{42}$$

In order to include the fuzzy membership functions in the mathematical modeling, the crisp parameters P_m^j, $\varepsilon_{ij}^{m(l)}$ and $\varepsilon_j^{m(u)}$ are replaced with their relevant fuzzy membership functions $f_m^j(P_m^j)$, $g_{ij}^m(\varepsilon_{ij}^{m(l)})$ and $h_j^m(\varepsilon_j^{m(u)})$ respectively. Hence, one may want to maximize joint membership function of all fuzzy variable. For this purpose, the joint membership function could be computed as the minimum value of all values, i.e., the max-min operator [30].

$$\mu_{ct}(x) = \max\left\{\min_{i,j,m}\left\{f_m^j(P_m^j), g_{ij}^m\left(\varepsilon_{ij}^{m(l)}\right), h_j^m\left(\varepsilon_j^{m(u)}\right)\right\}\right\} \tag{43}$$

For linearization, the above function can be rewritten as follows, where $\mu_{ct}(x)$ is the obtained membership function of the cycle time.

$$\mu_{ct}(x) = \max \varphi \tag{44}$$

Subject to:

$$f_m^j(P_m^j) \geq \varphi \quad j = 1,2,\ldots,|J| \quad m = 1,2,\ldots|M| \tag{45}$$

$$g_{ij}^m(\varepsilon_{ij}^{m(l)}) \geq \varphi \quad i = 1,2,\ldots,|J| \quad j = 1,2,\ldots,|J| \quad m = 1,2,\ldots|M| \tag{46}$$

$$h_j^m(\varepsilon_j^{m(u)}) \geq \varphi \quad j = 1,2,\ldots,|J| \quad m = 1,2,\ldots|M| \tag{47}$$

$$x \in X,\ P_m^j, \varepsilon_{ij}^{m(l)}, \varepsilon_j^{m(u)} \geq 0 \quad i = 1,2,\ldots,|J| \quad j = 1,2,\ldots,|J| \quad m = 1,2,\ldots|M| \tag{48}$$

As a result, one can include the fuzzy constraints in the crisp mathematical modeling, mentioned in the previous section. However, in the crisp model, the parameters are replaced with their relative membership functions. It should be noticed that the objective function of the fuzzy mathematical model is maximization of membership function of the cycle time, which could be represented as Fig. 14.

4 The Robotic Control System

This paper introduces two control approaches namely, the dynamic and static technique. In the dynamic approach state space equations are defined and utilized to present the behavior of the system at each unit of time. On the other hand, the static

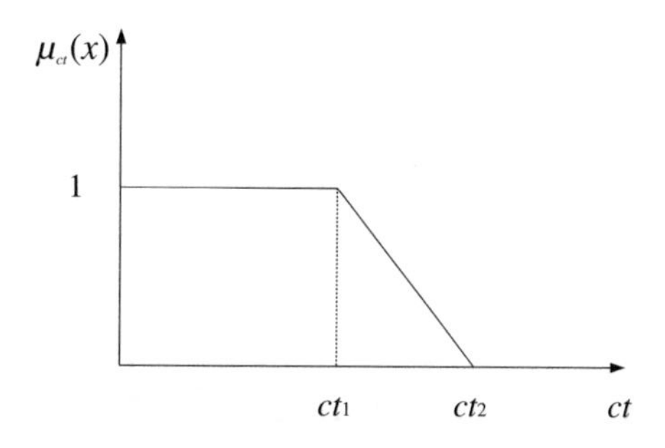

Fig. 14 Membership function of the cycle time

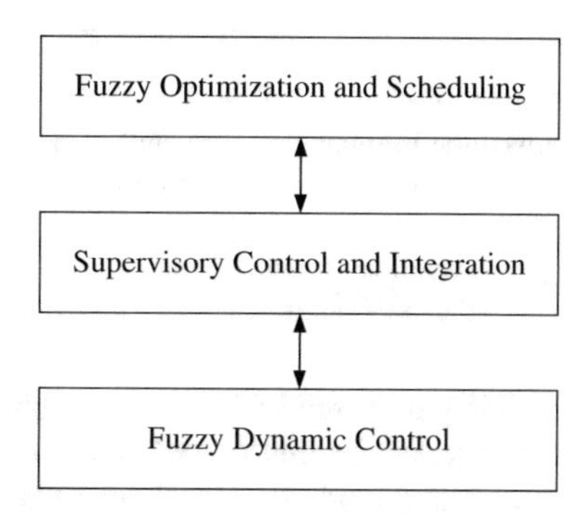

Fig. 15 The main framework of proposed integrated fuzzy optimization and control system

approach uses mathematical formulation in terms of an optimization problem that is solved periodically. Both approaches work with crisp data and parameters, however because of uncertainty, fuzzy set theory can be applied for both dynamic and static models.

Application of each approach individually may have some drawbacks. More precisely, when one aims to use dynamic equations and simulation, it may fail to manage jobs entrance to the shop to improve the productivity of the robot. When the static approach is used individually, one may lose the control of the system when some disturbances happen during the period. Hence, the combination of these approach may improve the performance of the system gradually. In what follows, the combined system is proposed for the robotic fuzzy optimization and control system. Figure 15 depicts the big picture of the integrated system.

The details of the proposed system is presented in Fig. 16, where j stands for the set of jobs entering to the shop.

Briefly, the structure of the proposed control system can be described as following steps.

Step 1. Find the optimal schedule of the jobs.
Step 2. Execute the scheduled plan and control the system at each unit of time.
Step 3. If the time of scheduling arrives, update the information and go to Step 1.

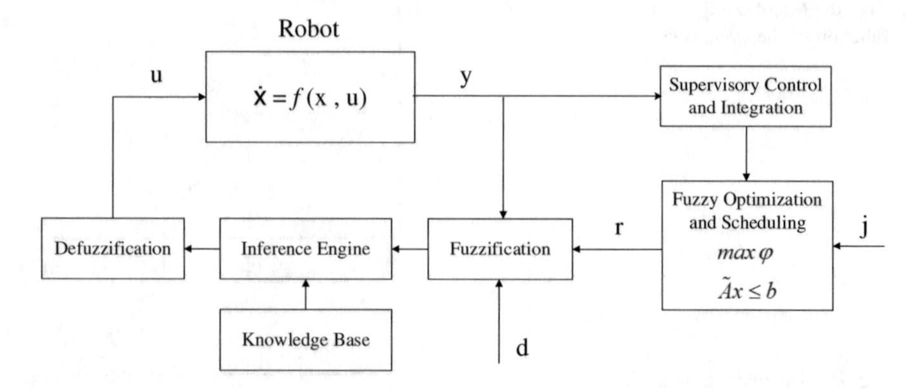

Fig. 16 The integrated fuzzy optimization on control system. The fuzzy mathematical model is solved the beginning of each shift or day. Then the fuzzy control system is utilized to take the control of the robot at each unit of time automatically

5 Conclusion

In this paper, the techniques of robotic control systems have been reviewed and investigated. In this regard, some conventional methods of modeling, namely the state space modeling, mathematical modeling and expert systems modeling have been presented specifically for a manipulator robot. The robot's dynamic formulations extremely depend on the kinematics and mechanical functioning of the robot, but a general state space modeling has been presented. Furthermore, a general mathematical scheduling model for an *m*-machine robotic cell system has been developed, by which the optimal sequence of jobs entering to the cell is obtained. On the other side, the concept of control theory and fuzzy control systems have been studied and explained. Moreover, the framework of fuzzy rules, knowledge-based approaches and their applications in control systems have been briefly discussed. Finally, an integrated optimization and control system has been proposed, in which all the three modelling approaches, i.e., the state space, the mathematical and the expert system models interact with each other on the basis of fuzzy logic. The proposed model can consider most aspects of real environment of every type of robot especially manipulator manufacturing robots. For further research, one may include stochastic approaches as well as robust control techniques.

References

1. Freeman, C.T., et al.: Iterative learning control in health care: electrical stimulation and robotic-assisted upper-limb stroke rehabilitation. Control Syst. IEEE **32**(1), 18–43 (2012)
2. Hussain, S., Xie, S.Q., Liu, G.: Robot assisted treadmill training: Mechanisms and training strategies. Med. Eng. Phys. **33**(5), 527–533 (2012)

3. Katić, D., Vukobratović, M.: Survey of intelligent control techniques for humanoid robots. J. Intell. Rob. Syst. **37**(2), 117–141 (2003)
4. Simorov, A., et al.: Review of surgical robotics user interface: what is the best way to control robotic surgery? Surg. Endosc. **26**(8), 2117–2125 (2012)
5. Liu, Y., Nejat, G.: Robotic urban search and rescue: a survey from the control perspective. J. Intell. Rob. Syst. **72**(2), 147–165 (2013)
6. Wen, L., et al.: Hydrodynamic investigation of a self-propelled robotic fish based on a force-feedback control method. Bioinspiration & Biomimentics **7** (2012)
7. Du, Y., et al.: Review on reliability in pipeline robotic control systems. Int. J. Comput. Appl. Technol. **49**(1), 12–21 (2014)
8. Lenarcic, J., Bajd, T., Stanisic, M.M.: Robot Mechanisms, Springer (2013)
9. Kim J.-H., et al.: Robot Intelligence Technology and Applications, vol. 2, Springer (2014)
10. Kim, J.-H., et al.: Robot Intelligence Technology and Applications, vol. 3, Springer (2015)
11. Christ, R.D., Wernli Sr R.L.: Chapter 19—Manipulators. In: Christ, R.D., Wernli, R.L. (ed.) The ROV Manual (Second Edition), pp. 503–534, Oxford, Butterworth-Heinemann (2014)
12. Sun, Y., Qian, H., Xu, Y.: Chapter 5.1—The state of the art in grasping and manipulation for household service. In: Wu, Y.X.Q. (ed.) Household Service Robotics, pp. 341–356, Oxford, Academic Press (2015)
13. Sethi, S.P., et al.: Sequencing of parts and robot moves in a robotic cell. Int. J. Flex. Manuf. Syst. **4**(3), 331–358 (1992)
14. Logendran, R., Sriskandarajah, C.: Sequencing of robot activities and parts in two-machine robotic cells. Int. J. Prod. Res. **34**(12), 3447–3463 (1996)
15. Chen, H., Chu, C., Proth, J.-M.: Sequencing of Parts in Robotic Cells. Int. J. Flex. Manuf. Syst. **9**(1), 81–104 (1997)
16. Sriskandarajah, C., Hall, N.G., Kamoun, H.: Scheduling large robotic cells without buffers. Ann. Oper. Res. **76**, 287–321 (1998)
17. Crama, Y., et al.: Cyclic scheduling in robotic flowshops. Ann. Oper. Res. **96**(1), 97–124 (2000)
18. Akturk, M.S., Gultekin, H., Karasan, O.E.: Robotic cell scheduling with operational flexibility. Discrete Appl. Math. **145**(3), 334–348 (2005)
19. Gultekin, H., Akturk, M.S., Karasan, O.E.: Scheduling in robotic cells: process flexibility and cell layout. Int. J. Prod. Res. **46**(8), 2105–2121 (2008)
20. Zarandi, Fazel, MHH, Mosadegh, Fattahi, M.: Two-machine robotic cell scheduling problem with sequence-dependent setup times. Comput. Oper. Res. **40**(5), 1420–1434 (2013)
21. Bagchi, T.P., Gupta, J.N.D., Sriskandarajah, C.: A review of TSP based approaches for flowshop scheduling. Eur. J. Oper. Res. **169**(3), 816–854 (2006)
22. Dawande, M., et al.: Sequencing and scheduling in robotic cells: recent developments. J. Sched. **8**(5), 387–426 (2005)
23. Cassandras Christos G., Stéphane, L.: Introduction to discrete event systems. Springer (2008)
24. Kandel, A., Langholz, G.: Fuzzy Control Systems. CRC Press (1993)
25. Sousa, J.M.C., Kaymak, U.: Fuzzy Decision Making in Modeling and Control 2002: World Scientific Publishing Co. Pte. Ltd
26. Peng, L., Peng-Yung, W.: Neural-fuzzy control system for robotic manipulators. Cont. Syst. IEEE **22**(1), 53–63 (2002)
27. Nanayakkara, T., Sahin, F., Jamshidi, M.: Intelligent control systems with an introduction to system of systems engineering. CRC Press (2010)
28. Al-Hadithi, B., Matía, F., Jiménez, A.: Fuzzy controller for robot manipulators. In: Melin, P., et al. (ed.) Foundations of Fuzzy Logic and Soft Computing, pp. 688–697, Springer, Berlin, Heidelberg
29. Siciliano, B., et al.: Advances in Control of Articulated and Mobile Robots. Springer (2004)
30. Zadeh, L.A.: Outline of a new approach to the analysis of complex systems and decision processes. IEEE Trans. Syst. Man Cybern. 1973. **SMC-3**(1), 28–44
31. Mamdani, E.H.: application of fuzzy logic to approximate reasoning using linguistic synthesis. IEEE Trans. Comput. **C-26**(12), 1182–1191 (1977)

32. Featherstone, R., Orin, D.: Robot dynamics: equations and algorithms. in Robotics and Automation. In: Proceedings IEEE International Conference on ICRA '00 (2000)
33. Spong, M.W., Vidyasagar, M.: Robot Dynamics and Control. Wiley (1989)
34. Durkin, J.: Expert systems: design and development (1994)

Fuzzy Sets in the Evaluation of Socio-Ecological Systems: An Interval-Valued Intuitionistic Fuzzy Multi-criteria Approach

Beyzanur Çayır Ervural, Bilal Ervural and Cengiz Kahraman

Abstract In recent days, the use of the fuzzy set theory to deal with social complexity has become more attractive as a research area to academics who wish to contribute to sustainable development. The socio-ecological systems are one of the well-known sub-disciplines of social science which is very critical issue to maintain sustainability and this concept basically concerns with the human-environment interactions. These systems involve various stakeholders with different levels of knowledge and experience from diverse social platforms. The basic characteristic of these systems is identified as having high level of uncertainty and incomplete information. The main purpose of this chapter is to demonstrate how to incorporate fuzzy sets theory into social sciences. In this chapter, an illustrative example which consists of a wide variety of social actors is used to evaluate sustainable management options utilizing the extended technique for order preference by similarity to ideal solution (TOPSIS) method for interval valued intuitionistic fuzzy multi-criteria group decision making.

Keywords Socio-ecological systems · Interval-valued intuitionistic fuzzy sets · Extended TOPSIS · Multi-criteria decision making

1 Introduction

The complexity of human relations and interactions requires to understand these systems deeply and to present efficient solutions extensively. It is obvious that the social complex systems demonstrate an ambiguous and uncertain structure since it consists of different stakeholders from various disciplines and try to satisfy all demands under conflicting objectives. At this point, the fuzzy set theory fulfills the need in an efficient way and overcomes the ambiguous nature of social complex

B.Ç. Ervural · B. Ervural · C. Kahraman (✉)
Department of Industrial Engineering, Istanbul Technical University, 34367 Macka, Istanbul, Turkey
e-mail: kahramanc@itu.edu.tr

C. Kahraman et al. (eds.), *Fuzzy Logic in Its 50th Year*,
Studies in Fuzziness and Soft Computing 341,
DOI 10.1007/978-3-319-31093-0_14

systems. Several studies in the literature have examined why the fuzzy sets theory are needed in the social sciences [14, 47, 52].

The social sciences consist of a variety of disciplines and methods. Environmental studies are one of the most popular sub-disciplines of social sciences. Scholars have handled the concept of socio-ecological systems to emphasize that the description between social systems and ecological systems which are related to each other and these issues couldn't consider individually [6]. Socio-ecological systems (SES) theory is relatively a new concept. Research in socio-ecological systems from a complex system perspective is an expanding interdisciplinary field which can be seen as an attempt to link different disciplines into a new concept to analyze some of the significant environmental problems [12]. Hence, different stakeholders from social, environmental, technical and economical platform take part in SES and they try to reach a decision or consensus under various and conflicting criteria in a multidimensional problem.

Social and ecological systems are not only related but also interconnected and show a change jointly in the temporal and spatial scales [6]. Ecological system dynamics is an essential discipline and it is really difficult to service while ignoring human factor in the system. Concentrating on the ecological aspect only does not reflect realities as a complete [49]. In order to maintain sustainability in a certain extent, all aspects of social and ecological systems must be considered, that affect decision making directly. The main reason of working on resilience is to consider social-ecological system all together. Investigating on research of these integrated systems is still proceeding to provide new perspectives for socio-ecological systems [17].

Due to human-environment interactions, socio-ecological systems are a kind of complex systems having vague and uncertain conditions. Additionally, these systems are also known as participatory systems since they involve many participants with different levels of knowledge and experience [40]. Participatory and social multi-criteria evaluation frameworks are developed to help this type of decision-making which consists of multi-dimensional framework [19].

The fuzzy set theory is a powerful tool in order to overcome ambiguous information. The fuzzy set theory was developed by Zadeh [57] whereas intuitionistic fuzzy sets (IFS) have been proposed by Atanassov [4]. IFSs are very effective to deal with uncertainty and vagueness. Therefore, the decision makers can utilize intuitionistic fuzzy sets, especially the interval-valued intuitionistic fuzzy sets (IVIFS) introduced by Atanassov and Gargov [3] to better express the information of the candidates under incomplete and uncertain information environment.

Complex socio-ecological systems (SESs) are difficult to analyze and model because of the nature of the complexity paradigm [27]. Owing to the uncertainty and vagueness of complex socio-ecological systems, we could not make decisions easily under such a large number of criteria and stakeholders. Besides, multi-criteria decision making with interval-valued intuitionistic fuzzy sets has received a great deal of attention from researchers recently. The aim of this study is to use fuzzy logic in complex socio-ecological systems. In this study, we have presented an illustrative example which evaluates the sustainable management options of a district by using an extended TOPSIS method with interval-valued intuitionistic fuzzy sets. The remainder of the study is organized as follows. Section 2 briefly gives the current literature review about complex socio-ecological systems. Section 3 reviews the basic concept about interval-valued intuitionistic fuzzy sets. Section 4 explains the methodology used in this chapter. Section 5 provides an illustrative example and finally some concluding remarks and future research directions are given in Sect. 6.

2 Socio-Ecological Systems: A Literature Review

In this section, firstly we give the general socio-ecological studies using multi criteria decision making and then discuss some specific applications of the fuzzy set theory in this complex system.

Until recent decades, many social science disciplines had disregarded environment with limiting their scope to humans and they had not considered humans as a part of the ecological studies [6]. However there are no natural systems without people, nor social systems without nature. Social and ecological systems are truly interdependent and constantly co-evolving [49].

Multi-criteria analysis is a mostly used method in ecological economics and which provides an option to combine different issues such as economic, ecologic and social criteria [24]. Multi-criteria analysis has been applied in different domains with specific focus such as forest management [2], river alteration projects [42], or bioenergy solutions [8]. Owing to these studies involve case-specific criteria and objectives, the results obtained solutions couldn't be adjusted to another cases easily [18]. Malekmohammadi and Rahimi Blouchi [38] considered a process of ecological risk assessment to determine stress factors and responses under ecosystem-based approach. The objective of the study is to provide zoning of wetland ecosystems and a systematic methodology for risk assessment. According to experts' opinions, these risks are determined with a MCDM method. Chen and Yang [11] proposed information based network environmental analysis for ecological risk assessment in wetland. In the literature, several scholars have developed a multi-criteria evaluation model based on Geographical Information System (GIS) in socio-ecological systems [5, 29, 30, 32, 37, 48]. The concept of systematic regional planning is another approach that can be used to integrate different tools such as GIS and spatial multi-criteria analyses for an efficient local management action that maximizes environmental benefits and minimizes economic costs [7]. Ferretti and Pomarico [16] analyzed the development of a Multi-criteria-Spatial Decision Support Systems (MC-SDSS) approach for studying the ecological connectivity of a region in Italy. The MC-SDSS model takes into account ecological and environmental spatial indicators which are combined by integrating the Analytic Network Process (ANP) and the Ordered Weighted Average (OWA) approach.

Some studies in the social science systems are discussed in specific applications of the fuzzy set theory: Castanon-Puga et al. [9] used type-2 fuzzy logic and multi-agent system to represent social systems since these are constitute of different parts of elements and these elements seem unrelated when considered individually. The nature of social systems demonstrates an uncertain behavior so fuzzy logic provides a good way to represent complex systems. del Acebo and de la Rosa [13] also considered a fuzzy system based approach to social modeling in multi-agent systems. The aim of this study is to show the necessity of multi-agent environments for a social model to determine the trust they can put in the assertions made by the other agents in the environment. And they have also proposed a mechanism based on fuzzy systems theory which allows them to satisfy that necessity.

Jeng and Tzeng [28] addressed the use of expert systems from a social science perspective. Their study investigated if social influence affects medical professionals' behavioral intention while introducing a new clinical decision support system. The fuzzy DEMATEL technique has been applied to address the vagueness of human judgments and has been used to explore cause and effect relationships from a top-down approach. Chatterjee et al. [10] applied fuzzy Analytical Network Process (FANP) for sensitivity level and ecological degradation analysis of the Keoladeo National Park which falls under the category of risk-prone wetland ecosystems in India.

Liwu et al. [59] considered the fuzzy decision system for an ecological distribution of area, which was established by using different factors and data. Because of the complexity of the system, L-R fuzzy numbers have been used in decision system. Taghizadeh et al. [50] evaluated the social responsibility activities related to the organizations and they used an expert system based fuzzy logic to measure the social responsibility activities of organizations. Hence, they evaluated the score of corporate social responsibility for the part-making companies in Iran. Márquez et al. [39] proposed a methodology to provide an option for the analysis of social problems. Owing to many interactions in social systems, modelling and analyzing them is not easy. The authorities utilized fuzzy logic, data mining theory and distributed agencies methods to explain these systems in a multidimensional perspective. Lin et al. [34] tried to explore the application of fuzzy logic control in the long term management of an ecological system. The results show that fuzzy logic control is better than with state feedback control or no control.

In the following works, it is focused on fuzzy cognitive maps (FCMs) as a tool to model and analyze causality in qualitative systems. Gray et al. [22] constructed a fuzzy logic cognitive map in order to evaluate differences and similarities in the structural and functional characteristics of stakeholder groups (managers, scientists, environmental NGOs) mental models about social–ecological systems. Gray et al. [21] gave a case study to examine the usefulness of a Fuzzy Cognitive Mapping (FCM) method to understand social-ecological systems (SESs). Vanwindekens et al. [53] introduced a fuzzy cognitive mapping approach in order to study the dynamic behavior of managers' system of practices. In the case study, they evaluated farmers' systems of practices based on their own conceptions. Ginis [20] considered a concept of modeling of complex systems. Fuzzy cognitive maps are proposed to analyze these complex systems which are difficult to construct and link between social and economic systems under uncertain environment. Elsawah et al. [15] proposed a methodology for integrating perceptions of stakeholders (qualitative) into formal simulation models (quantitative). Their methodology combines cognitive mapping and agent based modelling. The aim of their study is to contribute in order to develop understanding and communication about decision making in complex socio-ecological systems. Olazabal and Pascual [43] applied a participatory FCM approach in the context of urban low carbon energy transition planning for the city of Bilbao, Basque Country. They have drawn from previous work on fuzzy cognitive modelling applied to the resilience management of social–ecological systems [31] and on the potential of fuzzy thinking for addressing complex urban problems [23]. Mago et al. [36] employed a fuzzy cognitive map to model the interplay between high-level concepts, and cellular automata (CA) to model the low-level interactions between individual actors in environment which is characterized as a complex system. A FCM describes different aspects in the behavior of a complex system in terms of representing features of the system and interacting with each other.

Wozniak et al. [55] suggested a fuzzy inference system for the best possible surgical team for managing hospital schedules and surgical teams. The complex nature of social relationships requires fuzzy logic methods and they also presented a new usage of fuzzy logic methods in real life applications. Ocampo-Duque et al. [41] proposed a neuro fuzzy model to assess the ecological hazards in river basins. The proposed method combines with self-organizing maps and fuzzy inference system in order to give more reliable and powerful evaluation for aquatic ecosystem protection. Huang et al. [25] proposed an intelligent system which is an adaptive fuzzy control approach to investigate Kolmogorov's ecological system without any mathematical model. Lee et al. [33] examined the application of the fuzzy inference mechanism to develop a fuzzy expert system to determine stream water quality classification from ecological information under fuzziness.

It is evident from the previous studies in the literature, there has been no work using IVIF model to solve this problem in fuzzy environment. The contribution of the chapter lies in providing the extended TOPSIS with IVIF numbers to represent the implicit knowledge about socio-ecological decisions in terms of fuzzy rules which can be processed quantitatively to make reliable decisions.

3 Preliminaries of Interval-Valued Intuitionistic Fuzzy Sets

The fuzzy set theory was proposed by Zadeh [57] and has been applied in diverse fields. In the fuzzy set theory, the membership of an element to a fuzzy set (FS) is only a single value between zero and one [54]. Intuitionistic fuzzy sets (IFS), interval-valued fuzzy sets (IVFS) and interval-valued intuitionistic fuzzy sets (IVIFS) are extensions of the fuzzy set theory and they tackle with imprecision and uncertainty in different ways [45]. The concept of IFS has been introduced by Atanassov [4] as a generalization concept of FS. The theory of intuitionistic fuzzy sets is characterized by a membership degree, a non-membership degree, and a hesitation degree [56]. Atanassov and Gargov [3] introduced the concept of IVIFS as a further generalization of the fuzzy set theory. The basic characteristic of the IVFS and IVIFS is that the values of its membership and nonmembership functions are intervals rather than exact numbers [1]. Some basic concepts of fuzzy sets are as follows.

Definition 3.1. (*FS*) Let *X* is a set (space), with generic element of *X* denoted by *x*, that is $X = \{x\}$. Then a FS is defined as Eq. (1).

$$A = \{\langle x, \mu_A(x)\rangle | x \in X\} \tag{1}$$

where $\mu_A : X \rightarrow [0, 1]$ is the membership function of the FS *A*, $\mu_A(x) \in [0, 1]$ is the degree of membership of the element x to the set A.

Definition 3.2. (*IFS*) For a set X, an IFS *A* in the sense of Atanassov is given by Eq. (2).

$$A = \{\langle x, \mu_A(x), v_A(x)\rangle | x \in X\} \tag{2}$$

where the functions $\mu_A : X \rightarrow [0, 1]$ and $v_A : X \rightarrow [0, 1]$, with the condition $0 \leq \mu_A(x) + v_A(x) \leq 1, \forall x \in X$.

The numbers, $\mu_A(x) \in [0, 1]$ and $v_A(x) \in [0, 1]$, denote the degree of membership and the degree of nonmembership of the element x to the set A, respectively. For each IFS *A* in X, the amount $\pi_A(x) = 1 - (\mu_A(x) + v_A(x))$ is called the degree of indeterminacy (hesitation part) [1].

Definition 3.3. (*IVFS*) Let [I] be the set of all closed subintervals of the interval [0, 1] and $\mu = [\mu_L, \mu_U] \in [I]$, where μ_L and μ_U are the lower extreme and the upper extreme, respectively. For a set *X*, an IVFS A is given by Eq. (3).

$$A = \{\langle x, \mu_A(x)\rangle | x \in X\} \tag{3}$$

where the function $\mu_A : X \rightarrow$ [I] defines the degree of membership of an element x to A, and $\mu_A(x) = [\mu_{AL}(x), \mu_{AU}(x)]$ is called an interval-valued fuzzy number.

Definition 3.4. (*IVIFS*) For a set X, an IVIFS A is an object having the form Eq. (4).

$$A = \{\langle x, \mu_A(x), v_A(x)\rangle | x \in X\} \tag{4}$$

where $\mu_A\colon X \to [I]$ and $N_A\colon X \to [I]$ represent the degree of membership and non-membership, $0 \leq sup(\mu_A(x)) + sup(v_A(x)) \leq 1, \forall x \in X$.

$$\mu_A(x) = [\mu_{AL}(x), \mu_{AU}(x)]\, and\, v_A(x) = [v_{AL}(x), v_{AU}(x)] so$$

$A = ([\mu_{AL}(x), \mu_{AU}(x)], [v_{AL}(x), v_{AU}(x)])$. And the hesitation degree of an interval-valued intuitionistic fuzzy number x to set A can be defined as

$$\pi_A(x) = [1 - \mu_{AU}(x) - v_{AU}(x), 1 - \mu_{AL}(x) - v_{AL}(x)].$$

4 Interval Valued Intuitionistic Fuzzy TOPSIS

Technique for Order Performance by Similarity to Ideal Solution (TOPSIS) is one of the most classical methods developed by Hwang and Yoon [26] for solving MCDM problems. The underlying logic of TOPSIS method is to define the positive-ideal solution (PIS) and the negative-ideal solution (NIS). The TOPSIS was extended to various situations in which the decision data are fuzzy, interval-valued, or intuitionistic fuzzy sets.

In the literature, a number of fuzzy TOPSIS methods and applications with interval-valued intuitionistic fuzzy sets have been developed in recent years. The details of extended TOPSIS methodology can be seen extensively in [35, 46, 51, 56, 58].

The steps of interval-valued intuitionistic fuzzy TOPSIS method are given as follows [56]:

Step 1: Identify the evaluation attributes and alternatives

$X' = \{X'_i | i = 1, \ldots, n\}$ a finite set of possible alternatives and $A' = \{A'_j | j = 1, \ldots, m\}$ a finite set of attributes

Step 2: Construct the decision matrix

The decision makers cannot easily estimate an exact value to alternative with respect to attribute. Therefore, the decision makers k (k = 1, 2, …, K) utilize an interval-valued intuitionistic fuzzy number $\tilde{a}^k_{ij}$ to estimate their judgment on alternative with respect to attribute.

$\tilde{a}^k_{ij} = \langle \left[a^k_{ij}, b^k_{ij}\right], \left[c^k_{ij}, d^k_{ij}\right] \rangle$, where $\left[a^k_{ij}, b^k_{ij}\right]$ and $\left[c^k_{ij}, d^k_{ij}\right]$ denote the degree of membership and the degree of non-membership of the alternative X'_i with respect to the attribute A'_i.

Step 3: Determine PIS and NIS for each decision maker

First, we should identify PIS and NIS for each decision maker by using formulas (5) and (6).

$$\widetilde{R^{k+}} = \left(\widetilde{r_1^{k+}}, \widetilde{r_2^{k+}}, \ldots, \widetilde{r_m^{k+}}\right)$$
$$= \left(\left\langle\left[\widetilde{a_1^{k+}}, \widetilde{b_1^{k+}}\right], \left[\widetilde{c_1^{k+}}, \widetilde{d_1^{k+}}\right]\right\rangle, \left\langle\left[\widetilde{a_2^{k+}}, \widetilde{b_2^{k+}}\right], \left[\widetilde{c_2^{k+}}, \widetilde{d_2^{k+}}\right]\right\rangle, \ldots, \left\langle\left[\widetilde{a_m^{k+}}, \widetilde{b_m^{k+}}\right], \left[\widetilde{c_m^{k+}}, \widetilde{d_m^{k+}}\right]\right\rangle\right) \quad (5)$$

$$\widetilde{R^{k-}} = \left(\widetilde{r_1^{k-}}, \widetilde{r_2^{k-}}, \ldots, \widetilde{r_m^{k-}}\right)$$
$$= \left(\left\langle\left[\widetilde{a_1^{k-}}, \widetilde{b_1^{k-}}\right], \left[\widetilde{c_1^{k-}}, \widetilde{d_1^{k-}}\right]\right\rangle \left\langle\left[\widetilde{a_2^{k-}}, \widetilde{b_2^{k-}}\right], \left[\widetilde{c_2^{k-}}, \widetilde{d_2^{k-}}\right]\right\rangle, \ldots, \left\langle\left[\widetilde{a_m^{k-}}, \widetilde{b_m^{k-}}\right], \left[\widetilde{c_m^{k-}}, \widetilde{d_m^{k-}}\right]\right\rangle\right) \quad (6)$$

where

$$\widetilde{r^{k+}} = \left\langle\left[\widetilde{a_j^{k+}}, \widetilde{b_j^{k+}}\right], \left[\widetilde{c_j^{k+}}, \widetilde{d_j^{k+}}\right]\right\rangle = \left\langle\left[\max_i a_{ij}^k, \max_i b_{ij}^k\right], \left[\min_i c_{ij}^k, \min_i d_{ij}^k\right]\right\rangle,$$
$$\mathrm{i} = 1, \ldots, \mathrm{n}, \quad \mathrm{j} = 1, \ldots, \mathrm{m}, \quad \mathrm{k} = 1, \ldots, \mathrm{K}$$
$$\widetilde{r^{k-}} = \left\langle\left[\widetilde{a_j^{k-}}, \widetilde{b_j^{k-}}\right], \left[\widetilde{c_j^{k-}}, \widetilde{d_j^{k-}}\right]\right\rangle = \left\langle\left[\min_i a_{ij}^k, \min_i b_{ij}^k\right], \left[\max_i c_{ij}^k, \max_i d_{ij}^k\right]\right\rangle,$$
$$\mathrm{i} = 1, \ldots, \mathrm{n}, \quad \mathrm{j} = 1, \ldots, \mathrm{m}, \quad \mathrm{k} = 1, \ldots, \mathrm{K}$$

Step 4: Calculate separation measures
We can define the separation measure degree between alternative X_i' and the ideal solutions (PIS /NIS) for each decision maker by using formula (7) and (8).

$$D_i^{k+} = \sqrt{\frac{1}{2}\sum_{j=1}^{m} w_j\left\{\left(a_{ij}^k - \widetilde{a_j^{k+}}\right)^2 + \left(b_{ij}^k - \widetilde{b_j^{k+}}\right)^2 + \left(c_{ij}^k - \widetilde{c_j^{k+}}\right)^2 + \left(d_{ij}^k - \widetilde{d_j^{k+}}\right)^2 + \left(\pi_{ijL}^k - \widetilde{\pi_{jL}^{k+}}\right)^2 + \left(\pi_{ijU}^k - \widetilde{\pi_{jU}^{k+}}\right)^2\right\}} \quad (7)$$

where $\pi_{ijL}^k = 1 - b_{ij}^k - d_{ij}^k$, $\pi_{ijU}^k = 1 - a_{ij}^k - c_{ij}^k$, $\widetilde{\pi_{jL}^{k+}} = 1 - \widetilde{b_j^{k+}} - \widetilde{d_j^{k+}}$, $\widetilde{\pi_{jU}^{k+}} = 1 - \widetilde{a_j^{k+}} - \widetilde{c_j^{k+}}$, i = 1, …, n, j = 1, …, m, k = 1,…,K.

$$D_i^{k-} = \sqrt{\frac{1}{2}\sum_{j=1}^{m} w_j \left\{ \left(a_{ij}^k - \widetilde{a_j^{k-}}\right)^2 + \left(b_{ij}^k - \widetilde{b_j^{k-}}\right)^2 + \left(c_{ij}^k - \widetilde{c_j^{k-}}\right)^2 + \left(d_{ij}^k - \widetilde{d_j^{k-}}\right)^2 + \left(\pi_{ijL}^k - \widetilde{\pi_{jL}^{k-}}\right)^2 + \left(\pi_{ijU}^k - \widetilde{\pi_{jU}^{k-}}\right)^2 \right\}} \quad (8)$$

where $\pi_{ijL}^k = 1 - b_{ij}^k - d_{ij}^k$, $\pi_{ijU}^k = 1 - a_{ij}^k - c_{ij}^k$, $\widetilde{\pi_{jL}^{k-}} = 1 - \widetilde{b_j^{k-}} - \widetilde{d_j^{k-}}$, $\widetilde{\pi_{jU}^{k-}} = 1 - \widetilde{a_j^{k-}} - \widetilde{c_j^{k-}}$, i = 1, ..., n, j = 1, ..., m, k = 1, ..., K.

Step 5: Aggregate the separation measures

We can aggregate the separation measures for the group by using formulas (9) and (10).

$$D_i^+ = \sum_{k=1}^{K} \left(\lambda_k D_i^{k+}\right) \quad \text{for alternative i} \quad (9)$$

$$D_i^- = \sum_{k=1}^{K} \left(\lambda_k D_i^{k-}\right) \quad \text{for alternative i} \quad (10)$$

where i = 1, 2, ..., n; k = 1, 2, ..., K, and λ_k is the weight of decision maker k.

Step 6: Calculate the closeness coefficients and rank the alternatives

The closeness coefficient of the candidate $X_i^{'}$ is defined as:

$$U_i = \frac{D_i^-}{D_i^- + D_i^+}, \quad \mathrm{i} = 1, 2, \ldots, \mathrm{n} \quad (11)$$

Finally we can determine the ranking order of all alternatives, according to closeness coefficients. Finally we can select best option.

5 An Illustrative Example

In this section, we consider a multi attribute group decision making problem to illustrate the use of the fuzzy set theory in social sciences. Illustrative example is adapted from case study of Garmendia and Gamboa [19].

The illustrative example focuses on the sustainable management of a district, which is a part of the World Network of Biosphere Reserves by UNESCO owing to its natural and cultural value. A variety of stakeholders (tourism, fishing, industry, agriculture etc.) coexist in the district. In complex socio-ecological systems incorporating many stakeholders, decision makers are faced with serious difficulties. Coexistence of the various interest groups in the district is not always easy, and many conflicts have risen in recent years.

Firstly, actors from various social groups which include sociologists, economists, engineers, ecologists, and local actors are determined to evaluate different sustainable management options for the district. A committee comprised of five decision-makers D^k (k = 1, 2, …, 5) which are shipyard labour union (D^1); environmental organizations (D^2); shipyard (D^3); local council (D^4); fishers (D^5), respectively, from each strategic decision area has been set up to provide assessment information on the criteria.

In order to select the best environment option, a committee is set up to provide evaluations based on six criteria $x_i (i = 1, 2, \ldots, 6)$. These are summarized as follows:

1. Employment (x_1)—with the revival of trade, local economic activities and employment are enhanced, job stability is guaranteed.
2. Concord between socio-ecological activities (x_2)—compatibility between fishing, tourism, surfing, industry, and conservation should be ensured.
3. Cost of implementation (x_3)—budget constraints must be taken into account.
4. Environmental degeneration (x_4)—environmental quality of the area should be conserved.
5. Impact on ecosystem (x_5)—impact over habitat and fauna should be reduced.
6. Reversibility (x_6)—potential and the dynamics of the area should be maintained for the future

The criteria are independent and their weights are determined using pairwise comparisons by social actors. The weights of criteria are shown in Table 1.

In this example, four alternatives that represent sustainable management options for the district are determined $A = (A_1, A_2, A_3, A_4)$. These are summarized as follows:

(A_1)—do not allow dredging.
(A_2)—do not allow dredging and direct all the public resources into conservation measures for the district.
(A_3)—Minimum dredging according to the system's situation.
(A_4)—Satisfy demand from industry (shipyard) with a maximum dredging.

Table 1 Weights of criteria according to social actors' preferences

Social actors	Criteria					
	Employment	Socio-ecological concord	Cost of implementation	Environmental degeneration	Impact on ecosystem	Reversibility
D^1	0.32	0.05	0.23	0.23	0.03	0.14
D^2	0.20	0.23	0.05	0.22	0.25	0.05
D^3	0.27	0.23	0.05	0.18	0.18	0.09
D^4	0.13	0.20	0.20	0.13	0.13	0.20
D^5	0.18	0.21	0.06	0.15	0.21	0.21

Table 2 Linguistic scale

Linguistic terms	IVIFNs
Absolutely low (AL)	([0.0, 0.2], [0.5, 0.8])
Very low (VL)	([0.1, 0.3], [0.4, 0.7])
Low (L)	([0.2, 0.4], [0.3, 0.6])
Medium low (ML)	([0.3, 0.5], [0.2, 0.5])
Approximately equal (E)	([0.4, 0.6], [0.2, 0.4])
Medium high (MH)	([0.5, 0.7], [0.1, 0.3])
High (H)	([0.6, 0.8], [0.0, 0.2])
Very high (VH)	([0.7, 0.9], [0.0, 0.1])
Absolutely high (AH)	([0.8, 1.0], [0.0, 0.0])

In this example, in order to evaluate alternatives the scale proposed by Onar et al. [44] is used. The scale in Table 2 can be used by decision-makers to state their individual opinions and then, their lingual judgments are transformed into the corresponding IVIFNs.

The linguistic judgments given by the decision makers to four alternatives are shown in Table 3. Decision-makers state their individual opinions by using linguistic scale in Table 1. Then their linguistic judgments are transformed into the corresponding IVIFNs.

By using formulas (5) and (6), we can determine the PIS and NIS for each decision maker. Table 4 includes the values of PIS and NIS. Then, we can calculate the separation measure from the positive ideal and the negative ideal solutions, D_j^{k+} and D_j^{k-} respectively, for each decision maker by using formula (7) and (8). Calculated separation measure values are shown in Table 5.

By using formulas (9) and (10), separation measures for the group are aggregated. Finally, the aggregated relative closeness coefficients of all alternatives are calculated by using formula (11). Closeness coefficients and ranking of alternatives with different DM weights are shown in Table 6.

According to the results, A_2 is the best candidate in all of the cases. Moreover, Table 7 presents the rankings of options of sustainable management according to the preferences of each social group (DM). As shown in Table 7, Decision Makers 1, 3, 4, and 5, would prefer to restrict dredging activities and enhance conservation measures through ecosystem recovery plans, while improving the quality of the environment (alternative A_2).

6 Conclusion

Nowadays, the application of the fuzzy sets to social sciences is getting larger interest since the fuzzy logic is very appropriate to be used in real life problems. The fuzzy set theory presents an efficient way in order to overcome these complex systems' vagueness to determine the human-environment interactions clearly.

Table 3 The decision matrix with interval-valued intuitionistic fuzzy numbers for each DM

Social actors (DMs)	Alternatives	Criteria					
		Employment	Socio-ecological concord	Cost of implementation	Environmental degeneration	Impact on ecosystem	Reversibility
D^1	A_1	AH	VH	VH	VH	H	H
	A_2	VH	AH	VH	AH	AH	VH
	A_3	MH	AH	MH	H	MH	H
	A_4	E	H	ML	MH	MH	MH
D^2	A_1	VH	VH	H	VH	VH	VH
	A_2	H	VH	H	AH	H	VH
	A_3	MH	H	MH	H	MH	H
	A_4	E	MH	H	MH	MH	H
D^3	A_1	AH	VH	H	VH	H	VH
	A_2	VH	AH	VH	VH	VH	AH
	A_3	H	AH	H	AH	H	H
	A_4	AH	VH	VH	H	VH	VH
D^4	A_1	E	H	H	MH	H	AH
	A_2	H	VH	VH	VH	VH	AH
	A_3	MH	VH	AH	H	VH	VH
	A_4	ML	MH	MH	MH	H	H
D^5	A_1	VH	VH	VH	MH	H	E
	A_2	AH	AH	VH	VH	VH	H
	A_3	VH	VH	H	H	VH	H
	A_4	VH	VH	MH	E	H	MH

Table 4 The positive ideal solution (PIS) and negative ideal solution (NIS) for each DM

DM	Ideal solution	Criteria					
		Employment	Socio-ecological concord	Cost of implementation	Environmental degeneration	Impact on ecosystem	Reversibility
D^1	PIS	([0.8, 1.0], [0.0, 0.0])	([0.8, 1.0], [0.0, 0.0])	([0.7, 0.9], [0.0, 0.1])	([0.8, 1.0], [0.0, 0.0])	([0.8, 1.0], [0.0, 0.0])	([0.7, 0.9], [0.0, 0.1])
	NIS	([0.4, 0.6], [0.2, 0.4])	([0.6, 0.8], [0.0, 0.2])	([0.3, 0.5], [0.2, 0.5])	([0.5, 0.7], [0.1, 0.3])	([0.5, 0.7], [0.1, 0.3])	([0.5, 0.7], [0.1, 0.3])
D^2	PIS	([0.7, 0.9], [0.0, 0.1])	([0.7, 0.9], [0.0, 0.1])	([0.6, 0.8], [0.0, 0.2])	([0.8, 1.0], [0.0, 0.0])	([0.7, 0.9], [0.0, 0.1])	([0.7, 0.9], [0.0, 0.1])
	NIS	([0.4, 0.6], [0.2, 0.4])	([0.5, 0.7], [0.1, 0.3])	([0.5, 0.7], [0.1, 0.3])	([0.5, 0.7], [0.1, 0.3])	([0.5, 0.7], [0.1, 0.3])	([0.6, 0.8], [0.0, 0.2])
D^3	PIS	([0.8, 1.0], [0.0, 0.0])	([0.8, 1.0], [0.0, 0.0])	([0.7, 0.9], [0.0, 0.1])	([0.8, 1.0], [0.0, 0.0])	([0.7, 0.9], [0.0, 0.1])	([0.8, 1.0], [0.0, 0.0])
	NIS	([0.6, 0.8], [0.0, 0.2])	([0.7, 0.9], [0.0, 0.1])	([0.6, 0.8], [0.0, 0.2])	([0.6, 0.8], [0.0, 0.2])	([0.6, 0.8], [0.0, 0.2])	([0.6, 0.8], [0.0, 0.2])
D^4	PIS	([0.6, 0.8], [0.0, 0.2])	([0.7, 0.9], [0.0, 0.1])	([0.8, 1.0], [0.0, 0.0])	([0.7, 0.9], [0.0, 0.1])	([0.7, 0.9], [0.0, 0.1])	([0.8, 1.0], [0.0, 0.0])
	NIS	([0.3, 0.5], [0.2, 0.5])	([0.5, 0.7], [0.1, 0.3])	([0.5, 0.7], [0.1, 0.3])	([0.5, 0.7], [0.1, 0.3])	([0.6, 0.8], [0.0, 0.2])	([0.6, 0.8], [0.0, 0.2])
D^5	PIS	([0.8, 1.0], [0.0, 0.0])	([0.8, 1.0], [0.0, 0.0])	([0.7, 0.9], [0.0, 0.1])	([0.7, 0.9], [0.0, 0.1])	([0.7, 0.9], [0.0, 0.1])	([0.6, 0.8], [0.0, 0.2])
	NIS	([0.7, 0.9], [0.0, 0.1])	([0.7, 0.9], [0.0, 0.1])	([0.5, 0.7], [0.1, 0.3])	([0.4, 0.6], [0.2, 0.4])	([0.6, 0.8], [0.0, 0.2])	([0.4, 0.6], [0.2, 0.4])

Table 5 Separation measures

DMs	Alt.	Separation measures	
D^1	A_1	$D_1^{1+} = 0.1028$	$D_1^{1-} = 0.4195$
	A_2	$D_2^{1+} = 0.0806$	$D_2^{1-} = 0.4142$
	A_3	$D_3^{1+} = 0.3064$	$D_3^{1-} = 0.1837$
	A_4	$D_4^{1+} = 0.4588$	$D_4^{1-} = 0.0000$
D^2	A_1	$D_1^{2+} = 0.0663$	$D_1^{2-} = 0.2882$
	A_2	$D_2^{2+} = 0.0949$	$D_2^{2-} = 0.2335$
	A_3	$D_3^{2+} = 0.2361$	$D_3^{2-} = 0.1141$
	A_4	$D_4^{2+} = 0.3192$	$D_4^{2-} = 0.0314$
D^3	A_1	$D_1^{3+} = 0.1206$	$D_1^{3-} = 0.1652$
	A_2	$D_2^{3+} = 0.0953$	$D_2^{3-} = 0.1596$
	A_3	$D_3^{3+} = 0.1835$	$D_3^{3-} = 0.1381$
	A_4	$D_4^{3+} = 0.1446$	$D_4^{3-} = 0.1679$
D^4	A_1	$D_1^{4+} = 0.2066$	$D_1^{4-} = 0.1632$
	A_2	$D_2^{4+} = 0.0632$	$D_2^{4-} = 0.2781$
	A_3	$D_3^{4+} = 0.0967$	$D_3^{4-} = 0.2542$
	A_4	$D_4^{4+} = 0.3087$	$D_4^{4-} = 0.0000$
D^5	A_1	$D_1^{5+} = 0.1766$	$D_1^{5-} = 0.1260$
	A_2	$D_2^{5+} = 0.0000$	$D_2^{5-} = 0.2364$
	A_3	$D_3^{5+} = 0.1085$	$D_3^{5-} = 0.1831$
	A_4	$D_4^{5+} = 0.2086$	$D_4^{5-} = 0.0641$

Table 6 Closeness coefficients and ranking of alternatives with different DM weights

Case	Weights	Alternatives	Aggregated positive separation measure	Aggregated negative separation measure	U_i	Ranking
Case 1	D^1 ($\lambda_1 = 0.20$)	A_1	0.1346	0.2324	0.6333	2
	D^2 ($\lambda_2 = 0.20$)	A_2	0.0668	0.2722	0.8030	1
	D^3 ($\lambda_3 = 0.20$)	A_3	0.1862	0.1746	0.4839	3
	D^4 ($\lambda_4 = 0.20$)	A_4	0.2880	0.0527	0.1547	4
	D^5 ($\lambda_5 = 0.20$)					
Case 2	D^1 ($\lambda_1 = 0.10$)	A_1	0.1575	0.1924	0.5499	3
	D^2 ($\lambda_2 = 0.10$)	A_2	0.0587	0.2570	0.8140	1
	D^3 ($\lambda_3 = 0.20$)	A_3	0.1519	0.1921	0.5585	2
	D^4 ($\lambda_4 = 0.35$)	A_4	0.2669	0.0528	0.1650	4
	D^5 ($\lambda_5 = 0.25$)					
Case 3	D^1 ($\lambda_1 = 0.30$)	A_1	0.1413	0.2493	0.6383	2
	D^2 ($\lambda_2 = 0.10$)	A_2	0.0717	0.2905	0.8021	1
	D^3 ($\lambda_3 = 0.20$)	A_3	0.1921	0.1887	0.4956	3
	D^4 ($\lambda_4 = 0.30$)	A_4	0.3119	0.0431	0.1215	4
	D^5 ($\lambda_5 = 0.10$)					

Table 7 Ranking of alternatives according to individual preferences

	Ranking			
	First	Second	Third	Fourth
D^1	A_2	A_1	A_3	A_4
D^2	A_1	A_2	A_3	A_4
D^3	A_2	A_1	A_4	A_3
D^4	A_2	A_3	A_1	A_4
D^5	A_2	A_3	A_1	A_4

Various stakeholders from social, environmental, technical and economical platform take part in socio-ecological systems and they try to reach a decision under various and conflicting criteria in a multidimensional problem. It is often a challenge for decision makers to make a decision which will satisfy all interest group from diverse fields. Multi-criteria decision making tools are very effective to respond from different dimension demand.

In this chapter the literature on socio-ecological systems has been reviewed and the lack of socio-ecological studies under fuzziness has been observed. With a case study, we have demonstrated how to incorporate socio-ecological considerations into sustainable management options which consist of a wide variety of social actors from an integrated and inclusive perspective using the interval-valued intuitionistic fuzzy TOPSIS method.

For further research, we suggest that other extensions of ordinary fuzzy sets such as hesitant fuzzy sets or type-2 fuzzy sets to be used in the proposed fuzzy TOPSIS method. The obtained results through these extensions should be compared with the ones we obtained in this paper. Besides, the handled problem can be expanded to a social, economical and ecological problem to consider economical factors as well.

References

1. Ahn, J., Han, K., Oh, S., Lee, C.: An application of interval-valued intuitionistic fuzzy sets for medical diagnosis of headache. Int. J. Innov. Comput. Inf. Control, 2755–2762 (2011)
2. Ananda, J., Herath, G.: A critical review of multi-criteria decision making methods with special reference to forest management and planning. Ecol. Econ. **68**, 2535–2548 (2009). doi:10.1016/j.ecolecon.2009.05.010
3. Atanassov, K., Gargov, G.: Interval valued intuitionistic fuzzy sets. Fuzzy Sets Syst. **31**, 343–349 (1989). doi:10.1016/0165-0114(89)90205-4
4. Atanassov, K.T.: Intuitionistic fuzzy sets. Fuzzy Sets Syst. **20**, 87–96 (1986). doi:10.1016/S0165-0114(86)80034-3
5. Bell, N., Schuurman, N., Hayes, M.V.: Using GIS-based methods of multicriteria analysis to construct socio-economic deprivation indices. Int. J. Health Geogr. **6**, 17 (2007). doi:10.1186/1476-072X-6-17
6. Berkes, F., Folke, C. (eds.): Linking Social and Ecological Systems: Management Practices and Social Mechanisms for Building Resilience, Transferred to Digital Printing. Cambridge Univ. Press, Cambridge (2002)

7. Bryan, B.A., Crossman, N.D.: Systematic regional planning for multiple objective natural resource management. J. Environ. Manage. **88**, 1175–1189 (2008). doi:10.1016/j.jenvman.2007.06.003
8. Buchholz, T., Rametsteiner, E., Volk, T.A., Luzadis, V.A.: Multi Criteria Analysis for bioenergy systems assessments. Energy Policy **37**, 484–495 (2009). doi:10.1016/j.enpol.2008.09.054
9. Castanon-Puga, M., Gaxiola-Pacheco, C., Castro, J.R., Martinez, R.J., Flores, D.-L.: Towards A Multi-Dimensional Modelling Of Complex Social Systems Using Data Mining And Type-2 Neuro-Fuzzy System: Religious Affiliation Case Of Study, pp. 136–142. ECMS (2012) doi:10.7148/2012-0136-0142
10. Chatterjee, K., Bandyopadhyay, A., Ghosh, A., Kar, S.: Assessment of environmental factors causing wetland degradation, using fuzzy analytic network process: a case study on Keoladeo National Park. India. Ecol. Model. **316**, 1–13 (2015). doi:10.1016/j.ecolmodel.2015.07.029
11. Chen, Z., Yang, W.: A new multiple attribute group decision making method in intuitionistic fuzzy setting. Appl. Math. Model. **35**, 4424–4437 (2011). doi:10.1016/j.apm.2011.03.015
12. Cumming, G.S.: Spatial Resilience in Social-Ecological Systems. Springer, Dordrecht (2011)
13. del Acebo, E., de la Rosa, J.L.: A Fuzzy System Based Approach to Social Modeling in Multi-agent Systems, p. 463. ACM Press (2002). doi:10.1145/544741.544850
14. Dimitrov, V., Hodge, B.: Why does fuzzy logic need the challenge of social complexity? In: Dimitrov, V., Korotkich, V. (eds.) Fuzzy Logic, pp. 27–44. Physica-Verlag HD, Heidelberg (2002)
15. Elsawah, S., Guillaume, J.H.A., Filatova, T., Rook, J., Jakeman, A.J.: A methodology for eliciting, representing, and analysing stakeholder knowledge for decision making on complex socio-ecological systems: from cognitive maps to agent-based models. J. Environ. Manage. **151**, 500–516 (2015). doi:10.1016/j.jenvman.2014.11.028
16. Ferretti, V., Pomarico, S.: Ecological land suitability analysis through spatial indicators: an application of the analytic network process technique and ordered weighted average approach. Ecol. Indic. **34**, 507–519 (2013). doi:10.1016/j.ecolind.2013.06.005
17. Folke, C.: Resilience: the emergence of a perspective for social–ecological systems analyses. Glob. Environ. Change **16**, 253–267 (2006). doi:10.1016/j.gloenvcha.2006.04.002
18. Fontana, V., Radtke, A., Bossi Fedrigotti, V., Tappeiner, U., Tasser, E., Zerbe, S., Buchholz, T.: Comparing land-use alternatives: using the ecosystem services concept to define a multi-criteria decision analysis. Ecol. Econ. **93**, 128–136 (2013). doi:10.1016/j.ecolecon.2013.05.007
19. Garmendia, E., Gamboa, G.: Weighting social preferences in participatory multi-criteria evaluations: a case study on sustainable natural resource management. Ecol. Econ. **84**, 110–120 (2012). doi:10.1016/j.ecolecon.2012.09.004
20. Ginis, L.A.: The use of fuzzy cognitive maps for the analysis of structure of social and economic system for the purpose of its sustainable development. Mediterr. J. Soc. Sci. (2015). doi:10.5901/mjss.2015.v6n3s5p113
21. Gray, S.A., Gray, S., De Kok, J.L., Helfgott, A.E.R., O'Dwyer, B., Jordan, R., Nyaki, A.: Using fuzzy cognitive mapping as a participatory approach to analyze change, preferred states, and perceived resilience of social-ecological systems. Ecol. Soc. **20** (2015). doi:10.5751/ES-07396-200211
22. Gray, S., Chan, A., Clark, D., Jordan, R.: Modeling the integration of stakeholder knowledge in social–ecological decision-making: benefits and limitations to knowledge diversity. Ecol. Model. **229**, 88–96 (2012). doi:10.1016/j.ecolmodel.2011.09.011
23. Habib, F., Shokoohi, A.: Classification and resolving urban problems by means of fuzzy approach. World Acad. Sci. Eng. Technol. Int. Sci. Index 36 **3**(12), 774–781 (2009)
24. Huang, I.B., Keisler, J., Linkov, I.: Multi-criteria decision analysis in environmental sciences: Ten years of applications and trends. Sci. Total Environ. **409**, 3578–3594 (2011). doi:10.1016/j.scitotenv.2011.06.022

25. Huang, Z., Chen, S., Xia, Y.: Incorporate intelligence into an ecological system: An adaptive fuzzy control approach. Appl. Math. Comput. **177**, 243–250 (2006). doi:10.1016/j.amc.2005.11.004
26. Hwang, C.-L., Yoon, K.: Multiple attribute decision making. Lecture Notes in Economics and Mathematical Systems. Springer, Berlin, Heidelberg (1981)
27. Janssen, M.: Complexity and Ecosystem Management: The Theory and Practice of Multi-agent Systems. Edward Elgar Pub, Chelteham, UK, Northhampton, MA (2002)
28. Jeng, D. J.-F., Tzeng, G.-H.: Social influence on the use of Clinical Decision Support Systems: Revising the Unified Theory of Acceptance and Use of Technology by the fuzzy DEMATEL technique. Comput. Ind. Eng. **62**, 819–828 (2012)
29. Joerin, F., Thériault, M., Musy, A.: Using GIS and outranking multicriteria analysis for land-use suitability assessment. Int. J. Geogr. Inf. Sci. **15**, 153–174 (2001). doi:10.1080/13658810051030487
30. Karnatak, H.C., Saran, S., Bhatia, K., Roy, P.S.: Multicriteria spatial decision analysis in web GIS environment. GeoInformatica **11**, 407–429 (2007). doi:10.1007/s10707-006-0014-8
31. Kok, K.: The potential of fuzzy cognitive Maps for semi-quantitative scenario development, with an example from Brazil. Glob. Environ. Change **19**, 122–133 (2009). doi:10.1016/j.gloenvcha.2008.08.003
32. Lachassagne, P., Wyns, R., Bérard, P., Bruel, T., Chéry, L., Coutand, T., Desprats, J.-F., Strat, P.: Exploitation of high-yields in hard-rock aquifers: downscaling methodology combining GIS and multicriteria analysis to delineate field prospecting zones. Ground Water **39**, 568–581 (2001). doi:10.1111/j.1745-6584.2001.tb02345.x
33. Lee, H., Oh, K., Park, D., Jung, J., Yoon, S.: Fuzzy expert system to determine stream water quality classification from ecological information. Water Sci. Technol. **36**, 199–206 (1997). doi:10.1016/S0273-1223(97)00732-4
34. Lin, C.-M., Mon, Y.-J., Maa, J.-H.: Ecological systems control by fuzzy logic controller. Asian J. Control **2**, 274–280 (2008). doi:10.1111/j.1934-6093.2000.tb00032.x
35. Li, W., Guo, G., Yue, C., Zhao, Y.: Dynamic programming methodology for multi-criteria group decision-making under ordinal preferences. J. Syst. Eng. Electron. **21**, 975–980 (2010). doi:10.3969/j.issn.1004-4132.2010.06.008
36. Mago, V.K., Bakker, L., Papageorgiou, E.I., Alimadad, A., Borwein, P., Dabbaghian, V.: Fuzzy cognitive maps and cellular automata: an evolutionary approach for social systems modelling. Appl. Soft Comput. **12**, 3771–3784 (2012). doi:10.1016/j.asoc.2012.02.020
37. Malczewski, J.: Ordered weighted averaging with fuzzy quantifiers: GIS-based multicriteria evaluation for land-use suitability analysis. Int. J. Appl. Earth Obs. Geoinformation **8**, 270–277 (2006). doi:10.1016/j.jag.2006.01.003
38. Malekmohammadi, B., Rahimi Blouchi, L.: Ecological risk assessment of wetland ecosystems using multi criteria decision making and geographic information system. Ecol. Indic. **41**, 133–144 (2014). doi:10.1016/j.ecolind.2014.01.038
39. Márquez, B.Y., Castañon-Puga, M., Castro, J.R., Suarez, D.: Methodology for the Modeling of Complex Social System Using Neuro-Fuzzy and Distributed Agencies. J. Sel. Areas Softw. Eng, JSSE (2011)
40. Obiedat, M., Samarasinghe, S.: Fuzzy representation and aggregation of fuzzy cognitive maps. In: 20th International Congress Modeling Simulation (2013)
41. Ocampo-Duque, W., Juraske, R., Kumar, V., Nadal, M., Domingo, J.L., Schuhmacher, M.: A concurrent neuro-fuzzy inference system for screening the ecological risk in rivers. Environ. Sci. Pollut. Res. **19**, 983–999 (2012). doi:10.1007/s11356-011-0595-0
42. Oikonomou, V., Dimitrakopoulos, P.G., Troumbis, A.Y.: Incorporating ecosystem function concept in environmental planning and decision making by means of multi-criteria evaluation: the case-study of Kalloni, Lesbos. Greece. Environ. Manage. **47**, 77–92 (2011). doi:10.1007/s00267-010-9575-2
43. Olazabal, M., Pascual, U.: Use of fuzzy cognitive maps to study urban resilience and transformation. Innov. Soc. Transit, Environ (2015). doi:10.1016/j.eist.2015.06.006

44. Onar, S.C., Oztaysi, B., Otay, İ., Kahraman, C.: Multi-expert wind energy technology selection using interval-valued intuitionistic fuzzy sets. Energy (2015). doi:10.1016/j.energy.2015.06.086
45. Park, J.H., Lim, K.M., Park, J.S., Kwun, Y.C.: Distances between interval-valued intuitionistic fuzzy sets. J. Phys: Conf. Ser. **96**, 012089 (2008). doi:10.1088/1742-6596/96/1/012089
46. Park, J.H., Park, I.Y., Kwun, Y.C., Tan, X.: Extension of the TOPSIS method for decision making problems under interval-valued intuitionistic fuzzy environment. Appl. Math. Model. **35**, 2544–2556 (2011). doi:10.1016/j.apm.2010.11.025
47. Ragin, C.C.: Fuzzy sets and social research. Sociol. Methods Res. **33**, 423–430 (2005). doi:10.1177/0049124105274499
48. Roetter, R.P., Hoanh, C.T., Laborte, A.G., Van Keulen, H., Van Ittersum, M.K., Dreiser, C., Van Diepen, C.A., De Ridder, N., Van Laar, H.H.: Integration of systems network (SysNet) tools for regional land use scenario analysis in Asia. Environ. Model Softw. **20**, 291–307 (2005). doi:10.1016/j.envsoft.2004.01.001
49. Social-ecological systems—Stockholm Resilience Centre [WWW Document], 2007. URL http://stockholmresilience.org/21/research/what-is-resilience/research-background/research-framework/social-ecological-systems.html. Accessed 15 Sept 2015
50. Taghizadeh, H., Fasghandis, G.S., Zeinalzadeh, A.: Evaluation of corporate social responsibility using fuzzy expert system. Res. J. Appl. Sci. Eng. Technol. **6**, 3047–3053 (2013)
51. Tan, C.: A multi-criteria interval-valued intuitionistic fuzzy group decision making with Choquet integral-based TOPSIS. Expert Syst. Appl. **38**, 3023–3033 (2011). doi:10.1016/j.eswa.2010.08.092
52. Vaisey, S.: Fuzzy set theory: applications in the social sciences. Sage. Sociol. Methods Res. **37**, 455–457 (2009) (Book Review: Smithson, M., Verkuilen, J.: Thousand Oaks, CA, 2006). doi:10.1177/0049124107306675
53. Vanwindekens, F.M., Stilmant, D., Baret, P.V.: The relevance of fuzzy cognitive mapping approaches for assessing adaptive capacity and resilience in social-ecological systems. In: Papadopoulos, H., Andreou, A.S., Iliadis, L., Maglogiannis, I. (eds.) Artificial Intelligence Applications and Innovations, pp. 587–596. Springer, Berlin, Heidelberg (2013)
54. Wang, W., Xin, X.: Distance measure between intuitionistic fuzzy sets. Pattern Recognit. Lett. **26**, 2063–2069 (2005). doi:10.1016/j.patrec.2005.03.018
55. Wozniak, P., Jaworski, T., Fiderek, P., Kucharski, J., Romanowski, A.: clinical activity and schedule management with a fuzzy social preference system. In: Nguyen, N.T., Trawiński, B., Katarzyniak, R., Jo, G.-S. (eds.) Advanced Methods for Computational Collective Intelligence, pp. 345–354. Springer, Berlin, Heidelberg (2013)
56. Ye, F.: An extended TOPSIS method with interval-valued intuitionistic fuzzy numbers for virtual enterprise partner selection. Expert Syst. Appl. **37**, 7050–7055 (2010). doi:10.1016/j.eswa.2010.03.013
57. Zadeh, L.A.: Fuzzy sets. Inf. Control **8**, 338–353 (1965). doi:10.1016/S0019-9958(65)90241-X
58. Zhang, H., Yu, L.: MADM method based on cross-entropy and extended TOPSIS with interval-valued intuitionistic fuzzy sets. Knowl. Based Syst. **30**, 115–120 (2012). doi:10.1016/j.knosys.2012.01.003
59. Liwu, Z.H.U., Shaowen, J.B.L.I., Juanjuan, K.O.N.G.: Fuzzy decision system for ecological distribution of citrus in north-cultivated-marginal area. Chin. J. Appl, Ecol (2003). 502

A Survey on Models and Methods for Solving Fuzzy Linear Programming Problems

Ali Ebrahimnejad and José L. Verdegay

Abstract Fuzzy Linear Programming (FLP), as one of the main branches of operation research, is concerned with the optimal allocation of limited resources to several competing activities on the basis of given criteria of optimality in fuzzy environment. Numerous researchers have studied various properties of FLP problems and proposed different approaches for solving them. This work presents a survey on models and methods for solving FLP problems. The solution approaches are divided into four areas: (1) Linear Programming (LP) problems with fuzzy inequalities and crisp objective function, (2) LP problems with crisp inequalities and fuzzy objective function, (3) LP problems with fuzzy inequalities and fuzzy objective function and (4) LP problems with fuzzy parameters. In the first area, the imprecise right-hand-side of the constraints is specified with a fuzzy set and a monotone membership function expressing the individual satisfaction of the decision maker. In the second area, a fuzzy goal and corresponding tolerance is defined for each coefficient of decision variables in the objective function. In the third area, both the coefficients of decision variables in the objective function and the right-hand-side of the constraints are specified by fuzzy goals and corresponding tolerances. In the fourth area, some or all parameters of the LP problem are represented in terms of fuzzy numbers. In this contribution, some of the most common models and procedures for solving FLP problems are analyzed. The solution approaches are illustrated with numerical examples.

Keywords Fuzzy mathematical programming · Simplex algorithm · Optimization

A. Ebrahimnejad (✉)
Department of Mathematics, Qaemshahr Branch,
Islamic Azad University, Qaemshahr, Iran
e-mail: aemarzoun@gmail.com; a.ebrahimnejad@qaemiau.ac.ir

J.L. Verdegay
Department of Computer Science and Artificial Intelligence,
University of Granada, 18014 Granada, Spain
e-mail: verdegay@decsai.ugr.es

C. Kahraman et al. (eds.), *Fuzzy Logic in Its 50th Year*,
Studies in Fuzziness and Soft Computing 341,
DOI 10.1007/978-3-319-31093-0_15

1 Introduction

Linear programming (LP) is concerned with the optimal allocation of limited resources to several competing activities on the basis of given criteria of optimality. In conventional LP problems the observed data or parameters are usually imprecise because of incomplete information. Imprecise evaluations are primarily the result of unquantifiable, incomplete and non-obtainable information. Fuzzy Sets theory could be used to represent such imprecise data in LP by generalizing the notion of membership in a set, and a fuzzy LP (FLP) problem appears in a natural way.

In this context, Tanaka et al. [35] first proposed the theory of fuzzy mathematical programming based on the fuzzy decision framework of Bellman and Zadeh [2]. The first formulation of FLP to address the impreciseness of the parameters in LP problems with fuzzy constraints and objective functions has been introduced by Zimmerman [43]. Tanaka and Asai [34] proposed a possibilistic LP formulation where the coefficients of the decision variables were crisp while the decision variables were fuzzy numbers. Verdegay [38] presented the concept of a fuzzy objective based on the fuzzification principle and used this concept to solve FLP problems. Herrera et al. [17] examined the fuzzified version of the mathematical problem assuming that the coefficients are represented in terms of fuzzy numbers and the relations in the definition of the feasible set are also fuzzy. Zhang et al. [42] proposed an FLP where the coefficients of the objective function are assumed to be fuzzy. They showed how to convert the FLP problems into multi-objective optimization problems with four objective functions. Gupta and Mehlawat [15] studied a pair of fuzzy primal-dual LP problems and calculated duality results using an aspiration level approach. Hatami-Marbini and Tavana [16] proposed a new method for solving LP problems with fuzzy parameters. Their method provides an optimal solution that is not subject to specific restrictive conditions and supports the interactive participation of the decision maker in all steps of the decision-making process. Kheirfam and Verdegay [20] carried out a sensitivity analysis on the right-hand-side of the constraints and the coefficients of the objective function for a fuzzy quadratic programming problem. Wan and Dong [39] developed a new possibility LP with trapezoidal fuzzy numbers. They proposed the auxiliary multi-objective programming with four objectives to solve the corresponding possibility LP.

Numerous researchers have studied various properties of FLP problems and proposed different approaches for solving them. Because of existing different assumptions and sources of fuzziness in the parameters, the definition of FLP problem is not unique. Bector and Chandra [1] have classified FLP problems into four following Categories:

- Type 1: LP problems with fuzzy inequalities and crisp objective function
- Type 2: LP problems with crisp inequalities and fuzzy objective function
- Type 3: LP problems with fuzzy inequalities and fuzzy objective function
- Type 4: LP problems with fuzzy parameters

In the FLP problems of Type 1 is assumed that the fuzziness of the available resources is characterized by the membership function over the tolerance range. Verdegay [37] proved that the optimal solution of a problem of this type can be found by use of solving an equivalent crisp parametric LP problem assuming that the objective function is crisp. Werners [40], on the other hands, considered that the objective function should be fuzzy because of fuzzy inequality constraints and proposed a non-symmetric model for solving the FLP problem of Type 1.

In the FLP problems of Type 2 is assumed the coefficient of decision variables in the objective function cannot be precisely determined. For solving a problem of Type 2, Verdegay [37] proposed an equivalent cost parametric LP problem. In addition, Verdegay [38] proved that for a given problem of Type 2 there always exist a problem of Type 1 and has a same solution.

The FLP problem of Type 3 is a combination of Type 1 and Type 2 problems. This means that not only a goal and its corresponding tolerance are considered for the objective function, but also a tolerance for each fuzzy constraint is known. Two important approaches for solving this kind of FLP problems are Zimmerman's approach [44] and Chanas's approach [3]. In the proposed approach by Zimmerman is assumed that the goal and its corresponding tolerance of the objective function are given initially. According to this approach, the optimal solution of such problem can be found by solving a crisp LP problem. On the other hand, in the proposed approach by Chanas [3], the goal and its corresponding tolerance are determined by solving two LP problems. Then based on the obtained results and the optimal solution of a crisp parametric LP problem, the optimal solution of such FLP problem is determined.

In the FLP problems of Type 4, the some or all parameters of the problems under consideration may be fuzzy numbers and the inequalities may be interpreted in terms of fuzzy rankings. The solution approaches for solving the problems of Type 4 can be classified into two general groups: Simplex based approach and Non-simplex based approach. In the simplex based approach, the classical simplex algorithms such as primal simplex algorithm, dual simplex algorithm and primal-dual simplex algorithm are generalized for solving LP problems with fuzzy parameters. In this approach, the comparison of fuzzy numbers is done by use of linear ranking functions. On the other hand, in the non-simplex based approach, such FLP problems are first converted into equivalent crisp problems, which are then solved by the standard methods.

Maleki et al. [29] consider a kind of FLP problem in which only the coefficients of decision variables in the objective function are represented in terms of trapezoidal fuzzy numbers. They first proved the fuzzy analogues of some important theorems of LP and then extended the primal simplex algorithm in fuzzy sense to obtain the optimal solution of the LP problems with fuzzy cost coefficients. After that, Nasseri and Ebrahimnejad [30] generalized the dual simplex algorithm in fuzzy sense for solving the same problem based on duality results developed by Mahdavi-Amiri and Nasseri [27]. Another approach namely primal-dual simplex algorithm in fuzzy sense has been proposed by Ebrahimnejad [7] that unlike the dual simplex algorithm does not require a dual feasible solution to be basic. Using these algorithms the sensitivity analysis for the same problem was discussed in [6].

Maleki et al. [29] used the crisp solution of the LP problem with fuzzy cost coefficients as an auxiliary problem, for finding the fuzzy solution of fuzzy variable linear programming (FVLP) problems in which the right-hand-side vectors and decision variables are represented by trapezoidal fuzzy numbers. Mahdavi-Amiri and Nasseri [28] showed that the auxiliary problem is indeed the dual of the FVLP problem and proved duality results by a natural extension of the results of crisp LP. Using these results, Mahdavi-Amiri and Nasseri [28] and Ebrahimnejad et al. [13] developed two new methods in fuzzy sense, namely the fuzzy dual simplex algorithm and the fuzzy primal-dual simplex algorithm, respectively, for solving the FVLP problem directly and without any need of an auxiliary problem. Ebrahimnejad and Verdegay [12] and Ebrahimnejad [8] generalized the bounded simplex algorithms in fuzzy sense for solving that kind of FVLP problems in which some or all variables are restricted to lie within fuzzy lower and fuzzy upper.

Ganesan and Veeramani [14] introduced a type of fuzzy arithmetic for symmetric trapezoidal fuzzy numbers and then proposed a primal simplex method for solving FLP problems in which the coefficients of the constraints are represented by real numbers and all the other parameters as well as the variables are represented by symmetric trapezoidal fuzzy numbers. Nasseri et al. [31] discussed a concept of duality for the same FLP problems and derived the weak and strong duality theorems. Based on these duality results, Ebrahimnejad and Nasseri [11] proposed a duality approach for the same problem for situation in which a fuzzy dual feasible solution is at hand. Kumar and Kaur [23] proposed an alternative method for solving the existing symmetric FLP problem. Ebrahimnejad and Tavana [10] proposed a novel method for solving the same FLP problem which is simpler and computationally more efficient than the above mentioned algorithms.

Kheirfam and Verdegay [19] formulated a new type of FLP problems, namely symmetric fully fuzzy linear programming (SFFLP) problem in which all parameters as well as the decision variables were represented by symmetric trapezoidal fuzzy numbers. Then they established the duality and complementary slackness for the same problems and from the obtained results presented a fuzzy dual simplex for solving SFFLP problems.

For solving the LP problems with fuzzy cost coefficients according to non-simplex based approach some auxiliary programming problems are first defined. The three approaches proposed by Lai and Hwang [25], Rommenlfanger et al. [33] and Delgado et al. [4] are the main approaches for solving this kind of FLP problems.

The proposed approaches by Ramik and Ramanek [32], Tanaka et al. [36] and Dubois and Prade [5] belonging to non-simplex based approach can be used to solve the FLP problems involving fuzzy numbers for the coefficients of the decision variables in the constraints and the right-hand-side of the constraints. Ramik and Ramanek [32] first defined the concept of fuzzy inequality and then converted each fuzzy constraint into three or four crisp constraints depending on triangular and trapezoidal membership functions, respectively. Tanaka et al. [36], based on the concept of α-cut, proposed an auxiliary problem for solving this kind of FLP problems by assuming symmetric triangular membership functions for the fuzzy parameters. Dubois and Prade [5] provided two cases of "soft" and "hard" equivalent constraints for the fuzzy constraints of the FLP problem under consideration. Then they expressed the soft and hard constraints by means of α-weak feasibility and α-hard feasibility, respectively, leading to two separate auxiliary problems.

The FLP problems involving fuzzy numbers for the decision variables, the coefficients of the decision variables in the objective function, the coefficients of the decision variables in the constraints and the right-hand-side of the constraints, is called fully FLP (FFLP) problems. Hosseinzadeh Lotfi et al. [18] considered the FFLP problems where all the parameters and variables were triangular fuzzy numbers. Kumar et al. [24] using arithmetic operations and definition of fuzzy equality, first converted each fuzzy equality constraint into several crisp constraints and then optimized the rank of fuzzy objective function over the obtained crisp feasible space. In contrast to most existing approaches which provide crisp solution, their method not only gives fuzzy optimal solution but also preserve the form of non-negative fuzzy optimal solution and optimal objective function.

The purpose of this chapter is to shortly describe models and methods following the above four types of FLP problems, in order to provide the reader a clear view of the main results produced along the last 50 years on this key research area.

2 Preliminaries

In this section, some necessary concepts and backgrounds on fuzzy arithmetic are reviewed [6, 14, 22, 28, 29].

Definition 1 The characteristic function μ_A of a crisp set A assigns a value of either one or zero to each individual in the universal set X. This function can be generalized to a function $\mu_{\tilde{A}}$ such that the values assigned to the element of the universal set X fall within a specified range i.e., $\mu_{\tilde{A}} : X \to [0, 1]$. The assigned value indicates the membership grade of the element in the set $\tilde{A}$. Larger values denote the higher degrees of set membership.

The function $\mu_{\tilde{A}}$ is called membership function and the set $\tilde{A} = \{(x, \mu_{\tilde{A}}(x)) | x \in X\}$ defined by $\mu_{\tilde{A}}$ for each $x \in X$ is called a fuzzy set.

Definition 2 Given a fuzzy set $\tilde{A}$ defined on universal set of real numbers X and any number $\alpha \in [0, 1]$, the α-cut of is defined as $[\tilde{A}]_\alpha = \{x \in X; \mu_{\tilde{A}}(x) \geq \alpha\}$.

Definition 3 A fuzzy set $\tilde{A}$, defined on universal set of real numbers $\mathbb{R}$, is said to be a fuzzy number if its membership function has the following characteristics:

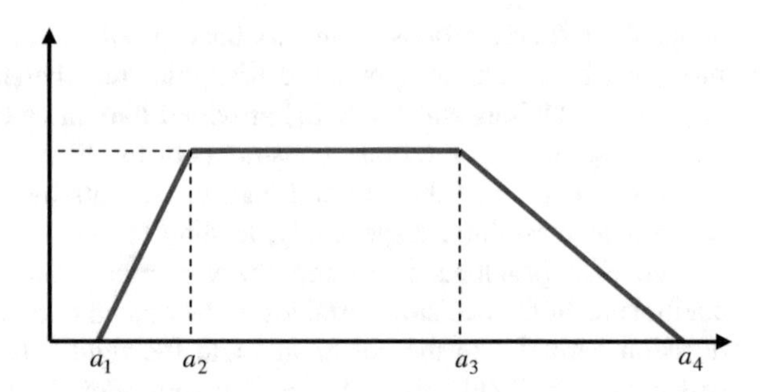

Fig. 1 A trapezoidal fuzzy number $\tilde{a} = (a_1, a_2, a_3, a_4)$

- $\tilde{A}$ is convex, i.e.

$$\forall x, y \in \mathbb{R}, \forall \lambda \in [0,1], \mu_{\tilde{A}}(\lambda x + (1-\lambda)y) \geq \min\{\mu_{\tilde{A}}(x), \mu_{\tilde{A}}(y)\},$$

- $\tilde{A}$ is normal, i.e., $\exists \bar{x} \in \mathbb{R}; \mu_{\tilde{A}}(\bar{x}) = 1$,
- $\mu_{\tilde{A}}$ is piecewise continues.

Definition 4 A fuzzy number $\tilde{A}$ is said to be a positive (negative) fuzzy number if for each $x \leq 0 (x \geq 0), \mu_{\tilde{A}}(x) = 0$.

Definition 5 A fuzzy number $\tilde{a} = (a_1, a_2, a_3, a_4)$ is said to be a trapezoidal fuzzy number if its membership function is given by (see Fig. 1).

$$\mu_{\tilde{a}}(x) = \begin{cases} \frac{x-a_1}{a_2-a_1}, & a_1 \leq x \leq a_2, \\ 1, & a_2 \leq x \leq a_3, \\ \frac{a_4-x}{a_4-a_3}, & a_3 \leq x \leq a_4. \end{cases}$$

Definition 6 A trapezoidal fuzzy number $\tilde{a} = (a_1, a_2, a_3, a_4)$ is said to be a symmetric trapezoidal fuzzy number if $a_2 - a_1 = a_4 - a_3$. Assuming $a_2 - a_1 = a_4 - a_3 = \alpha$, a symmetric trapezoidal fuzzy number is denoted by $\tilde{a} = (a_2 - \alpha, a_2, a_3, a_3 + \alpha)$.

Definition 7 Two trapezoidal fuzzy numbers $\tilde{a} = (a_1, a_2, a_3, a_4)$ and $\tilde{b} = (b_1, b_2, b_3, b_4)$ are said to be equal, i.e. $\tilde{a} = \tilde{b}$ if and only if $a_1 = b_1, a_2 = b_2, a_3 = b_3$ and $a_4 = b_4$.

Definition 8 A trapezoidal fuzzy number $\tilde{a} = (a_1, a_2, a_3, a_4)$ is said to be a non-negative trapezoidal fuzzy number if and only if $a_1 \geq 0$.

Definition 9 A fuzzy number $\tilde{a} = (a_1, a_2, a_3)$ is said to be a triangular fuzzy number if its membership function is given by

$$\mu_{\tilde{a}}(x) = \begin{cases} \frac{x-a_1}{a_2-a_1}, & a_1 \le x \le a_2, \\ \frac{a_3-x}{a_3-a_2}, & a_2 \le x \le a_3. \end{cases}$$

Definition 10 Let $\tilde{a} = (a_1, a_2, a_3, a_4)$ and $\tilde{b} = (b_1, b_2, b_3, b_4)$ be two trapezoidal fuzzy numbers. Then the arithmetic operations on $\tilde{a}$ and $\tilde{b}$ are given by:

- $\tilde{a} + \tilde{b} = (a_1 + b_1, a_2 + b_2, a_3 + b_3, a_4 + b_4)$
- $\tilde{a} \otimes \tilde{b} = (m_1, m_2, m_3, m_4)$, where

$$m_1 = \min\{a_1b_1, a_1b_4, a_4b_1, a_4b_4\}, \quad m_2 = \min\{a_2b_2, a_2b_3, a_3b_2, a_3b_3\},$$
$$m_3 = \max\{a_1b_1, a_1b_4, a_4b_1, a_4b_4\}, \quad m_4 = \max\{a_2b_2, a_2b_3, a_3b_2, a_3b_3\},$$

- $k > 0, k\tilde{a} = (ka_1, ka_2, ka_3, ka_4)$,
- $k < 0, k\tilde{a} = (ka_4, ka_3, ka_2, ka_1)$.

Definition 11 For any two fuzzy numbers $\tilde{a}$ and $\tilde{b}$, the operations MIN and MAX are defined as follows:

$$\mathrm{MIN}(\tilde{a}, \tilde{b})(z) = \sup_{z=\min(x,y)} \min\left[\mu_{\tilde{a}}(x), \mu_{\tilde{b}}(y)\right]$$
$$\mathrm{MAX}(\tilde{a}, \tilde{b})(z) = \sup_{z=\max(x,y)} \min\left[\mu_{\tilde{a}}(x), \mu_{\tilde{b}}(y)\right]$$

Theorem 1 *Let* MIN *and* MAX *be binary operations on the set of all fuzzy numbers* $F(\mathbb{R})$ *defined in* Definition 11. *Then, the triple* $\langle F(\mathbb{R}), \mathrm{MIN}, \mathrm{MAX}\rangle$ *is a distributive lattice.*

Remark 1 The lattice $\langle F(\mathbb{R}), \mathrm{MIN}, \mathrm{MAX}\rangle$ can also be expressed as the pair $\langle F(\mathbb{R}), \preceq\rangle$, where $\preceq$ is a partial ordering defined as:

$$\tilde{a} \preceq \tilde{b} \Leftrightarrow \mathrm{MIN}(\tilde{a}, \tilde{b}) = \tilde{a} \text{ or, alternatively,}$$
$$\tilde{a} \preceq \tilde{b} \Leftrightarrow \mathrm{MAX}(\tilde{a}, \tilde{b}) = \tilde{b}.$$

Remark 2 For any two trapezoidal fuzzy numbers $\tilde{a} = (a_1, a_2, a_3, a_4)$ and $\tilde{b} = (b_1, b_2, b_3, b_4)$, $\mathrm{MAX}(\tilde{a}, \tilde{b}) = \tilde{b}$ or $\tilde{a} \preceq \tilde{b}$ if and only if $a_1 \le b_1, a_2 \le b_2, a_3 \le b_3$ and $a_4 \le b_4$.

Definition 12 A ranking function is a function $\Re: F(\mathbb{R}) \to \mathbb{R}$ that maps each fuzzy number into the real line, where a natural order exists. A ranking function $\Re$ is said to be a linear ranking function if $\Re(k\tilde{a}_1 + \tilde{a}_2) = k\Re(\tilde{a}_1) + \Re(\tilde{a}_2)$ for any $k \in \mathbb{R}$.

Remark 3 For a trapezoidal fuzzy number $\tilde{a} = (a_1, a_2, a_3, a_4)$, one of the most linear ranking functions introduced by Yager [41] is $\Re(\tilde{a}_1) = \frac{a_1 + a_2 + a_3 + a_4}{4}$.

Remark 4 For a triangular fuzzy number $\tilde{a} = (a_1, a_2, a_3)$, Yager's linear ranking function is reduced to $\Re(\tilde{a}_1) = \frac{a_1 + 2a_2 + a_3}{4}$.

3 LP with Fuzzy Inequalities and Crisp Objective Function

The general model of LP problems with fuzzy inequalities and crisp objective function is formulated as follows:

$$\begin{aligned} &\max z = \sum_{j=1}^{n} c_j x_j \\ &s.t. \;\; g_i(x) = \sum_{j=1}^{n} a_{ij} x_j \lesssim b_i, \quad i = 1, 2, \ldots, m, \\ &\qquad x_j \geq 0, \quad j = 1, 2, \ldots, n. \end{aligned} \tag{1}$$

In model (1), $\lesssim$ is called "less than or equal to" and it is assumed that tolerance p_i for each constraint is given. This means that the decision maker can accept a violation of each constraint up to degree p_i. In this case, constraints $g_i(x) \lesssim b_i$, are equivalent to $g_i(x) \leq b_i + \theta p_i$, $(i = 1, 2, \ldots, m)$, where $\theta \in [0, 1]$.

Verdegay [37] and Werners [40] considered two solution methods namely nonsysmmetric method and symmetric method, respectively, for solving the FLP problem (1). In what follows, theses solution approaches are discussed.

3.1 Verdegay's Approach

Verdegay [37] proved that the FLP problem (1) is equivalent to crisp parametric LP problem when the membership functions of the fuzzy constraints are continues and non-increasing functions. According to his approach, the membership functions of the fuzzy constraints of problem (1) can be modeled as follows:

$$\mu_i(g_i(x)) = \begin{cases} 1, & g_i(x) < b_i, \\ 1 - \frac{g_i(x) - b_i}{p_i}, & b_i \le g_i(x) \le b_i + p_i, \\ 0, & g_i(x) > b_i + p_i. \end{cases} \tag{2}$$

In this case, FLP problem (1) is equivalent to:

$$\begin{aligned} &\max z = \sum_{j=1}^{n} c_j x_j \\ &s.t.\ \ \mu_i(g_i(x)) \ge \alpha, \quad i = 1, 2, \ldots, m, \\ &\qquad \alpha \in [0, 1], \quad x_j \ge 0, \quad j = 1, 2, \ldots, n. \end{aligned} \tag{3}$$

Now, by substituting membership functions (2) into problem (3), the following crisp parametric LP problem is obtained:

$$\begin{aligned} &\max z = \sum_{j=1}^{n} c_j x_j \\ &s.t.\ \ g_i(x) = \sum_{j=1}^{n} a_{ij} x_j \le b_i + (1 - \alpha) p_i, \quad i = 1, 2, \ldots, m, \\ &\alpha \in [0, 1], \quad x_j \ge 0, \quad j = 1, 2, \ldots, n. \end{aligned} \tag{4}$$

It should be note that for each $\alpha \in [0, 1]$, an optimal solution is obtained. This indicates that the solution with α grade of membership function is actually fuzzy.

Example 1 [26] Consider the following FLP problem:

$$\begin{aligned} &\max z = 4x_1 + 5x_2 + 9x_3 + 11x_4 \\ &s.t.\ \ g_1(x) = x_1 + x_2 + x_3 + x_4 \le 15, \\ &\quad g_2(x) = 7x_1 + 5x_2 + 3x_3 + 2x_4 \le 120, \\ &\quad g_3(x) = 3x_1 + 5x_2 + 10x_3 + 15x_4 \le 100, \\ &\quad x_1, x_2, x_3, x_4 \ge 0. \end{aligned} \tag{5}$$

Assume that the first and third constrains are imprecise and their maximum tolerances are 3 and 20, respectively, i.e. $p_1 = 3$ and $p_3 = 20$. Then, according to (4), the following crisp parametric LP problem is solved:

$$\begin{aligned} &\max z = 4x_1 + 5x_2 + 9x_3 + 11x_4 \\ &s.t.\ \ g_1(x) = x_1 + x_2 + x_3 + x_4 \le 15 + 3(1 - \alpha), \\ &\quad g_2(x) = 7x_1 + 5x_2 + 3x_3 + 2x_4 \le 120, \\ &\quad g_3(x) = 3x_1 + 5x_2 + 10x_3 + 15x_4 \le 100 + 20(1 - \alpha), \\ &\quad x_1, x_2, x_3, x_4 \ge 0. \end{aligned} \tag{6}$$

By use of the parametric technique the optimal solution and the optimal objective value are obtained as follows:

$$x^* = (7.14 + 1.43(1-\alpha), 0, 7.86 + 1.57(1-\alpha), 0),$$
$$z^* = 99.29 + 19.86(1-\alpha) \tag{7}$$

3.2 *Werners's Approach*

In the symmetric method proposed by Werners [40], it is assumed that the objective function of problem (1) should be fuzzy because of fuzzy inequality constraints. For doing this, he calculated the lower and upper bounds of the optimal values by solving the standard LP problems (8) and (9), respectively, as follows:

$$\begin{aligned} z^l = \max z &= \sum_{j=1}^{n} c_j x_j \\ s.t. \quad &\sum_{j=1}^{n} a_{ij} x_j \leq b_i, \quad i = 1, 2, \ldots, m, \\ &x_j \geq 0, \quad j = 1, 2, \ldots, n. \end{aligned} \tag{8}$$

$$\begin{aligned} z^u = \max z &= \sum_{j=1}^{n} c_j x_j \\ s.t. \quad &\sum_{j=1}^{n} a_{ij} x_j \leq b_i + p_i, \quad i = 1, 2, \ldots, m, \\ &x_j \geq 0, \quad j = 1, 2, \ldots, n. \end{aligned} \tag{9}$$

In this case, the membership function of the objective function is defined as follows:

$$\mu_0(z) = \begin{cases} 1, & z > z^u, \\ 1 - \frac{z^u - z}{z^u - z^l}, & z^l \leq z \leq z^u, \\ 0, & z < z^l. \end{cases} \tag{10}$$

Now, problem (1) can be solved by solving the following problem, a problem of finding a solution that satisfies the constraints and goal with the maximum degree:

$$\begin{aligned} &\max z = \alpha \\ &s.t. \quad \mu_0(z) \geq \alpha, \\ &\qquad \mu_i(g_i(x)) \geq \alpha, \quad i = 1, 2, \ldots, m, \\ &\qquad \alpha \in [0, 1], \quad x_j \geq 0, \quad j = 1, 2, \ldots, n. \end{aligned} \tag{11}$$

By substituting the membership functions of the constraints given in (2) and the membership function of the objective function given in (10) into problem (11), the following crisp LP problem is obtained:

$$
\begin{aligned}
&\max \alpha \\
&s.t. \ \sum_{j=1}^{n} c_j x_j \geq z^l - (1-\alpha)(z^u - z^l), \\
&\quad g_i(x) = \sum_{j=1}^{n} a_{ij} x_j \leq b_i + (1-\alpha) p_i, \quad i = 1, 2, \ldots, m, \\
&\quad \alpha \in [0, 1], \quad x_j \geq 0, \quad j = 1, 2, \ldots, n.
\end{aligned} \tag{12}
$$

Example 2 [22] Consider the following FLP problem:

$$
\begin{aligned}
&\max z = 0.4x_1 + 0.3x_2 \\
&s.t. \ g_1(x) = x_1 + x_2 \leq \tilde{400}, \\
&\quad g_2(x) = 2x_1 + x_2 \leq \tilde{500}, \\
&\quad x_1, x_2 \geq 0.
\end{aligned} \tag{13}
$$

where the membership functions of the fuzzy constraints are defined by:

$$
\mu_1(x) = \begin{cases} 1, & g_1(x) < 400, \\ 1 - \frac{g_1(x) - 400}{100}, & 400 \leq g_1(x) \leq 500, \\ 0, & g_1(x) > 500. \end{cases} \tag{14}
$$

$$
\mu_2(x) = \begin{cases} 1, & g_2(x) < 500, \\ 1 - \frac{g_2(x) - 500}{100}, & 500 \leq g_2(x) \leq 600, \\ 0, & g_2(x) > 600. \end{cases} \tag{15}
$$

Then, the lower and upper bounds of the objective function are calculated by solving the following two crisp LP problems (16) and (17), respectively:

$$
\begin{aligned}
&z^l = \max z = 0.4x_1 + 0.3x_2 \\
&s.t. \ g_1(x) = x_1 + x_2 \leq 400, \\
&\quad g_2(x) = 2x_1 + x_2 \leq 500, \\
&\quad x_1, x_2 \geq 0.
\end{aligned} \tag{16}
$$

$$\begin{aligned} z^u &= \max z = 0.4x_1 + 0.3x_2 \\ s.t.\ \ & g_1(x) = x_1 + x_2 \leq 500, \\ & g_2(x) = 2x_1 + x_2 \leq 600, \\ & x_1, x_2 \geq 0. \end{aligned} \tag{17}$$

Solving two problems (16) and (17) give $z^l = 130$ and $z^u = 160$, respectively. Then, the membership function μ_0 of the objective function is defined as follows:

$$\mu_0(z) = \begin{cases} 1, & z > 160, \\ 1 - \frac{160-z}{30}, & 130 \leq z \leq 160, \\ 0, & z < 130. \end{cases} \tag{18}$$

Then, according to problem (12), the FLP problem (13) becomes:

$$\begin{aligned} & \max \alpha \\ s.t.\ \ & 0.4x_1 + 0.3x_2 \geq 130 - 30(1-\alpha), \\ & g_1(x) = x_1 + x_2 \leq 400 + 100(1-\alpha), \\ & g_2(x) = 2x_1 + x_2 \leq 500 + 100(1-\alpha), \\ & \alpha \in [0,1], \quad x_1, x_2 \geq 0, \quad j = 1, 2, \ldots, n. \end{aligned} \tag{19}$$

The optimal solution of the problem (19) is $x^* = (x_1^*, x_2^*) = (100, 350), \alpha^* = 0.5$. The optimal value of the objective function is then calculated by

$$z^* = 0.4x_1^* + 0.3x_2^* = 145.$$

4 LP with Crisp Inequalities and Fuzzy Objective Function

The general from of LP problems with crisp inequalities and fuzzy objective function can be formulated as follows:

$$\begin{aligned} \widetilde{\max}\, z &= \sum_{j=1}^{n} c_j x_j \\ s.t.\ \ & g_i(x) = \sum_{j=1}^{n} a_{ij} x_j \leq b_i, \quad i = 1, 2, \ldots, m, \\ & x_j \geq 0, \quad j = 1, 2, \ldots, n. \end{aligned} \tag{20}$$

Assume that the membership function of the fuzzy objective be given by

$$\phi(c) = \inf\{\phi_1(c_1), \phi_2(c_2), \ldots, \phi_n(c_n)\} \tag{21}$$

In this case, Verdegay [38] proved that if the membership function $\phi_j(c_j) : \mathbb{R} \to [0, 1], (j = 1, 2, \ldots, n)$, are continues and strictly monotone, then the fuzzy optimal solution of problem (20) can be obtained by solving the following parametric LP problem:

$$\begin{aligned}
&\max z = \sum_{j=1}^{n} \phi_j^{-1}(1-\alpha)x_j \\
&s.t.\ \ g_i(x) = \sum_{j=1}^{n} a_{ij}x_j \leq b_i, \quad i = 1, 2, \ldots, m, \\
&\quad \alpha \in [0, 1], \quad x_j \geq 0, \quad j = 1, 2, \ldots, n.
\end{aligned} \tag{22}$$

Example 3 [38] Consider the following FLP problem:

$$\begin{aligned}
&\widetilde{\max}\, z = c_1x_1 + 75x_2 \\
&s.t.\ \ g_1(x) = 3x_1 - x_2 \leq 2, \\
&\quad g_2(x) = x_1 + 2x_2 \leq 3, \\
&\quad x_1, x_2 \geq 0.
\end{aligned} \tag{23}$$

where the membership function ϕ_1 for c_1 is given by:

$$\phi_1(c_1) = \begin{cases} 1, & c_1 > 115, \\ \frac{(c_1-40)^2}{5625}, & 40 \leq c_1 \leq 115, \\ 0, & c_1 < 40. \end{cases} \tag{24}$$

Then, we have $\phi_1^{-1}(t) = 40 + 75\sqrt{t}$. Now, according to model (20) we have to solve the following problem:

$$\begin{aligned}
&\max z = \left(40 + 75\sqrt{1-\alpha}\right)x_1 + 75x_2 \\
&s.t.\ \ g_1(x) = 3x_1 - x_2 \leq 2, \\
&\quad g_2(x) = x_1 + 2x_2 \leq 3, \\
&\quad \alpha \in [0, 1], \quad x_1, x_2 \geq 0.
\end{aligned} \tag{25}$$

By use of the parametric technique, the optimal solution of problem (25) is $x^* = (x_1^*, x_2^*) = (1, 1)$. The fuzzy set of the objective function values of the objective function is then obtained as follows:

$$\left\{(75 + c_1), \frac{(c_1 - 40)^2}{5625}\right\}, \quad 40 \leq c_1 \leq 115.$$

5 LP with Fuzzy Inequalities and Fuzzy Objective Function

The general from of LP problems with fuzzy inequalities and fuzzy objective function can be formulated as follows:

$$\begin{aligned} &\widetilde{\max}\, z = \sum_{j=1}^{n} c_j x_j \\ &s.t.\ \ g_i(x) = \sum_{j=1}^{n} a_{ij} x_j \precsim b_i, \quad i = 1, 2, \ldots, m, \\ &\qquad x_j \geq 0, \quad j = 1, 2, \ldots, n. \end{aligned} \tag{26}$$

In what follows, two solution approaches for solving FLP of type (26) are explored.

5.1 *Zimmerman's Approach [43]*

According to this approach, the membership functions $\mu_i (i = 1, 2, \ldots, m)$ of the fuzzy constraints are given by (2). Also, it is required to have an aspiration level z_0 for the objective function. In this case, problem (26) can be described as follows:

$$\begin{aligned} &\textit{Find } x \\ &s.t.\ \ z = \sum_{j=1}^{n} c_j x_j \succsim z_0 \\ &\qquad g_i(x) = \sum_{j=1}^{n} a_{ij} x_j \precsim b_i, \quad i = 1, 2, \ldots, m, \\ &\qquad x_j \geq 0, \quad j = 1, 2, \ldots, n. \end{aligned} \tag{27}$$

Assuming p_0 be the permissible tolerance for the objective function, the membership function μ_0 of the fuzzy objective is given by

$$\mu_0(z) = \begin{cases} 1, & z > z_0, \\ 1 - \dfrac{z_0 - z}{p_0}, & z_0 - p_0 \le z \le z_0, \\ 0, & z < z_0 - p_0. \end{cases} \tag{28}$$

In this case, to solve the fuzzy system of inequalities (27) corresponding to problem (26), the following crisp LP problem is solved:

$$\begin{aligned} & \max \alpha \\ & s.t.\ \ \mu_0(z) \ge \alpha, \\ & \qquad \mu_i(g_i(x)) \ge \alpha, \quad i = 1, 2, \ldots, m, \\ & \qquad \alpha \in [0, 1], \quad x_j \ge 0, \quad j = 1, 2, \ldots, n. \end{aligned} \tag{29}$$

By substituting the membership functions of the fuzzy constraints given in (2) and the membership function of the objective function given in (28) into problem (29), the following crisp LP problem is obtained:

$$\begin{aligned} & \max \alpha \\ & s.t.\ \ \sum_{j=1}^{n} c_j x_j \ge z_0 - (1 - \alpha) p_0, \\ & \qquad \sum_{j=1}^{n} a_{ij} x_j \le b_i + (1 - \alpha) p_0, \quad i = 1, 2, \ldots, m, \\ & \qquad \alpha \in [0, 1], \quad x_j \ge 0, \quad j = 1, 2, \ldots, n. \end{aligned} \tag{30}$$

It needs to point out that if (x^*, α^*) is an optimal solution of the crisp LP problem (30), then it is said to be an optimal solution of the FLP problem (26) with α^* as the degree up to which the aspiration level z_0 of the decision maker is satisfied.

Example 4 [43] Consider the following FLP problem:

$$\begin{aligned} & \widetilde{\max}\, z = x_1 + x_2 \\ & s.t.\ \ g_1(x) = -x_1 + 3x_2 \lesssim 21, \\ & \qquad g_2(x) = x_1 + 3x_2 \lesssim 27, \\ & \qquad g_3(x) = 4x_1 + 3x_2 \lesssim 45, \\ & \qquad g_4(x) = 3x_1 + x_2 \le 30, \\ & \qquad x_1, x_2 \ge 0. \end{aligned} \tag{31}$$

Assume that $z_0 = 14.5, p_0 = 2, p_1 = 3, p_2 = 6,$ and $p_3 = 6$. To obtain the optimal solution of problem (31), we have to solve the following crisp LP problem with regard to problem (30):

$$\begin{aligned} &\max \alpha \\ &s.t.\ \ z = x_1 + x_2 \geq 14.5 - 2(1-\alpha) \\ &\qquad g_1(x) = -x_1 + 3x_2 \leq 21 - 3(1-\alpha), \\ &\qquad g_2(x) = x_1 + 3x_2 \leq 27 - 6(1-\alpha), \\ &\qquad g_3(x) = 4x_1 + 3x_2 \leq 45 - 6(1-\alpha), \\ &\qquad g_4(x) = 3x_1 + x_2 \leq 30, \\ &\qquad x_1, x_2 \geq 0. \end{aligned} \tag{32}$$

The optimal solution of the crisp LP problem (32) is $(x_1^*, x_2^*, \alpha^*) = (6, 7.75, 0.625)$ with the optimal objective function value $z^* = 13.75$. Then, $(x_1^*, x_2^*) = (6, 7.75)$ is the optimal solution of the FLP problem (31).

5.2 *Chanas's Approach*

In this approach, it is assumed that the decision maker cannot provide initially the goal and its tolerance for the objective function. Chanas [3] therefore proposed first to solve the following problem and then present the results to the decision maker to estimate the goal and the tolerance of the objective function:

$$\begin{aligned} &\max z = \sum_{j=1}^{n} c_j x_j \\ &s.t.\ \ g_i(x) = \sum_{j=1}^{n} a_{ij} x_j \lesssim b_i, \quad i = 1, 2, \ldots, m, \\ &\qquad x_j \geq 0, \quad j = 1, 2, \ldots, n. \end{aligned} \tag{33}$$

In this problem, the membership functions of the fuzzy constraints are assumed as Eq. (2). Therefore, following the Verdegay's approach [38] and assuming $\theta = 1 - \alpha$ the problem (33) becomes:

$$\begin{aligned} &\max z = \sum_{j=1}^{n} c_j x_j \\ &s.t.\ \ g_i(x) = \sum_{j=1}^{n} a_{ij} x_j \leq b_i + \theta p_i, \quad i = 1, 2, \ldots, m, \\ &\qquad \theta \in [0, 1], \quad x_j \geq 0, \quad j = 1, 2, \ldots, n. \end{aligned} \tag{34}$$

The optimal solution of the parametric LP problem (34) can be found by use of the parametric techniques. For a given θ, assume that $x^*(\theta)$ and $z^*(\theta)$ are, respectively an optimal solution and the corresponding optimal objective value of the parametric LP problem (34). Then, the following constrains are hold:

$$\mu_i(g_i(x^*(\theta))) \geq 1-\theta, \quad i=1,2,\ldots,m, \tag{35}$$

On the other hand, in the case of $p_i > 0$, for every non-zero basic solution, there is at least one i such that $\mu_i(g_i(x^*(\theta))) = 1-\theta$. This follows that the common degree of the constraints satisfaction is $\mu_c(g(x^*(\theta))) = \min_{1\leq i\leq m} \mu_i(g_i(x^*(\theta))) = 1-\theta$. This means that for every θ, a solution can be obtained which satisfies all the constraints with degree $1-\theta$. Now, such solution is presented to the decision maker to determine the goal z_0 of the objective function and its tolerance p_0. Therefore, the membership function μ_0 of the objective function becomes:

$$\mu_0(z^*(\theta)) = \begin{cases} 1, & z^*(\theta) > z_0, \\ 1-\frac{z_0 - z^*(\theta)}{p_0}, & z_0 - p_0 \leq z^*(\theta) \leq z_0, \\ 0, & z^*(\theta) < z_0. \end{cases} \tag{36}$$

In this case, the final optimal solution $x^*(\theta^*)$ of the FLP problem (26) will exist at:

$$\mu_D(\theta^*) = \max\{\mu_D(\theta)\} = \mu_c(g(x^*(\theta))) = \max\{\min\{\mu_o(\theta), \mu_c(\theta)\}\}$$

This approach can be summarized as follows:

Step 1: Solve the parametric LP problem (34) to determine $x^*(\theta)$ and $z^*(\theta)$.
Step 2: Determine an appropriate value θ to get z_0 and choose p_0.
Step 3: Obtain the membership function $\mu_0(z^*(\theta))$ as (36).
Step 4: Set $\mu_c = \mu_0 = 1-\theta$ to find θ^*.
Step 5: The optimal solution of the FLP problem (26) is $x^*(\theta^*)$ with the optimal objective value $z^*(\theta^*)$.

Example 5 Again, consider the FLP problem (31) with $p_1 = 3$, $p_2 = 6$, and $p_3 = 6$.

The solution procedure is illustrated as follows:

Step 1: According to problem (34), the following parametric LP problem should be solved:

$$\begin{aligned} &\max z = x_1 + x_2 \\ &s.t. \;\; g_1(x) = -x_1 + 3x_2 \leq 21 + 3\theta, \\ &\qquad g_2(x) = x_1 + 3x_2 \leq 27 + 6\theta, \\ &\qquad g_3(x) = 4x_1 + 3x_2 \leq 45 + 6\theta, \\ &\qquad g_4(x) = 3x_1 + x_2 \leq 30, \\ &\qquad \theta \in [0,1], \quad x_1, x_2 \geq 0. \end{aligned} \tag{37}$$

The optimal solution is $x^*(\theta) = \left(x_1^*(\theta), x_2^*(\theta)\right) = (6, 7+2\theta)$ with $z^*(\theta) = 13+2\theta$.

Step 2: Assume $z_0 = 14$ and choose $p_0 = 1$.

Step 3: Now, the membership function $\mu_0(z^*(\theta))$ is then given by

$$\mu_0(z^*(\theta)) = \begin{cases} 1, & 13+2\theta > 14, \\ 1 - \frac{14-(13+2\theta)}{1} = 2\theta, & 13 \leq 13+2\theta \leq 14, . \\ 0, & 13+2\theta < 14. \end{cases} \quad (38)$$

Step 4: In order to find θ^*, we set $\mu_c = \mu_0 = 1 - \theta$. We obtain $2\theta = 1 - \theta$ and then $\theta^* = \frac{1}{3}$.

Step 5: The optimal solution is then $x^*\left(\frac{1}{3}\right) = \left(x_1^*\left(\frac{1}{3}\right), x_2^*\left(\frac{1}{3}\right)\right) = \left(6, \frac{23}{3}\right)$ with $z^*\left(\frac{1}{3}\right) = 13 + \frac{2}{3} = 13.667$.

6 LP with Fuzzy Parameters

In this section, different solution approaches for solving several kinds of LP problems with fuzzy parameters are explored.

The solution approaches are classified into two general groups: Simplex based approach and Non-simplex based approach. In the simplex based approach, the classical simplex algorithms such as primal simplex algorithm and dual simplex algorithm are generalized for solving LP problems with fuzzy parameters. In the non-simplex based approach, such FLP problems are first converted into equivalent crisp problems, which are then solved by the standard methods. In what follows, these approaches are explored to find the optimal solution of LP problems with fuzzy parameters.

6.1 The Simplex Based Approach

In this section, the simplex based approach for solving several kinds of LP problems with fuzzy parameters is explored. In such approach, the comparison of fuzzy numbers is done by use of linear ranking functions.

6.1.1 LP with Fuzzy Cost Coefficients

The general from of LP problems with fuzzy cost coefficients can be formulated as follows:

$$\begin{aligned} &\max \tilde{z} \approx \sum_{j=1}^{n} \tilde{c}_j x_j \\ &s.t. \ \sum_{j=1}^{n} a_{ij} x_j \leq b_i, \quad i = 1, 2, \ldots, m, \\ &x_j \geq 0, \quad j = 1, 2, \ldots, n. \end{aligned} \tag{39}$$

This problem can be rewritten as follows:

$$\begin{aligned} &\max \tilde{z} \approx \tilde{c}x \\ &s.t. \ Ax \leq b, \\ &x \geq 0. \end{aligned} \tag{40}$$

where $b = (b_1, b_2, \ldots, b_m) \in \mathbb{R}^m$, $\tilde{c} = (\tilde{c}_1, \tilde{c}_2, \ldots, \tilde{c}_n) \in F(\mathbb{R})^n$, $A = (a_{ij})_{m \times n} \in \mathbb{R}^{m \times n}$ and $x = (x_1, x_2, \ldots, x_n) \in \mathbb{R}^n$.

In what follows, the solution approach proposed by Maleki et al. [29] for solving the FLP problem (40) is explored. The other approaches can be found in [7, 30].

Consider the system of equality constraints of (40) where A is a matrix of order $(m \times n)$ and $rank(A) = m$. Therefore, A can be partitioned as $[B, \quad N]$, where $B_{m \times m}$ is a nonsingular matrix with $rank(B) = m$. Let y_j is the solution of $By_j = a_j$. Moreover, the basic solution $x = (x_B, x_N) = (B^{-1}b, 0)$ is a solution of $Ax = b$. This basic solution is feasible, whenever $x_B \geq 0$. Furthermore, the corresponding fuzzy objective function value is obtained as $\tilde{z} \approx \tilde{c}_B x_B, \tilde{c}_B = (\tilde{c}_{B_1}, \tilde{c}_{B_2}, \ldots, \tilde{c}_{B_m})$.

Suppose J_N is the set of indices associated with the current non-basic variables. For each non-basic variable $x_j, j \in J_N$ the fuzzy variable $\tilde{z}_j$ is defined as $\tilde{z}_j \approx \tilde{c}_B B^{-1} a_j = \tilde{c}_B y_j$.

Maleki et al. [29] stated the following important results in order to improve a feasible solution and to provide unbounded criteria and the optimality conditions for the FLP problem (40) with equality constraints.

Theorem 1 *If for a basic feasible solution with basis B and objective value* $\tilde{z}, \tilde{z}_k \prec \tilde{c}_k$ *for some non-basic variable* x_k *while* $y_k \nleq 0$, *then a new feasible solution can be obtained as follows with objective value* $\tilde{z}_{new} = \tilde{z} - (\tilde{z}_k - \tilde{\tilde{c}}_k)x_k$, *such that* $\tilde{z} \preceq \tilde{z}_{new}$.

$$\begin{aligned} &x_{B_r} = x_k = \theta = \frac{\bar{b}_r}{y_{rk}} = \min_{1 \leq i \leq m} \left\{ \frac{\bar{b}_i}{y_{ik}} | y_{ik} > 0 \right\}, \\ &x_{B_i} = \bar{b}_i - y_{ik} \frac{\bar{b}_r}{y_{rk}}, \quad i = 1, 2, \ldots, m, \quad i \neq r, \\ &x_j = 0, \quad j \in J_N, \quad j \neq k. \end{aligned}$$

Theorem 2 *If for a basic feasible solution with basis B and objective value* $\tilde{z}, \tilde{z}_k \prec \tilde{c}_k$ *for some non-basic variable* x_k *while* $y_k \leq 0$, *then the optimal solution of the FLP problem* (40) *is unbounded.*

Theorem 3 *The basic solution* $x = (x_B, x_N) = (B^{-1}b, 0)$ *is an optimal solution for the FLP problem* (40) *if* $\tilde{z}_j \succeq \tilde{c}_j$ *for all* $j \in J_N$.

Now we are in a position to summarize the fuzzy primal simplex algorithm [29] for solving the FLP problem (40).

Algorithm 1: Fuzzy Primal Simplex Algorithm (Maximization Problem)

Initialization step

Choose a starting feasible basic solution with basis B.
Main steps

1. Solve the system $Bx_B = b$. Let $x_B = B^{-1}b, x_N = 0$, and $\tilde{z} = \tilde{c}_B x_B$.
2. Solve the system $\tilde{w}B = \tilde{c}_B$. Let $\tilde{w} = \tilde{c}_B B^{-1}$.
3. Calculate $\tilde{z}_j = \tilde{c}_B B^{-1} a_j = \tilde{w} a_j$ for all $j \in J_N$. Find the rank of $\tilde{z}_j - \tilde{c}_j$ for all $j \in J_N$ based on ranking function $\Re$ given in Remark 3 and let $\Re(\tilde{z}_k - \tilde{c}_k) = \min_{j \in J_N} \{\Re(\tilde{z}_j - \tilde{c}_j)\}$. If $\Re(\tilde{z}_k - \tilde{c}_k) \geq 0$, then stop with the current basic solution as an optimal solution.
4. Solve the system $By_k = a_k$ and let $y_k = B^{-1}a_k$. If $y_k \leq 0$ then stop with the conclusion that the problem is unbounded.
   If $y_k \nleq 0$, then x_k enters the basis and x_{Br} leaves the basis providing that

$$\frac{\bar{b}_r}{y_{rk}} = \min_{1 \leq i \leq m} \left\{ \frac{\bar{b}_i}{y_{ik}} | y_{rk} > 0 \right\}.$$

5. Update the basic B where a_k replace a_{Br}, update the index set J_N and go to (1).

Example 6 [29] Consider the following FLP problem:

$$\begin{aligned} &\text{Max } \tilde{z} \approx (5, 8, 2, 5)x_1 + (6, 10, 2, 6)x_2 \\ &s.t. \;\; 2x_1 + 3x_2 \leq 6, \\ &\qquad 5x_1 + 4x_2 \leq 10, \\ &\qquad x_1, x_2 \geq 0. \end{aligned} \tag{41}$$

After introducing the slack variables x_3 and x_4 we obtain the following FLP problem:

$$\begin{aligned}
&\text{Max } \tilde{z} \approx (5,8,2,5)x_1 + (6,10,2,6)x_2 \\
&s.t. \;\; 2x_1 + 3x_2 + x_3 = 6, \\
&\qquad 5x_1 + 4x_2 + x_4 = 10, \\
&\qquad x_1, x_2, x_3, x_4 \geq 0.
\end{aligned} \tag{42}$$

The steps of Algorithm 1 for solving the FLP problem (42) are presented as follows:

Iteration 1

The starting feasible basic is $B = [a_3, \quad a_4] = \begin{bmatrix} 1 & 0 \\ 0 & 1 \end{bmatrix}$. Thus, the non-basic matrix N is $N = [a_1, \quad a_2] = \begin{bmatrix} 2 & 3 \\ 5 & 4 \end{bmatrix}$.

Step 1: We find the basic feasible solution by solving the system $Bx_B = b$:

$$\begin{bmatrix} 1 & 0 \\ 0 & 1 \end{bmatrix} \begin{bmatrix} x_3 \\ x_4 \end{bmatrix} = \begin{bmatrix} 6 \\ 10 \end{bmatrix}$$

Solving this system leads to $x_{B_1} = x_3 = 6$ and $x_{B_2} = x_4 = 10$. The non-basic variables are $x_1 = 0$, $x_2 = 0$ and the fuzzy objective value is $\tilde{z} = \tilde{c}_B x_B = 0$.

Step 2: We find $\tilde{w}$ by solving the fuzzy system $\tilde{w}B = \tilde{c}_B$:

$$(\tilde{w}_1, \tilde{w}_2) \begin{bmatrix} 1 & 0 \\ 0 & 1 \end{bmatrix} = \begin{bmatrix} (0,0,0,0) \\ (0,0,0,0) \end{bmatrix} \Rightarrow \tilde{w}_1 = \tilde{w}_2 = (0,0,0,0)$$

Step 3: We calculate $\tilde{z}_j - \tilde{c}_j = \tilde{w}a_j - \tilde{c}_j$ for all $j \in J_N = \{1, 2\}$:

$$\tilde{z}_1 - \tilde{c}_1 = \tilde{w}a_1 - \tilde{c}_1 = ((0,0,0,0),(0,0,0,0)) \begin{bmatrix} 2 \\ 5 \end{bmatrix} - (5,8,2,5) = (-8,-5,5,2)$$

$$\tilde{z}_2 - \tilde{c}_2 = \tilde{w}a_2 - \tilde{c}_2 = ((0,0,0,0),(0,0,0,0)) \begin{bmatrix} 3 \\ 4 \end{bmatrix} - (6,10,2,6) = (-10,-6,6,2)$$

Computing the rank of $\tilde{z}_1 - \tilde{c}_1$ and $\tilde{z}_2 - \tilde{c}_2$ based on ranking function given in Remark 3 leads to $\Re(\tilde{z}_1 - \tilde{c}_1) = -\frac{29}{4}$ and $\Re(\tilde{z}_2 - \tilde{c}_2) = -9$. Therefore,

$$\min \left\{ \Re(\tilde{z}_1 - \tilde{c}_1) = -\frac{29}{2}, \Re(\tilde{z}_2 - \tilde{c}_2) = -9 \right\} = \Re(\tilde{z}_2 - \tilde{c}_2) = -9$$

Step 4: We find y_2 by solving the system $By_2 = a_2$:

$$\begin{bmatrix} 1 & 0 \\ 0 & 1 \end{bmatrix} \begin{bmatrix} y_{12} \\ y_{22} \end{bmatrix} = \begin{bmatrix} 3 \\ 4 \end{bmatrix} \Rightarrow y_2 = \begin{bmatrix} y_{12} \\ y_{22} \end{bmatrix} = \begin{bmatrix} 3 \\ 4 \end{bmatrix}$$

Step 5: The variable x_{Br} leaving the basis is determined by the following test:

$$\min\left\{\frac{\bar{b}_1}{y_{12}}, \frac{\bar{b}_2}{y_{22}}\right\} = \min\left\{\frac{6}{3}, \frac{10}{4}\right\} = \frac{6}{3} = 2$$

Hence, $x_{B_1} = x_3$ leaves the basis.

Step 6: We update the basis B where a_2 replaces $a_{B_1} = a_3$:

$$B = [a_2, a_4] = \begin{bmatrix} 3 & 0 \\ 4 & 1 \end{bmatrix}$$

Also, we update $J_N = \{1, 2\}$ as $J_N = \{1, 3\}$.

This process is continued and after two more iterations the optimal basis $B = [a_2, a_1] = \begin{bmatrix} 3 & 2 \\ 4 & 5 \end{bmatrix}$ is found. The optimal solution and the optimal objective function value of the FLP problem (41) are therefore given by:

$$x^* = (x_1^*, x_2^*, x_3^*, x_4^*) = \left(\frac{6}{7}, \frac{10}{7}, 0, 0\right), \quad \tilde{z}^* = \left(\frac{90}{7}, \frac{148}{7}, \frac{32}{7}, \frac{96}{7}\right) \tag{43}$$

6.1.2 LP Problem with Fuzzy Decision Variables and Fuzzy Right-Hand-Side of the Constraints

The LP problem involving fuzzy numbers for the decision variables and the right-hand-side of the constraints is called fuzzy variables linear programming problem (FVLP) problem and is formulated as follows:

$$\begin{aligned} &\min \tilde{z} = c\tilde{x} \\ &s.t.\ A\tilde{x} \succeq \tilde{b}, \\ &\tilde{x} \succeq \tilde{0}. \end{aligned} \tag{44}$$

In what follows, two solution approaches proposed by Maleki et al. [29] and Mahdavi-Amiri and Nasseri [28] for solving the FVLP problem (44) are explored. The other approaches can be found in [12, 13].

6.1.3 Maleki et al.'s Approach

Maleki et al. [29] introduced an auxiliary problem, having only fuzzy cost coefficients, for an FVLP problem and then used its crisp solution for obtaining the fuzzy solution of the FVLP problem. In order to describe this approach, we reformulate the FVLP problem (44), with respect to the change of all the parameters and variables, as the following FVLP problem:

$$\begin{aligned} \min \tilde{u} &= \tilde{w}b \\ s.t.\ \ \tilde{w}A &\succeq \tilde{c}, \\ \tilde{w} &\succeq \tilde{0}. \end{aligned} \tag{45}$$

Maleki et al. [29] considered the FLP problem (40) as the fuzzy auxiliary problem for the FVLP problem (45) and studied the relation between the FVLP problem (45) and the fuzzy auxiliary problem (40). In fact, they proved that if $x_B = B^{-1}b$ is an optimal basic feasible solution of the auxiliary problem (40) then $\tilde{w} = \tilde{c}_B B^{-1}$ is a fuzzy optimal solution of the FVLP problem (45).

Example 7 [28, 29] Consider the following FVLP problem:

$$\begin{aligned} &\text{Min } \tilde{z} \approx 6\tilde{x}_1 + 10\tilde{x}_2 \\ &s.t.\ \ 2\tilde{x}_1 + 5\tilde{x}_2 \succeq (5, 8, 2, 5), \\ &\quad\ 3\tilde{x}_1 + 4\tilde{x}_2 \succeq (6, 10, 2, 6), \\ &\quad\ \tilde{x}_1, \tilde{x}_2 \succeq \tilde{0}. \end{aligned} \tag{46}$$

Now, we solve this problem using the solution approach proposed by Maleki et al. [29]. To do this, we rewrite the FVLP problem (46) with respect to the change of all the parameters and variables as follows:

$$\begin{aligned} &\text{Min }\ \tilde{u} \approx 6\tilde{w}_1 + 10\tilde{w}_2 \\ &s.t.\ \ 2\tilde{w}_1 + 5\tilde{w}_2 \succeq (5, 8, 2, 5), \\ &\quad\ 3\tilde{w}_1 + 4\tilde{w}_2 \succeq (6, 10, 2, 6), \\ &\quad\ \tilde{w}_1, \tilde{w}_2 \succeq \tilde{0} \end{aligned} \tag{47}$$

The fuzzy auxiliary problem corresponding to the FVLP problem (47) is defined as follows:

$$\begin{aligned} &\text{Max } \tilde{z} \approx (5, 8, 2, 5)x_1 + (6, 10, 2, 6)x_2 \\ &s.t.\ \ 2x_1 + 3x_2 \le 6, \\ &\quad\ 5x_1 + 4x_2 \le 10, \\ &\quad\ x_1, x_2 \ge 0. \end{aligned} \tag{48}$$

The fuzzy auxiliary problem (48) is exactly the LP problem with fuzzy costs given in (41). Thus, the optimal solution given in (43) is its optimal solution. Now, since the basis $B = [a_2, a_1] = \begin{bmatrix} 3 & 2 \\ 4 & 5 \end{bmatrix}$ is the optimal basis of the fuzzy auxiliary problem (48) or (41), the fuzzy optimal solution of the FVLP problem (47) is obtained as follows based on the formulation $\tilde{w} = \tilde{c}_B B^{-1}$:

$$\tilde{w}^* = \left(\tilde{w}_1^*, \tilde{w}_2^*\right) = ((6,10,2,6),(5,8,2,5)) \begin{bmatrix} \frac{5}{7} & \frac{-2}{7} \\ \frac{-4}{7} & \frac{3}{7} \end{bmatrix}$$
$$= \left(\left(\frac{-2}{7},\frac{30}{7},\frac{30}{7},\frac{38}{7}\right),\left(\frac{-5}{7},\frac{12}{7},\frac{18}{7},\frac{19}{7}\right)\right) \tag{49}$$

This means that the optimal solution of FVLP problem (46) is as follows:

$$\tilde{x}^* = \begin{bmatrix} \tilde{x}_1^* \\ \tilde{x}_2^* \end{bmatrix} = \begin{bmatrix} \left(\frac{-2}{7},\frac{30}{7},\frac{30}{7},\frac{38}{7}\right) \\ \left(\frac{-5}{7},\frac{12}{7},\frac{18}{7},\frac{19}{7}\right) \end{bmatrix} \tag{50}$$

Unlike the non-simplex based approach, this approach doesn't increase the number of the constraint of the primary FLP problem. Another advantage of this approach is to produce the fuzzy optimal solution that provides possible outcomes with certain degree of memberships to the decision maker. However, the fuzzy optimal solution by use of this approach may be not non-negative. For instance the value of $\tilde{x}_1^*$ in the fuzzy optimal solution (50) is non-negative as there is negative part in this fuzzy number.

Mahdavi-Amiri and Nasseri's Approach

Mahdavi-Amiri and Nasseri [28] proved that the fuzzy auxiliary problem introduced by Maleki et al. [29] is the dual of the FVLP problem. Hence, they introduced a fuzzy dual simplex algorithm for solving the FVLP problem directly on the primary problem and without solving any fuzzy auxiliary problem.

Mahdavi-Amiri and Nasseri [28] defined the dual problem of the FVLP problem (44) as follows:

$$\begin{aligned} \max \tilde{u} &= w\tilde{b} \\ s.t.\ \ wA &\leq c, \\ w &\geq 0. \end{aligned} \tag{51}$$

We see that only the coefficients of the objective function of this problem are fuzzy numbers. Thus, it plays the role of the auxiliary problem for the FVLP problem (44).

A summary of the fuzzy dual simplex algorithm proposed by Mahdavi-Amiri and Nasseri [28] for solving FVLP problem (44) is given as follows.

Algorithm 2: Fuzzy Dual Simplex Algorithm (Minimization Problem)

Initialization step

Choose a starting dual feasible basic solution with basis B.

Main steps

1. Solve the system $B\tilde{x}_B = \tilde{b}$. Let $\tilde{x}_B = B^{-1}\tilde{b} = \tilde{\bar{b}}$, and $\tilde{z} \approx c_B\tilde{x}_B$.
2. Find the rank of $\tilde{\bar{b}}_i$ for all $1 \le i \le m$ based on ranking function $\Re$ given in Remark 3 and let $\Re\left(\tilde{\bar{b}}_i\right) = \min_{1 \le i \le m}\left\{\Re\left(\tilde{\bar{b}}_i\right)\right\}$. If $\Re\left(\tilde{\bar{b}}_r\right) \ge 0$, then stop with the current basic solution as an optimal solution.
3. Solve the system $By_j = a_j$ for all $j \in J_N$ and let $y_j = B^{-1}a_j$. If $y_{rj} \ge 0$ for all $j \in J_N$ then stop with the conclusion that the FVLP problem is infeasible.
4. Solve the system $wB = c_B$. Let $w = c_B B^{-1}$. Calculate $z_j = wa_j$ for all $j \in J_N$. If $y_{rj} \ngeq 0$ for all $j \in J_N$, then $\tilde{x}_{B_r}$ leaves the basis and $\tilde{x}_k$ enters the basis providing that

$$\frac{z_k - c_k}{y_{rk}} = \min_{j \in J_N}\left\{\frac{z_j - c_j}{y_{rk}}\middle| y_{rk} < 0\right\}.$$

5. Update the basis B where a_k replaces a_{Br}, update the index set J_N and go to (1).

Example 8 Again, consider the FVLP problem (46). Now, we solve this problem using the solution approach proposed by Mahdavi-Amiri and Nasseri [28]. To do this, we rewrite the FVLP problem (46) as follows in order to obtain a dual feasible solution with identity basis where $\tilde{x}_3$ and $\tilde{x}_4$ are fuzzy surplus variables:

$$\begin{aligned} &\text{Min } \tilde{z} \approx 6\tilde{x}_1 + 10\tilde{x}_2 \\ &s.t. \ -2\tilde{x}_1 - 5\tilde{x}_2 + \tilde{x}_3 \approx (-8, -5, 5, 2), \\ &\quad -3\,\tilde{x}_1 - 4\tilde{x}_2 + \tilde{x}_4 \approx (-10, -6, 6, 2), \\ &\quad \tilde{x}_1, \tilde{x}_2, \tilde{x}_3, \tilde{x}_4 \succeq \tilde{0}. \end{aligned} \tag{52}$$

The steps of Algorithm 2 for solving the FVLP problem (52) are explored as follows:

Iteration 1 The starting dual feasible basic is $B = [a_3, a_4] = \begin{bmatrix} 1 & 0 \\ 0 & 1 \end{bmatrix}$. Thus, the non-basic matrix N is $N = [a_1, a_2] = \begin{bmatrix} -2 & -5 \\ -3 & -4 \end{bmatrix}$.

Step 1: We find the basic solution by solving the fuzzy system $B\tilde{x}_B = \tilde{b}$:

$$\begin{bmatrix} 1 & 0 \\ 0 & 1 \end{bmatrix}\begin{bmatrix} \tilde{x}_3 \\ \tilde{x}_4 \end{bmatrix} = \begin{bmatrix} (-8,-5,5,2) \\ (-10,-6,6,2) \end{bmatrix} \Rightarrow \begin{bmatrix} \tilde{x}_3 \\ \tilde{x}_4 \end{bmatrix} = \begin{bmatrix} (-8,-5,5,2) \\ (-10,-6,6,2) \end{bmatrix} = \begin{bmatrix} \bar{\tilde{b}}_1 \\ \bar{\tilde{b}}_2 \end{bmatrix}$$

The fuzzy objective function value is obtained based on the formulation $\tilde{z} \approx c_B\tilde{x}_B$:

$$\tilde{z} \approx (0,0)\begin{bmatrix} (-8,-5,5,2) \\ (-10,-6,6,2) \end{bmatrix} = (0,0,0,0)$$

Step 2: We find the rank of $\bar{\tilde{b}}_1$ and $\bar{\tilde{b}}_2$ by use of ranking function $\Re$ given in Remark 3. Hence we have $\Re\left(\bar{\tilde{b}}_1\right) = -\frac{29}{4}$ and $\Re\left(\bar{\tilde{b}}_2\right) = -9$. Therefore,

$$\min\left\{\Re\left(\bar{\tilde{b}}_1\right) = -\frac{29}{4}, \Re\left(\bar{\tilde{b}}_2\right) = -9\right\} = \Re(\bar{b}_2) = -9 \not\geq 0$$

Step 3: We find y_j for all $j \in J_N = \{1,2\}$ by solving the system $By_j = a_j$:

$$\begin{bmatrix} 1 & 0 \\ 0 & 1 \end{bmatrix}\begin{bmatrix} y_{11} \\ y_{21} \end{bmatrix} = \begin{bmatrix} -2 \\ -3 \end{bmatrix} \Rightarrow y_1 = \begin{bmatrix} y_{11} \\ y_{21} \end{bmatrix} = \begin{bmatrix} -2 \\ -3 \end{bmatrix}$$

$$\begin{bmatrix} 1 & 0 \\ 0 & 1 \end{bmatrix}\begin{bmatrix} y_{12} \\ y_{22} \end{bmatrix} = \begin{bmatrix} -5 \\ -4 \end{bmatrix} \Rightarrow y_2 = \begin{bmatrix} y_{12} \\ y_{22} \end{bmatrix} = \begin{bmatrix} -5 \\ -4 \end{bmatrix}$$

Step 4: We find w by solving the system $wB = c_B$:

$$(w_1, w_2)\begin{bmatrix} 1 & 0 \\ 0 & 1 \end{bmatrix} = \begin{bmatrix} 0 \\ 0 \end{bmatrix} \Rightarrow w_1 = w_2 = 0$$

Now, we calculate $z_j - c_j = wa_j - c_j$ for all $j \in J_N = \{1,2\}$:

$$z_1 - c_1 = wa_1 - c_1 = (0,0)\begin{bmatrix} -2 \\ -3 \end{bmatrix} - 6 = -6$$

$$z_2 - c_2 = wa_2 - c_2 = (0,0)\begin{bmatrix} -5 \\ -4 \end{bmatrix} - 10 = -10$$

The variable $\tilde{x}_{B_2} = \tilde{x}_4$ leaves the basis and the variable $\tilde{x}_k$ entering the basis is determined by the following test:

$$\min\left\{\frac{z_1 - c_1}{y_{21}}, \frac{z_2 - c_2}{y_{22}}\right\} = \min\left\{\frac{-6}{-3}, \frac{-10}{-4}\right\} = \frac{-6}{-3} = 2$$

Hence, $x_k = x_1$ enters the basis.

Step 5: We update the basis B where a_1 replaces $a_{B_2} = a_4$:

$$B = [a_3, a_1] = \begin{bmatrix} 1 & -2 \\ 0 & -3 \end{bmatrix}$$

Also, we update $J_N = \{1, 2\}$ as $J_N = \{2, 4\}$.
This process is continued and after two more iterations, the optimal solution of the FVLP problem (44) is given as follows, which is matched with that given in (50) obtained by Maleki et al.'s approach [29]:

$$\tilde{x}^* = \begin{bmatrix} \tilde{x}_1^* \\ \tilde{x}_2^* \end{bmatrix} = \begin{bmatrix} \left(\frac{-2}{7}, \frac{30}{7}, \frac{30}{7}, \frac{38}{7}\right) \\ \left(\frac{-5}{7}, \frac{12}{7}, \frac{18}{7}, \frac{19}{7}\right) \end{bmatrix}$$

It should be note that the advantages and disadvantages of this approach are same with the Maleki et al.'s approach [29].

6.1.4 LP Problem with Fuzzy Cost Coefficients, Fuzzy Decision Variables and Fuzzy Right-Hand-Side of the Constraints

Ganesan and Veeramani [14] formulated a LP problem in which the cots coefficients, decision variables and the values of the right-hand-side are represented by symmetric trapezoidal fuzzy numbers while the elements of the coefficient matrix are represented by real numbers. This problem can be formulated as follows:

$$\begin{aligned} &\max \tilde{z} \approx \tilde{c}\tilde{x} \\ &s.t. \ \ A\tilde{x} \approx \tilde{b}, \\ &\qquad \tilde{x} \succeq \tilde{0}. \end{aligned} \tag{54}$$

In what follows, two solution approaches are reviewed for solving FLP problem (54). The other approaches can be found in [10, 23].

Ganesan and Veeramani's Approach

Ganesan and Veeramani [14] introduced a new type of fuzzy multiplication and fuzzy ordering for symmetric trapezoidal fuzzy numbers and then proposed a simplex based method for solving SFFLP problem (55) without converting it into crisp LP problem.

Definition 13 Let $\tilde{a} = (a_2 - \alpha, a_2, a_3, a_3 + \alpha)$ and $\tilde{b} = (b_2 - \beta, b_2, b_3, b_3 + \beta)$ be two symmetric trapezoidal fuzzy numbers. The new type of multiplication on $\tilde{a}$ and $\tilde{b}$ has been defined in [14] as follows:

$$\tilde{a}\tilde{b} \approx (p - w - \gamma, p - w, p + w, p + w + \gamma)$$

where

$$p = \left(\frac{a_2 + a_3}{2}\right)\left(\frac{b_2 + b_3}{2}\right), \quad w = \frac{t_2 - t_1}{2}, \quad t_1 = \min\{a_2b_2, a_2b_3, a_3b_2, a_3b_3\},$$
$$t_2 = \max\{a_2b_2, a_2b_3, a_3b_2, a_3b_3\}, \quad \gamma = |a_3\beta + b_3\alpha|.$$

Definition 14 For any two symmetric trapezoidal fuzzy numbers $\tilde{a} = (a_2 - \alpha, a_2, a_3, a_3 + \alpha)$ and $\tilde{b} = (b_2 - \alpha, b_2, b_3, b_3 + \alpha)$ the relations $\preceq$ and $\approx$ have been defined in [14] as follows:

$$\tilde{a} \preceq b \Leftrightarrow \frac{a_2 + a_3}{2} \leq \frac{b_2 + b_3}{2}$$
$$\tilde{a} \approx b \Leftrightarrow \frac{a_2 + a_3}{2} \approx \frac{b_2 + b_3}{2}$$

Remark 5 It should be note that the ordering defined in Definition 14 is same as with the ordering defined in terms of linear ranking function in Remark 3. This means that the linear ranking function given in Remark 3 for any symmetric trapezoidal fuzzy numbers $\tilde{a} = (a_2 - \alpha, a_2, a_3, a_3 + \alpha)$ is reduced to

$$\Re(\tilde{a}) = \frac{a_2 - \alpha + a_2 + a_3 + a_3 + \alpha}{4} = \frac{a_2 + a_3}{2}$$

Now, consider a system of m simultaneous fuzzy linear equations involving symmetric trapezoidal fuzzy numbers in n unknowns $A\tilde{x} = \tilde{b}$. Let B be any matrix formed by m linear independent of A. In this case, the solution $\tilde{x} = (\tilde{x}_B, \tilde{x}_N) = (B^{-1}\tilde{b}, \tilde{0})$ is a fuzzy basic solution. Let y_j and $\tilde{w}$ be the solutions to $By_j = a_j$ and $\tilde{w}B = \tilde{c}_B$, respectively. Define $\tilde{z}_j = \tilde{c}_B B^{-1} a_j = \tilde{w}a_j$. Based on these results, Ganesan and Veeramani [14] proved fuzzy analogues of some important theorems of LP leading to a new method for solving FLP problem (54) without converting it into a crisp LP problem.

Algorithm 3: Symmetric fuzzy primal simplex algorithm (Maximization Problem)

Initialization step
Choose a starting dual feasible basic solution with basis B. Form the initial tableau as Table 1.

Main steps

1. Calculate $\tilde{z}_j - \tilde{c}_j$ for all $j \in J_N$ where J_N is the index set of non-basic variables. Suppose that $\tilde{z}_j - \tilde{c}_j = (h_j^2 - \alpha_j, h_j^2, h_j^3, h_j^3 + \alpha_j)$. Let

$$h_k^2 + h_k^3 = \max_{j \in J_N} \left\{ h_j^2 + h_j^3 \right\}.$$

   If $h_k^2 + h_k^3 \geq 0$, then stop with the current basic solution as an optimal solution.
2. Solve the system $By_k = a_k$ and let $y_k = B^{-1}a_k$. If $y_k \leq 0$ then stop with the conclusion that the problem is unbounded. Otherwise, suppose $\bar{\tilde{b}}_i = \left(\bar{b}_i^2 - \alpha_i, \bar{b}_i^2, \bar{b}_i^3, \bar{b}_i^3 + \alpha_j\right)$. In this case, then $\tilde{x}_k$ enters the basis and $\tilde{x}_{B_r}$ leaves the basis providing that

$$\frac{\bar{b}_r^2 + \bar{b}_r^3}{y_{rk}} = \min_{1 \leq i \leq m} \left\{ \frac{\bar{b}_i^2 + \bar{b}_i^3}{y_{ik}} \middle| y_{rk} > 0 \right\}$$

3. Update the tableau by pivoting on y_{rk} and go to (1).

Example 9 [11] Consider the following FLP problem:

$$\begin{aligned} &\text{Max } \tilde{z} \approx (11, 13, 15, 17)\tilde{x}_1 + (9, 12, 14, 17)\tilde{x}_2 \\ &s.t. \;\; 12\tilde{x}_1 + 13\tilde{x}_2 \preceq (469, 475, 505, 511), \\ &\qquad 12\tilde{x}_1 + 15\tilde{x}_2 \preceq (460, 465, 495, 500), \\ &\qquad \tilde{x}_1, \tilde{x}_2 \succeq \tilde{0}. \end{aligned} \tag{55}$$

Table 1 The initial simplex tableau

Basis	$\tilde{x}_1$	…	$\tilde{x}_r$	…	$\tilde{x}_m$	…	$\tilde{x}_j$	…	$\tilde{x}_k$	…	R.H.S	$\Re$
$\tilde{z}$	$\tilde{0}$	…	$\tilde{0}$	…	$\tilde{0}$	…	$\tilde{z}_j - \tilde{c}_j$	…	$\tilde{z}_k - \tilde{c}_k$	…	$\tilde{z} = \tilde{c}_B\bar{\tilde{b}}$	–
$\Re(\tilde{z})$	0	…	0	…	0	…	$h_j^2 + h_j^3$	…	$h_k^2 + h_k^3$	…	–	$z^2 + z^3$
$\tilde{x}_1$	1	…	0	…	0	…	y_{1j}	…	y_{1k}	…	$\bar{\tilde{b}}_1$	$\bar{b}_1^2 + \bar{b}_1^3$
⋮	⋮	…	⋮	…	⋮	…	⋮	…	⋮	…	⋮	
$\tilde{x}_r$	⋮	…	1	…	0	…	y_{rj}	…	y_{rk}	…	$\bar{\tilde{b}}_r$	$\bar{b}_r^2 + \bar{b}_r^3$
⋮	⋮	…	⋮	…	⋮	…	⋮	…	⋮	…	⋮	
$\tilde{x}_m$	0	…	0	…	1	…	y_{mj}	…	y_{mk}	…	$\bar{\tilde{b}}_m$	$\bar{b}_m^2 + \bar{b}_m^3$

After introducing the slack variables $\tilde{x}_3$ and $\tilde{x}_4$ we obtain the initial fuzzy primal simplex tableau as Table 2.

By considering $\tilde{x}_1$ and $\tilde{x}_4$ as the entering variable the leaving variable, respectively, and then by pivoting on $y_{21} = 12$, we obtain the next tableau as Table 3.

Since $h_2^2 + h_2^3 \geq 0$ and $h_4^2 + h_4^3 \geq 0$, the algorithm stops and Table 3 is the optimal tableau.

Ebrahimnejad and Nasseri's Approach

In the proposed approach by Ganesan and Veeramani [14] it is assumed that a fuzzy primal feasible basic solution of the problem (54) is at hand. For situation in which a primal fuzzy feasible basic solution is not at hand, but there is a fuzzy dual feasible basic solution, Ebrahimnejad and Nasseri [9] generalized the fuzzy analogue of the dual simplex algorithm of LP to the FLP problem under consideration. A summary of their method for a minimization problem in tableau format is given as follows.

Algorithm 4: Symmetric Fuzzy Dual Simplex Algorithm (Minimization Problem)

Initialization step

Choose a starting primal feasible basic solution with basis B. Form the initial tableau as Table 1. Suppose $\tilde{z}_j - \tilde{c}_j = (h_j^2 - \alpha_j, h_j^2, h_j^3, h_j^3 + \alpha_j)$, so we have $h_j^2 + h_j^3 \geq 0$ for all j.

Main steps

1. Suppose $\tilde{\bar{b}} = B^{-1}\tilde{b}$. If $\tilde{\bar{b}} = B^{-1}\tilde{b}$ then stop; the current fuzzy solution is optimal. Else, suppose $\tilde{\bar{b}}_i = \left(\bar{b}_i^2 - \alpha_i, \bar{b}_i^2, \bar{b}_i^3, \bar{b}_i^3 + \alpha_j\right)$ and let

Table 2 The initial simplex tableau of Example 8

Basis	$\tilde{x}_1$	$\tilde{x}_2$	$\tilde{x}_3$	$\tilde{x}_4$	R.H.S	$\Re$
$\tilde{z}$	$(-17, -15, -13, -11)$	$(-17, -14, -12, -9)$	$\tilde{0}$	$\tilde{0}$	$\tilde{0}$	–
$\Re(\tilde{z})$	−28	−26	0	0	–	0
$\tilde{x}_3$	12	13	1	0	$(469, 475, 505, 511)$	980
$\tilde{x}_4$	12	15	0	1	$(460, 465, 495, 500)$	960

Table 3 The optimal tableau of Example 8

Basis	$\tilde{x}_1$	$\tilde{x}_2$	$\tilde{x}_3$	$\tilde{x}_4$	R.H.S	$\Re$
$\tilde{z}$	$\tilde{0}$	$\left(\frac{-13}{4}, \frac{9}{4}, \frac{27}{4}, \frac{49}{4}\right)$	$\tilde{0}$	$\left(\frac{11}{12}, \frac{13}{12}, \frac{15}{12}, \frac{17}{12}\right)$	$\left(\frac{4965}{12}, \frac{1005}{2}, \frac{1235}{2}, \frac{8475}{12}\right)$	–
$\Re(\tilde{z})$	0	9	0	$\frac{7}{3}$	–	1120
$\tilde{x}_3$	0	−2	1	−1	$(-31, -20, 40, 51)$	20
$\tilde{x}_1$	1	$\frac{5}{4}$	0	$\frac{1}{12}$	$\left(\frac{460}{12}, \frac{465}{12}, \frac{495}{12}, \frac{500}{12}\right)$	80

$$\bar{b}_r^2 + \bar{b}_r^3 = \min_{1 \le i \le m} \{\bar{b}_i^2 + \bar{b}_i^3\}$$

2. If $y_{rj} \geq 0$ for all j, then stop; the FLP problem under consideration is infeasible. Else, $\tilde{x}_{B_r}$ leaves the basis and $\tilde{x}_k$ enters to the basis providing that

$$\frac{h_k^2 + h_k^3}{y_{rk}} = \min_{1 \le j \le n} \left\{ \frac{h_j^2 + h_j^3}{y_{rj}} \middle| y_{rj} < 0 \right\}$$

3. Update the tableau by pivoting on y_{rk} and go to (1).

Example 9 [10] Consider the following FLP problem:

$$\begin{aligned} &\text{Min } \tilde{z} \approx (1,4,8,13)\tilde{x}_1 + (2,4,6,8)\tilde{x}_2 + (1,2,4,5)\tilde{x}_3 \\ &s.t. \ \ 3\tilde{x}_1 + 4\tilde{x}_2 + 2\tilde{x}_3 \succeq (3,6,10,13), \\ &\quad 4\tilde{x}_1 + 2\tilde{x}_2 + \tilde{x}_3 \succeq (2,4,6,8), \\ &\quad 2\tilde{x}_1 + \tilde{x}_2 + \tilde{x}_3 \succeq (1,2,6,7), \\ &\quad \tilde{x}_1, \tilde{x}_2, \tilde{x}_3 \succeq \tilde{0}. \end{aligned} \tag{56}$$

After introducing the surplus variables $\tilde{x}_4, \tilde{x}_5$ and $\tilde{x}_6$ the initial fuzzy dual simplex tableau is given as Table 4.

After two iterations of the Algorithm 4, the optimal tableau is obtained as Table 5.

In contrast to the FVLP problem considered in [28, 29], in the FLP problem introduced by Ganesan and Veeramini [14], not only the decision variables and the right-hand-side of the constraints are fuzzy, but also the coefficients of decision variables in objective function are fuzzy. However, the proposed approaches for solving FLP problem considered in [14] are valid only for situation in which the parameters are represented by symmetric trapezoidal fuzzy numbers. But, the proposed methods [28, 29] for solving FVLP problems are applicable to both symmetric and non-symmetric trapezoidal fuzzy numbers. Also, unlike the non-simplex based approach and similar to the existing approaches [28, 29] in solving FVLP problems, this approach doesn't increase the number of the

Table 4 The initial tableau of Example 9

Basis	$\tilde{x}_1$	$\tilde{x}_2$	$\tilde{x}_3$	$\tilde{x}_4$	$\tilde{x}_5$	$\tilde{x}_6$	R.H.S	$\Re$
$\tilde{z}$	$(-13,-8,-4,-1)$	$(-8,-6,-4,-2)$	$(-5,-4,-2,-1)$	$\tilde{0}$	$\tilde{0}$	$\tilde{0}$	$\tilde{0}$	–
$\Re(\tilde{z})$	−12	−10	−6	0	0	0	–	0
$\tilde{x}_4$	−3	−4	−2	1	0	0	$(-13,-10,-6,-3)$	−16
$\tilde{x}_5$	−4	−2	−1	0	1	0	$(-8,-6,-4,-2)$	−10
$\tilde{x}_6$	−2	−1	−1	0	0	1	$(-7,-6,-2,-1)$	−8

Table 5 The optimal tableau of Example 9

Basis	$\tilde{x}_1$	$\tilde{x}_2$	$\tilde{x}_3$	$\tilde{x}_4$	$\tilde{x}_5$	$\tilde{x}_6$	R.H.S	$\Re$
$\tilde{z}$	$\tilde{0}$	$\tilde{0}$	$\tilde{0}$	$\tilde{0}$	$\tilde{0}$	$\tilde{0}$	$\tilde{0}$	–
$\Re(\tilde{z})$	0	0	0	$\left(\frac{-14}{5}, \frac{-19}{10}, \frac{3}{10}, \frac{11}{5}\right)$	$\left(\frac{-22}{5}, \frac{-11}{5}, \frac{3}{5}, \frac{14}{5}\right)$	$\left(\frac{-6}{5}, \frac{-4}{5}, \frac{2}{5}, \frac{4}{5}\right)$	$\left(\frac{-75}{5}, \frac{-2}{5}, \frac{114}{5}, \frac{187}{5}\right)$	112
$\tilde{x}_2$	1	0	0	$\frac{-2}{5}$	$\frac{1}{5}$	$\frac{1}{5}$	$\left(\frac{-9}{5}, 0, \frac{14}{5}, \frac{23}{5}\right)$	$\frac{14}{5}$
$\tilde{x}_1$	0	1	0	$\frac{1}{5}$	$\frac{-2}{5}$	0	$\left(\frac{-9}{5}, \frac{-2}{5}, \frac{6}{5}, \frac{13}{5}\right)$	$\frac{4}{5}$
$\tilde{x}_3$	0	0	1	0	$\frac{1}{5}$	$\frac{-2}{5}$	$\left(\frac{-11}{5}, \frac{-4}{5}, \frac{10}{5}, \frac{17}{5}\right)$	$\frac{6}{5}$

constraint of the primary FLP problem. In addition, although these approaches [9, 14] give fuzzy optimal solution, but, the fuzzy optimal solution by use of these approaches may be not non-negative, too. For instance the value of $\tilde{x}_3^* = (-31, -20, 40, 51)$ in the fuzzy optimal solution of Example 8 (see Table 3) and the value of $\tilde{x}_1^* = \left(\frac{-9}{5}, \frac{-2}{5}, \frac{6}{5}, \frac{13}{5}\right)$ in the optimal solution of Example 9 (see Table 5) are non-negative as there are negative parts in these fuzzy solutions.

6.1.5 LP Problem with Fuzzy Cost Coefficients, Fuzzy Decision Variables, Fuzzy Coefficients Matrix and Fuzzy Right-Hand-Side of the Constraints

Kheirfam and Verdegay [19] defined a new type of FLP problems in which all parameters and decision variables were represented by symmetric trapezoidal fuzzy numbers. This problem, named as symmetric fully fuzzy linear programming (SFFLP), can be represented with the following model:

$$\begin{aligned} \max \tilde{z} &= \tilde{c}\tilde{x} \\ s.t. \ \tilde{A}\tilde{x} &\approx \tilde{b}, \\ \tilde{x} &\succeq \tilde{0}. \end{aligned} \tag{57}$$

Kheirfam and Verdegay [19] introduced a new type of fuzzy inverse and division for symmetric trapezoidal fuzzy numbers and then proposed a simplex based method for solving SFFLP problem (57) without converting it to crisp LP problem.

Definition 15 Let $\tilde{a} = (a_2 - \alpha, a_2, a_3, a_3 + \alpha)$ and $\tilde{b} = (b_2 - \beta, b_2, b_3, b_3 + \beta)$ be symmetric trapezoidal fuzzy number and symmetric non zero trapezoidal fuzzy number, respectively. Two new types of arithmetic operations on $\tilde{a}$ and $\tilde{b}$ have been defined in [19] as follows:

- $\frac{1}{\tilde{b}} = \tilde{b}^{-1}$
$$\approx \left(\left[\frac{2}{b_2 + b_3} - w \right] - \beta, \left[\frac{2}{b_2 + b_3} - w \right], \left[\frac{2}{b_2 + b_3} + w \right], \left[\frac{2}{b_2 + b_3} + w \right] + \beta \right)$$
where $w = \frac{t_2 - t_1}{2}, t_2 = \max_{j=2,3} \left\{ \frac{1}{b_j} \right\}, t_1 = \min_{j=2,3} \left\{ \frac{1}{b_j} \right\}, \frac{1}{b_j} = \begin{cases} \frac{1}{b_j}, & b_j \neq 0, \\ 0, & b_j = 0. \end{cases}$

- $\frac{\tilde{a}}{\tilde{b}} \approx \left(\left[\frac{a_2 + a_3}{b_2 + b_3} - w \right] - \gamma, \left[\frac{a_2 + a_3}{b_2 + b_3} - w \right], \left[\frac{a_2 + a_3}{b_2 + b_3} + w \right], \left[\frac{a_2 + a_3}{b_2 + b_3} + w \right] + \gamma \right)$

where $w = \frac{t_2 - t_1}{2}, t_2 = \max_{i,j=2,3} \left\{ \frac{a_i}{b_j} \right\}, t_1 = \min_{i,j=2,3} \left\{ \frac{a_i}{b_j} \right\}, \gamma = |a_3 \beta + b_3 \alpha|.$

A symmetric trapezoidal fuzzy matrix $\widetilde{A} = [\tilde{a}_{ij}]_{m \times n}$ is any rectangular array of symmetric trapezoidal fuzzy numbers. For any values of indices i and j, the ijth minor of the square matrix $\tilde{A} = [\tilde{a}_{ij}]_{n \times n}$, denoted by $\tilde{A}_{ij}$, is the $(n-1) \times (n-1)$ sub-matrix of $\tilde{A}$ obtained by deleting the ith row and the jth column of $\tilde{A}$. In this case, determinant $\tilde{A}$, denoted by $|\tilde{A}|$, is computed as follows [19]:

- For $n = 1$, $|\tilde{A}| = \tilde{a}_{11}$.
- For $n = 2$, $|\tilde{A}| = \tilde{a}_{11}\tilde{a}_{22} - \tilde{a}_{12}\tilde{a}_{21}$.
- For $n > 2$, $|\tilde{A}| = \sum_{j=1}^{n} (-1)^{i+j} \tilde{a}_{ij} |\tilde{A}_{ij}|$, for any value of index $i = 1, 2, \ldots, n$.

A square matrix $\tilde{A} = [\tilde{a}_{ij}]_{n \times n}$, is called singular if $|\tilde{A}| = \tilde{0}$. In other case, it said to be non-singular fuzzy matrix and its inverse is calculated by $\tilde{A}^{-1} = \frac{(1,1,0,0)}{|\tilde{A}|} [(-1)^{i+j} |\tilde{A}_{ij}|]_{n \times n}$.

Now, we are in a position to summary the proposed method by Kheirfam and Verdegay [19] solving the SFFLP problem (57).

Suppose a fuzzy basic feasible solution of problem (57) with basis $\tilde{B}$ is at hand. Let $\tilde{y}_j$ and $\tilde{w}$ be the solutions to $\tilde{B}\tilde{y}_j = \tilde{a}_j$ and $\tilde{w}\tilde{B} = \tilde{c}_B$, respectively. Define $\tilde{z}_j = \tilde{c}_B \tilde{B}^{-1} \tilde{a}_j = \tilde{w}\tilde{a}_j$. Using these notations, Kheirfam and Verdegay [19] developed and presented for first time an original dual simplex algorithm for solving the SFFLP problem (57).

Algorithm 5: Symmetric Fully Fuzzy Dual Simplex Algorithm (Maximization Problem)

Initialization step

Let $\tilde{B}$ be a basis for the SFFLP problem (57) such that $\tilde{z}_j - \tilde{c}_j = \left(h_j^2 - \alpha_j, h_j^2, h_j^3, h_j^3 + \alpha_j \right) \succeq \tilde{0}$ for all j, i.e. $h_j^2 + h_j^3 \geq 0$. Form the initial tableau as Table 6.

Table 6 Tableau of the SFFLP problem

Basis	$\tilde{x}_1$	...	$\tilde{x}_r$	...	$\tilde{x}_m$	...	$\tilde{x}_j$	...	$\tilde{x}_k$	...	R.H.S	$\Re$
$\tilde{z}$	$\tilde{0}$	...	$\tilde{0}$	...	$\tilde{0}$	...	$\tilde{z}_j - \tilde{c}_j$	...	$\tilde{z}_k - \tilde{c}_k$	...	$\tilde{z} = \tilde{c}_B \bar{\tilde{b}}$	–
$\Re(\tilde{z})$	0	...	0	...	0	...	$h_j^2 + h_j^3$	...	$h_k^2 + h_k^3$	...	–	$z^2 + z^3$
$\tilde{x}_1$	1	...	0	...	0	...	$\tilde{y}_{1j}$	...	$\tilde{y}_{1k}$	...	$\bar{\tilde{b}}_1$	$\bar{b}_1^2 + \bar{b}_1^3$
$\vdots$	$\vdots$	...	$\vdots$	...	$\vdots$	...	$\vdots$	...	$\vdots$	...	$\vdots$	
$\tilde{x}_r$	$\vdots$	...	1	...	0	...	$\tilde{y}_{rj}$	...	$\tilde{y}_{rk}$	...	$\bar{\tilde{b}}_r$	$\bar{b}_r^2 + \bar{b}_r^3$
$\vdots$	$\vdots$	...	$\vdots$	...	$\vdots$	...	$\vdots$	...	$\vdots$	...	$\vdots$	
$\tilde{x}_m$	0	...	0	...	1	...	$\tilde{y}_{mj}$	...	$\tilde{y}_{mk}$	...	$\bar{\tilde{b}}_m$	$\bar{b}_m^2 + \bar{b}_m^3$

Main steps

1. Suppose $\bar{\tilde{b}} = \tilde{B}^{-1}\tilde{b}$. If $\bar{\tilde{b}} \succeq \tilde{0}$ then stop; the current fuzzy solution is optimal. Else, suppose $\bar{\tilde{b}}_i = \left(\bar{b}_i^2 - \alpha_i, \bar{b}_i^2, \bar{b}_i^3, \bar{b}_i^3 + \alpha_j\right)$ and let

$$\bar{b}_r^2 + \bar{b}_r^3 = \min_{1 \le i \le m} \left\{\bar{b}_i^2 + \bar{b}_i^3\right\}$$

2. If $\tilde{y}_{rj} = \left(y_{rj}^2 - \alpha_{rj}, y_{rj}^2, y_{rj}^3, y_{rj}^3 - \alpha_{rj}\right) \succeq \tilde{0}$ for all j, i.e. $y_{rj}^2 + y_{rj}^3 \ge 0$, then stop; the SFFLP problem (57) is infeasible.
Else, $\tilde{x}_{B_r}$ leaves the basis and $\tilde{x}_k$ enters the basis providing that

$$\frac{h_k^2 + h_k^3}{y_{rk}^2 + y_{rk}^3} = \max_{1 \le j \le n} \left\{ \frac{h_j^2 + h_j^3}{y_{rj}^2 + y_{rj}^3} \middle| y_{rj}^2 + y_{rj}^3 < 0 \right\}$$

3. Update the tableau by pivoting on $\tilde{y}_{rk}$ and go to (1).

Example 10 [19] Consider the following SFFLP problem:

$$\begin{aligned}
&\text{Max } \tilde{z} \approx (-4,-2,2,5)\tilde{x}_1 + (-3,-2,0,1)\tilde{x}_2 \\
&s.t. \ (1,2,4,5)\tilde{x}_1 + (-5,-3,-1,1)\tilde{x}_2 + (1,1,1,1)\tilde{x}_3 \approx (-6,-4,-4,2), \\
&\quad (-3,-2,0,1)\tilde{x}_1 + (-4,-2,4,6)\tilde{x}_2 + (1,1,1,1)\tilde{x}_4 \approx (-4,-1,5,8), \\
&\quad \tilde{x}_1, \tilde{x}_2, \tilde{x}_3, \tilde{x}_4 \succeq \tilde{0}.
\end{aligned} \tag{59}$$

The first dual feasible simplex tableau is showed in Table 7.

Since $\bar{b}_1^2 + \bar{b}_1^3 < 0$, thus $\tilde{x}_{B_1} = \tilde{x}_3$ leaves the basis and according to test given in step (2) of Algorithm 5, $\tilde{x}_2$ is the entering variable. By pivoting $\tilde{y}_{12} = (-5,-3,-1,1)$ on the new tableau shown in Table 8 is obtained.

Table 7 The initial simplex tableau of Example 10

Basis	$\tilde{x}_1$	$\tilde{x}_2$	$\tilde{x}_3$	$\tilde{x}_4$	R.H.S	$\Re$
$\tilde{z}$	$(-4,-2,2,5)$	$(-3,-2,0,1)$	$\tilde{0}$	$\tilde{0}$	$\tilde{0}$	–
$\Re(\tilde{z})$	0	−2	0	0	–	0
$\tilde{x}_3$	$(1,2,4,5)$	$(-5,-3,-1,1)$	$(1,1,1,1)$	$\tilde{0}$	$(-6,-4,-4,2)$	−8
$\tilde{x}_4$	$(-3,-2,0,1)$	$(-4,-2,4,6)$	$\tilde{0}$	$(1,1,1,1)$	$(-4,-1,5,8)$	4

Table 8 The optimal tableau of Example 8

Basis	$\tilde{x}_1$	$\tilde{x}_2$	$\tilde{x}_3$	$\tilde{x}_4$	R.H.S	$\Re$
$\tilde{z}$	$\left(-20,\frac{-23}{6},\frac{41}{6},\frac{138}{6}\right)$	$\tilde{0}$	$\left(\frac{-25}{6},\frac{-1}{3},\frac{4}{3},\frac{31}{6}\right)$	$\tilde{0}$	$\left(\frac{-86}{6},\frac{-16}{3},\frac{4}{3},\frac{74}{6}\right)$	–
$\Re(\tilde{z})$	3	0	1	0	–	−4
$\tilde{x}_2$	$\left(\frac{-61}{6},\frac{-19}{6},\frac{1}{6},\frac{43}{6}\right)$	$(1,1,1,1)$	$\left(\frac{-17}{6},\frac{-5}{6},\frac{-1}{6},\frac{11}{6}\right)$	$\tilde{0}$	$\left(\frac{-28}{3},\frac{2}{3},\frac{10}{3},\frac{40}{3}\right)$	4
$\tilde{x}_4$	$\left(\frac{-383}{12},\frac{-29}{3},\frac{32}{3},\frac{484}{12}\right)$	$\tilde{0}$	$\tilde{0}$	$(1, 1,1,1)$	$\tilde{0}$	0

Since $\tilde{\bar{b}}_1 = \left(\frac{-28}{3},\frac{2}{3},\frac{10}{3},\frac{40}{3}\right) \succ \tilde{0}$ and $\tilde{\bar{b}}_2 = \left(\frac{-308}{3},-13,13,\frac{308}{3}\right) \approx \tilde{0}$, the current solution is optimal. Thus, the optimal solution of the problem (59) is

$$\tilde{x}^* = \begin{bmatrix} \tilde{x}_1^* \\ \tilde{x}_2^* \end{bmatrix} = \begin{bmatrix} (0,0,0,0) \\ \left(\frac{-28}{3},\frac{2}{3},\frac{10}{3},\frac{40}{3}\right) \end{bmatrix} \tag{60}$$

In contrast to the FVLP problem considered in [28, 29] and the FLP problem introduced by Ganesan and Veeramini [14], in the FLP problem considered by Kheirfam and Verdegay [19] all parameters and decision variable are specified in terms of fuzzy data. In addition, the solution approach proposed by Kheirfam and Verdegay [19], not only gives fuzzy optimal solution, but also doesn't increase the number of the constraint of the primary FLP problem. However, similar to the existing approaches in [9, 14, 28, 29], the fuzzy optimal solution given by use of this approach may be not non-negative. For instance the value of $\tilde{x}_2^* = \left(\frac{-28}{3},\frac{2}{3},\frac{10}{3},\frac{40}{3}\right)$ in the fuzzy optimal solution (60) is non-negative as there is negative part in this fuzzy solution.

6.2 *The Non-simplex Based Approach*

In this section, the non-simplex based approach for solving two kinds of LP problems with fuzzy parameters is explored. In such approach, the fuzzy constraints are first converted to crisp ones based on arithmetic operations on fuzzy numbers and then the standard method are used for solving the crisp problem. Such approach increases the number of functional constraints and thus directly effects the computational time of the simplex method.

6.2.1 LP Problem with Fuzzy Coefficients Matrix and Fuzzy Right-Hand-Side of the Constraints

The general form of LP problems in which the right-hand-side of the constraints and the coefficients of the constraint matrix are fuzzy numbers, can be formulated as follows [32]:

$$\begin{aligned} \max z &= \sum_{j=1}^{n} c_j x_j \\ s.t. \quad & \sum_{j=1}^{n} \tilde{a}_{ij} x_j \preceq \tilde{b}_i, \quad i = 1, 2, \ldots, m, \\ & x_j \geq 0, \quad j = 1, 2, \ldots, n. \end{aligned} \tag{61}$$

Let us assume that all fuzzy numbers are triangular. Thus, $\tilde{a}_{ij}$ and $\tilde{b}_i$ are represented as $\tilde{a}_{ij} = (a_{1,ij}, a_{2,ij}, a_{3,ij})$ and $\tilde{b}_i = (b_{1,i}, b_{2,i}, b_{3,i})$, respectively. Hence, FLP problem (61) can be reformulated as follows:

$$\begin{aligned} \max z &= \sum_{j=1}^{n} c_j x_j \\ s.t. \quad & \sum_{j=1}^{n} (a_{1,ij}, a_{2,ij}, a_{3,ij}) x_j \preceq (b_{1,i}, b_{2,i}, b_{3,i}), \quad i = 1, 2, \ldots, m, \\ & x_j \geq 0, \quad j = 1, 2, \ldots, n. \end{aligned} \tag{62}$$

Equivalently, with regard to Remarks 1 and 2, the FLP (62) can be rewritten as follows:

$$\begin{aligned} \max z &= \sum_{j=1}^{n} c_j x_j \\ s.t. \quad & \sum_{j=1}^{n} a_{1,ij} x_j \leq b_{1,i}, \quad i = 1, 2, \ldots, m, \\ & \sum_{j=1}^{n} a_{2,ij} x_j \leq b_{2,i}, \quad i = 1, 2, \ldots, m, \\ & \sum_{j=1}^{n} a_{3,ij} x_j \leq b_{3,i}, \quad i = 1, 2, \ldots, m, \\ & x_j \geq 0, \quad j = 1, 2, \ldots, n. \end{aligned} \tag{63}$$

The optimal solution of crisp LP problem (63) can be considered as the solution of the FLP problem (61) [32].

Example 11 [26] Consider the following FLP problem:

$$\begin{aligned} &\text{Max } z = 5x_1 + 4x_2 \\ &s.t.\ \ (2,4,5)x_1 + (2,5,6)x_2 \preceq (19,24,32), \\ &\qquad (3,4,6)x_1 + (4,5,6)x_2 \preceq (6,12,19) \\ &\qquad x_1, x_2 \geq 0. \end{aligned} \tag{64}$$

This problem is converted into the following crisp LP problem with regard to problem (63):

$$\begin{aligned} &\text{Max } z = 5x_1 + 4x_2 \\ &s.t.\ \ 2x_1 + 2x_2 \leq 19, \\ &\qquad 4x_1 + 5x_2 \leq 24, \\ &\qquad 5x_1 + 6x_2 \leq 32, \\ &\qquad 3x_1 + 4x_2 \leq 6, \\ &\qquad 4x_1 + 5x_2 \leq 12, \\ &\qquad 6x_1 + 6x_2 \leq 19, \\ &\qquad x_1, x_2 \geq 0. \end{aligned} \tag{65}$$

Solving this problem gives the optimal solution as $x^* = \left(x_1^*, x_1^*\right) = (1.5, 3), z^* = 19.5$.

The optimal solutions obtained by this approach are real numbers, which represent a compromise in terms of fuzzy numbers involved. In addition, this approach increases the number of functional constraints and thus this proposed need more computation cost than the simplex based approach.

6.2.2 Fully Fuzzy LP Problem [24]

The FLP problem in which all the parameters as well as the decision variables are represented by fuzzy numbers is called FFLP problem. The general form of FFLP problem can be formulated as follows:

$$\begin{aligned} &\max \tilde{z} \approx \sum_{j=1}^{n} \tilde{c}_j \otimes \tilde{x}_j \\ &s.t.\ \sum_{j=1}^{n} \tilde{a}_{ij} \otimes \tilde{x}_j \approx \tilde{b}_i, \quad i = 1, 2, \ldots, m, \\ &\qquad \tilde{x}_j \succeq \tilde{0}, \quad j = 1, 2, \ldots, n. \end{aligned} \tag{66}$$

Let us assume that all fuzzy numbers are triangular. Thus $\tilde{a}_{ij}, \tilde{x}_{ij}, \tilde{c}_j$ and $\tilde{b}_i$, are represented as $\tilde{a}_{ij} = (a_{1,ij}, a_{2,ij}, a_{3,ij})$ $\tilde{x}_j = (x_{1,j}, x_{2,j}, x_{3,j})$ $\tilde{b}_i = (b_{1,i}, b_{2,i}, b_{3,i})$ and $\tilde{b}_i = (b_{1,i}, b_{2,i}, b_{3,i})$, respectively. Hence, FLP problem (66) can be reformulated as follows:

$$\begin{aligned}
&\max \tilde{z} \approx \sum_{j=1}^{n} (c_{1,j}, c_{2,j}, c_{3,j}) \otimes (x_{1,j}, x_{2,j}, x_{3,j}) \\
&s.t. \sum_{j=1}^{n} (a_{1,ij}, a_{2,ij}, a_{3,ij}) \otimes (x_{1,j}, x_{2,j}, x_{3,j}) \approx (b_{1,i}, b_{2,i}, b_{3,i}), \quad i = 1, 2, \ldots, m, \\
&\quad (x_{1,j}, x_{2,j}, x_{3,j}) \text{ is a non negative trinagular fuzzy number}, \quad j = 1, 2, \ldots, n.
\end{aligned} \tag{67}$$

Assuming $(a_{1,ij}, a_{2,ij}, a_{3,ij}) \otimes (x_{1,ij}, x_{2,ij}, x_{3,ij}) = (m_{1,ij}, m_{2,ij}, m_{3,ij})$ and using ranking function given in Remark 4 for the objective function, the FLP (67) can be rewritten as follows:

$$\begin{aligned}
&\max \Re(\tilde{z}) = \Re\left(\sum_{j=1}^{n} (c_{1,j}, c_{2,j}, c_{3,j}) \otimes (x_{1,j}, x_{2,j}, x_{3,j})\right) \\
&s.t. \sum_{j=1}^{n} (m_{1,ij}, m_{2,ij}, m_{3,ij}) \approx (b_{1,i}, b_{2,i}, b_{3,i}), \quad i = 1, 2, \ldots, m, \\
&\quad (x_{1,j}, x_{2,j}, x_{3,j}) \text{ is a non negative trinagular fuzzy number}, \quad j = 1, 2, \ldots, n.
\end{aligned} \tag{68}$$

Using Definitions 10 and 7, the FLP problem (68) becomes:

$$\begin{aligned}
&\max \Re(\tilde{z}) = \Re\left(\sum_{j=1}^{n} (c_{1,j}, c_{2,j}, c_{3,j}) \otimes (x_{1,j}, x_{2,j}, x_{3,j})\right) \\
&s.t. \sum_{j=1}^{n} m_{1,ij} = b_{1,i}, \quad i = 1, 2, \ldots, m, \\
&\quad \sum_{j=1}^{n} m_{2,ij} = b_{2,i}, \quad i = 1, 2, \ldots, m, \\
&\quad \sum_{j=1}^{n} m_{3,ij} = b_{3,i}, \quad i = 1, 2, \ldots, m, \\
&\quad x_{1,j} \geq 0, \quad x_{2,j} - x_{1,j} \geq 0, \quad x_{3,j} - x_{2,j} \geq 0, \quad j = 1, 2, \ldots, n.
\end{aligned} \tag{69}$$

The optimal solution of the crisp LP problem (69) can be considered as the optimal solution of the FFLP problem (66).

Example 12 [24] Consider the following FFLP problem:

$$\begin{aligned}
&\text{Max } z = (1,6,9) \otimes (x_{1,1}, x_{1,2}, x_{1,3}) + (2,3,8) \otimes (x_{2,1}, x_{2,2}, x_{2,3}) \\
&s.t.\ \ (2,3,4)(x_{1,1}, x_{1,2}, x_{1,3}) + (1,2,3)(x_{2,1}, x_{2,2}, x_{2,3}) = (6,16,30), \\
&\quad\ (-1,1,2)(x_{1,1}, x_{1,2}, x_{1,3}) + (1,3,4)(x_{2,1}, x_{2,2}, x_{2,3}) = (1,17,30) \\
&\quad\ (x_{1,1}, x_{1,2}, x_{1,3}) \text{ and } (x_{2,1}, x_{2,2}, x_{2,3}) \text{ are non negative fuzzy numbers.}
\end{aligned} \tag{70}$$

This problem is converted into the following crisp LP problem with regard to problem (69):

$$\begin{aligned}
&\text{Max } \Re(\tilde{z}) = \frac{1}{4}\left[\left(x_{1,1} + 2x_{1,2} + 12x_{2,1} + 6x_{2,2} + 9x_{3,1} + 8x_{2,3}\right)\right] \\
&s.t.\ \ 2x_{1,1} + x_{2,1} = 6, \\
&\quad -x_{1,3} + x_{2,1} = 1, \\
&\quad 3x_{1,2} + 2x_{2,2} = 16, \\
&\quad x_{1,2} + 3x_{2,2} = 17, \\
&\quad 4x_{1,3} + 3x_{2,3} = 30, \\
&\quad 2x_{1,3} + 4x_{2,3} = 30, \\
&\quad x_{1,1} \geq 0, \quad x_{1,2} - x_{1,1} \geq 0, \quad x_{1,3} - x_{1,2} \geq 0, \\
&\quad x_{2,1} \geq 0, \quad x_{2,2} - x_{2,1} \geq 0, \quad x_{2,3} - x_{2,2} \geq 0.
\end{aligned} \tag{71}$$

The optimal solution of the crisp LP problem (71) is as follows:

$$x^*_{1,1} = 1, \quad x^*_{1,2} = 2, \quad x^*_{1,3} = 3, \quad x^*_{2,1} = 4, \quad x^*_{2,2} = 5, \quad x^*_{2,3} = 6 \tag{72}$$

Hence, the fuzzy optimal solution of the FFLP problem (70) is

$$\tilde{x}^* = (\tilde{x}^*_1, \tilde{x}^*_2) = \left(\left(x^*_{1,1}, x^*_{1,2}, x^*_{1,3}\right), \left(x^*_{2,1}, x^*_{2,2}, x^*_{2,3}\right)\right) = ((1,2,3),(4,5,6)) \tag{73}$$

Putting the fuzzy optimal solution (73) in the objective function of the problem (70) gives $\tilde{z}^* = (9, 27, 75)$.

In contrast to the proposed approaches in [32], Kumar et al.'s method [24] gives fuzzy optimal solution. In contrast to the simplex based approach [9–11, 14, 19, 21, 28–30], Kumar et al.'s method [24] gives non-negative fuzzy optimal solutions.

7 Conclusions

In the conventional LP problems it is assumed that the parameters of the problem are exactly known, but there may be some situations in formulating real-life problems where the parameters are not known in a precise way. A frequently used manner to represent the imprecision in parameters is by means of fuzzy numbers that enlarges the range of applications of LP. Many different techniques have been used in the literature to solve the LP problems in fuzzy environment. In this contribution, a taxonomy and review of some of these techniques for solving four categories of FLP problems was provided. The limitations and advantages of each category were also pointed out. Moreover, the solution approaches were illustrated with several numerical examples.

Acknowledgments Research supported in part under projects P11-TIC-8001 (VASCAS) from the Andalusian Government and TIN2014-55024-P (MODAS-3D) from the Spanish Ministry of Economy and Competitiveness (both including FEDER funds).

References

1. Bector, C.R., Chandra, S.: Fuzzy mathematical programming and fuzzy matrix games. In: Studies in Fuzziness and Soft Computing, vol. 169. Springer, Berlin (2005)
2. Bellman, R.E., Zadeh, L.A.: Decision making in a fuzzy environment. Manage. Sci. **17**(4), 141–164 (1970)
3. Chanas, S.: The use of parametric programming in fuzzy linear programming problems. Fuzzy Sets Syst. **11**, 243–251 (1983)
4. Delgado, M., Verdegay, J.L., Vila, M.A.: Relating different approaches to solve linear programming problems with imprecise costs. Fuzzy Sets Syst. **37**(1), 33–42 (1990)
5. Dubois, D., Prade, H.: A review of fuzzy set aggregation connectives. Inf. Sci. **36**(1–2), 85–121 (1985)
6. Ebrahimnejad, A.: Sensitivity analysis in fuzzy number linear programming problems. Math. Comput. Model. **53**(9–10), 1878–1888 (2011)
7. Ebrahimnejad, A.: A primal–dual method for solving linear programming problems with fuzzy cost coefficients based on linear ranking functions and its applications. Int. J. Ind. Syst. Eng. **12**(2), 119–140 (2012)
8. Ebrahimnejad, A.: A duality approach for solving bounded linear programming problems with fuzzy variables based on ranking functions and its application in bounded transportation problems. Int. J. Syst. Sci. **46**(11), 2048–2060 (2015)
9. Ebrahimnejad, A., Nasseri, S.H.: Linear programmes with trapezoidal fuzzy numbers: a duality approach. Int. J. Oper. Res. **13**(1), 67–89 (2012)
10. Ebrahimnejad, A., Tavana, M.: A novel method for solving linear programming problems with symmetric trapezoidal fuzzy numbers. Appl. Math. Model. **38**(17–18), 4388–4395 (2014)
11. Ebrahimnejad, A., Verdegay, J.L.: A novel approach for sensitivity analysis in linear programs with trapezoidal fuzzy numbers. J. Intell. Fuzzy Syst. **27**(1), 173–185 (2014)
12. Ebrahimnejad, A., Verdegay, J.L.: On solving bounded fuzzy variable linear program and its applications. J. Intell. Fuzzy Syst. **27**(5), 2265–2280 (2014)
13. Ebrahimnejad, A., Nasseri, S.H., HosseinzadehLotfi, F., Soltanifar, M.: A primal-dual method for linear programming problems with fuzzy variables. Eur. J. Ind. Eng. **4**(2), 189–209 (2010)

14. Ganesan, K., Veeramani, P.: Fuzzy linear programming with trapezoidal fuzzy numbers. Ann. Oper. Res. **143**(1), 305–315 (2006)
15. Gupta, P., Mehlawat, M.K.: Bector-Chandra type duality in fuzzy linear programming with exponential membership functions. Fuzzy Sets Syst. **160**(22), 3290–3308 (2009)
16. Hatami-Marbini, A., Tavana, M.: An extension of the linear programming method with fuzzy parameters. Int. J. Math. Oper. Res. **3**(1), 44–55 (2011)
17. Herrera, F., Kovacs, M., Verdegay, J.L.: Optimality for fuzzified mathematical programming problems: a parametric approach. Fuzzy Sets Syst. **54**(3), 279–284 (1993)
18. Hosseinzadeh Lotfi, F., Allahviranloo, T., Alimardani Jondabeh, M., Alizadeh, L.: Solving a full fuzzy linear programming using lexicography method and fuzzy approximate solution. Appl. Math. Model. **33**(7), 3151–3156 (2009)
19. Kheirfam, B., Verdegay, J.L.: Optimization and reoptimization in fuzzy linear programming problems. In: The 8th Conference of the European Society for Fuzzy Logic and Technology, pp. 527–533 (2013)
20. Kheirfam, B., Verdegay, J.L.: Strict sensitivity analysis in fuzzy quadratic programming. Fuzzy Sets Syst. **198**, 99–111 (2012)
21. Kheirfam, B., Verdegay, J.L.: The dual simplex method and sensitivity analysis for fuzzy linear programming with symmetric trapezoidal numbers. Fuzzy Optim. Decis. Making **12**(1), 171–189 (2013)
22. Klir, G.J., Yuan, B.: Fuzzy Sets and Fuzzy Logic, Theory and Applications. Prentice-Hall, PTR, Englewood Cliffs (1995)
23. Kumar, A., Kaur, J.: A new method for fuzzy linear programming programs with trapezoidal fuzzy numbers. J. Fuzzy Valued Syst. Anal. 1–12. Article ID jfsva-00102. doi:10.5899/2011/jfsva-00102
24. Kumar, A., Kaur, J., Singh, P.: A new method for solving fully fuzzy linear programming problems. Appl. Math. Model. **35**(2), 817–823 (2011)
25. Lai, Y.J., Hwang, C.L.: A new approach to some possibilistic linear programming problem. Fuzzy Sets Syst. **49**(2), 121–133 (1992)
26. Lai, Y.J., Hwang, C.L.: Fuzzy Mathematical Programming. Springer, Berlin (1992)
27. Mahdavi-Amiri, N., Nasseri, S.H.: Duality in fuzzy number linear programming by use of a certain linear ranking function. Appl. Math. Comput. **180**(1), 206–216 (2006)
28. Mahdavi-Amiri, N., Nasseri, S.H.: Duality results and a dual simplex method for linear programming problems with trapezoidal fuzzy variables. Fuzzy Sets Syst. **158**(17), 1961–1978 (2007)
29. Maleki, H.R., Tata, M., Mashinchi, M.: Linear programming with fuzzy variables. Fuzzy Sets Syst. **109**(1), 21–33 (2000)
30. Nasseri, S.H., Ebrahimnejad, A.: A fuzzy dual simplex method for fuzzy number linear programming problem. Adv. Fuzzy Sets Syst. **5**(2), 81–95 (2010)
31. Nasseri, S.H., Ebrahimnejad, A., Mizuno, S.: Duality in fuzzy linear programming with symmetric trapezoidal numbers. Appl. Appl. Math. **5**(10), 1467–1482 (2010)
32. Ramik, J., Rimanek, J.: Inequality relation between fuzzy numbers and its use in fuzzy optimization. Fuzzy Sets Syst. **16**(2), 123–138 (1085)
33. Rommelfanger, H., Hanuscheck, R., Wolf, J.: Linear programming with fuzzy objective. Fuzzy Sets Syst. **29**(1), 31–48 (1989)
34. Tanaka, H., Asai, K.: Fuzzy solution in fuzzy linear programming problems. IEEE Trans. Syst. Man Cybern. **14**(2), 325–328 (1984)
35. Tanaka, H., Okuda, T., Asai, K.: On fuzzy mathematical programming. J. Cybern. **3**(4), 37–46 (1974)
36. Tanaka, H., Ichihashi, H., Asai, K.: A formulation of fuzzy linear programming problems based on comparison of fuzzy numbers. Control Cybern. **13**, 186–194 (1984)
37. Verdegay, J.L.: Fuzzy mathematical programming. In: Gupta, M.M., Sanchez, E. (eds.) Fuzzy Information and Decision Processes. North Holland, Amsterdam (1982)
38. Verdegay, J.L.: A dual approach to solve the fuzzy linear programming problem. Fuzzy Sets Syst. **14**(2), 131–141 (1984)

39. Wan, S.P., Dong, J.Y.: Possibility linear programming with trapezoidal fuzzy numbers. Appl. Math. Model. **38**, 1660–1672 (2014)
40. Werners, B.: Interactive multiple objective programming subject to flexible constraints. Eur. J. Oper. Res. **31**, 342–349 (1987)
41. Yager, R.R.: A procedure for ordering fuzzy subsets of the unit interval. Inf. Sci. **24**, 143–161 (1981)
42. Zhang, G., Wu, Y.H., Remias, M., Lu, J.: Formulation of fuzzy linear programming problems as four-objective constrained optimization problems. Appl. Math. Comput. **139**(2–3), 383–399 (2003)
43. Zimmerman, H.J.: Fuzzy programming and linear programming with several objective functions. Fuzzy Sets Syst. **1**(1), 45–55 (1978)
44. Zimmermann, H.J.: Description and optimization of fuzzy Systems. Int. J. Gen. Syst. **2**, 209–215 (1976)

Applications of Fuzzy Mathematical Programming Approaches in Supply Chain Planning Problems

Mohammad Javad Naderi, Mir Saman Pishvaee and Seyed Ali Torabi

Abstract Supply chain planning includes numerous decision problems over strategic (i.e. long-term), tactical (i.e. mid-term) and operational (i.e. short-term) planning horizons in a supply chain. As most of supply chain planning problems deal with decision making in *real world* while configuring *future situations*, relevant data should be predicted and described for multiple time periods in the future. Such prediction and description involve imprecision and vagueness due to errors and absence of sharp boundaries in the subjective data and/or insufficient or unreliable objective data. If the uncertainty in supply chain planning problems is to be neglected by the decision maker, the plausible performance of supply chain in future conditions will be in doubt. This is why considerable body of the recent literature account for uncertainty through applying different uncertainty programming approaches with respect to the nature of uncertainty. This chapter aims to provide useful and updated information about different sources and types of uncertainty in supply chain planning problems and the strategies used to confront with uncertainty in such problems. A hyper methodological framework is proposed to cope with uncertainty in supply chain planning problems. Also, among the different uncertainty programming approaches, various fuzzy mathematical programming methods extended in the recent literature are introduced and a number of them are elaborated. Finally, a useful case study is illustrated to present the practicality of fuzzy programming methods in the area of supply chain planning.

Keywords Supply chain planning · Uncertainty · Fuzzy mathematical programming · Possibilistic programming · Flexible programming · Robust fuzzy programming

M.J. Naderi · M.S. Pishvaee
School of Industrial Engineering, Iran University of Science and Technology, Tehran, Iran
e-mail: mo_naderi@ind.iust.ac.ir

M.S. Pishvaee
e-mail: pishvaee@iust.ac.ir

S.A. Torabi (✉)
School of Industrial Engineering, College of Engineering, University of Tehran, Tehran, Iran
e-mail: satorabi@ut.ac.ir

C. Kahraman et al. (eds.), *Fuzzy Logic in Its 50th Year*,
Studies in Fuzziness and Soft Computing 341,
DOI 10.1007/978-3-319-31093-0_16

1 Introduction to Supply Chain Planning Under Uncertainty

Supply chain planning (SCP) consists of strategic, tactical and operational decision making problems of a supply chain network. Strategic decisions are related to supply chain design and configuration over a long term whilst tactical planning is related to those decisions affecting the efficient usage of the various resources. Also, operational planning is related to the detailed scheduling, sequencing, assigning loads, etc., in a short term planning horizon. The same facet of all above-mentioned problems is the decision making in a *real world* and with respect to *future situation*. In *real world*, vagueness may be arise with semantic description of events and phenomena, i.e., planning of supply chains in many cases may require human descriptions, judgments, statements and evaluations about the future conditions. These affairs are performed through the natural languages and in respect to this fact that in natural language the words are mostly appeared vaguely and imprecisely [98], the supply chain planning input data may be tainted with uncertainty. On the other hand, with respect to lack of information about the *future situations*, we need to predict the future, and in prediction of the future conditions, imprecision may be arise due to inherent errors and ambiguity. Therefore, supply chain planning problems are accompanied by imprecision and vagueness due to prediction errors and imprecise descriptions, respectively. Nevertheless, is it necessary to consider these uncertainties when planning of supply chains? According to Klibi et al. [32] any supply chain planning decision model, which is developed based on deterministic conditions has no assurance for acceptable performance in any plausible future situation. They also mentioned that in some cases it is not sufficient to only consider business as usual random variables such as demand, prices and exchange rate but one should include extreme and catastrophe events such as natural disasters and terrorist attacks. Therefore, it is necessary to have a specific strategy in facing with uncertainty in supply chain planning.

Accordingly, in this chapter, different types of uncertainty and their associated sources are first introduced and analyzed. Thereafter, various approaches used to cope with uncertainty in supply chain planning problems are discussed and a hyper conceptual framework is proposed for coping with uncertainty in SCP problems. Since the focus of this chapter is on the application of fuzzy mathematical programming in SCP area, the other sections of the chapter are dedicated to classification, review and description of fuzzy programming methods developed in the relevant literature.

1.1 Definition and Scope

According to Rosenhead et al. [71], certainty can be defined as the condition of decision making in which there is no element of chance intervene between decisions and their outcomes. Also, Zimmermann [98] defined the certainty as the situation that is definitely known, input parameters have predetermined values and there is no doubt in the occurrence of events. On the contrary, uncertainty is vagueness in reaching the target values and/or imperfect information about the input parameters or environmental conditions in which the decisions are made. Galbraith [19] also mentioned that uncertainty is the difference between the amount of information that is required to make the decision and the amount of available information. This definition implicitly unveils one of the important origins of uncertainty that is 'imperfect information'. There is also a distinction between uncertainty and risk. In the classical risk management, risk is defined as the product of probability and severity of an extreme event [23]. According to above-mentioned definitions, it can be concluded that uncertainty is a value neutral concept (i.e. includes both the chance for gain and the chance of damage), while risk is a negative concept and worthy of avoidance. As explained by Stewart [77], uncertainty gives rise to risk and in the risky situations undesirable outcomes could occur.

In an attempt for choosing an appropriate strategy for uncertainty management in supply chain planning area, it is necessary to understand the sources of uncertainty related to supply chain networks. According to Simangunsong et al. [76], identifying a full list of supply chain uncertainty sources is a precursor to develop appropriate portfolio of uncertainty management strategies.

An early classification was suggested by Davis [10] who mentioned three sources of uncertainty in the supply chain planning context including the demand, supply and process uncertainties. Later, this classification was supported by other researchers including van der Vaart et al. [85] and Gupta and Maranas [22]. Afterwards, Mason-Jones and Towill [49] added a fourth source to those suggested by Davis [10] called control uncertainty, which is related to the capability of a supply chain to transform the orders of customers to production flows by means of information flows. They denominated their model as the circle model which involves four uncertainty sources: (1) demand side uncertainty, (2) supply side uncertainty, (3) control uncertainty and (4) process uncertainty. The model advices that reducing uncertainty in each of these sources can lead to cost reduction.

Another taxonomy is proposed by Christopher and Peck [9], which classifies the risk sources into three categories as follows: (1) internal risk (process and control), (2) external risk and (3) network dependent risk (supply and demand). They also extended a framework for managing and mitigating the risks related to each source. Klibi et al. [32] identified uncertainty sources from the supply chain perspective and in a more detailed view. Table 1 integrates different prominent classifications for sources of uncertainty available in the literature. In the review paper of Simangunsong et al. [76], fourteen sources of uncertainty extracted from the literature are introduced. Simangunsong et al. [76] classify uncertainty management strategies into two broad categories: (1) reducing uncertainty, (2) coping with uncertainty. Their classification is summarized in Fig. 1. The first strategy (i.e. reducing uncertainty) disables or reduces uncertainty at its sources such as using the better forecasting approaches for demand predictions or integrating the organizations across the supply chain to achieve better collaboration between them. The

Table 1 Different classifications for sources of uncertainty

Klibi et al. [32]		Ho [26]	Davis [10]
Production centers	Endogenous assets	System	Process/manufacturing
Distribution centers			
Recovery centers			
Service centers			
External providers	Supply chain partners	Environment	Supply
Demand zones			
Nature	Exogenous geographical factors		Demand
Public infrastructures			
Socio-economic-political factors			

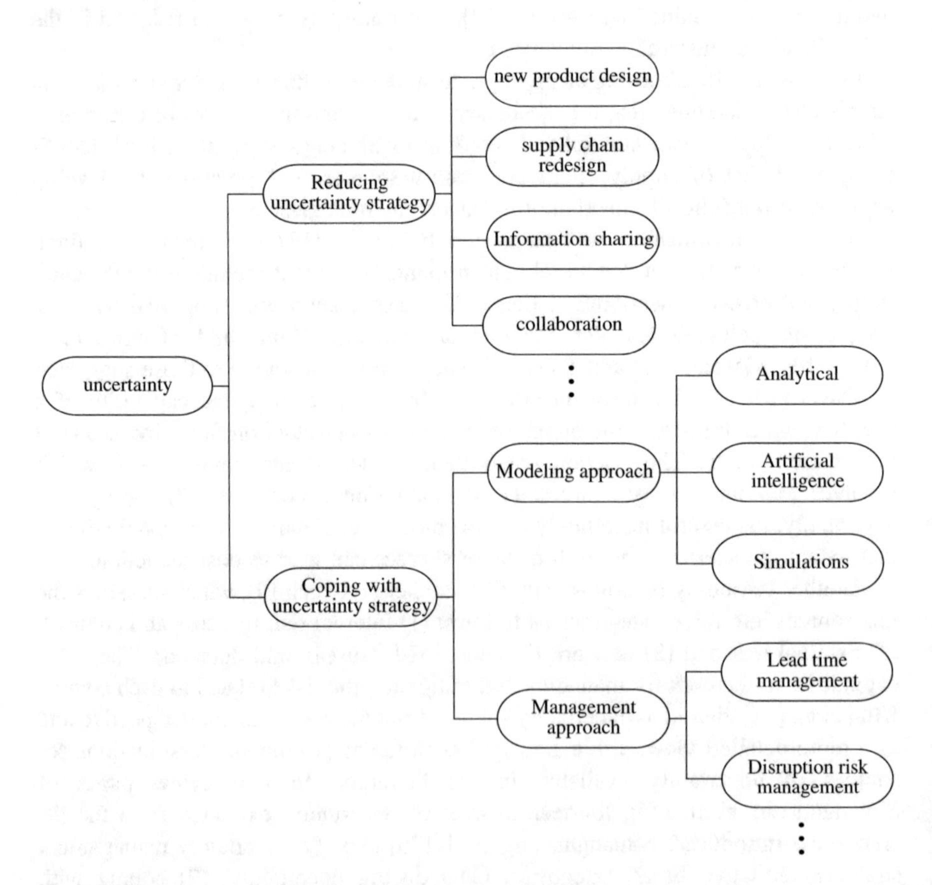

Fig. 1 Different strategies in facing with uncertainty along with some examples

latter strategy refers to one which tries to minimize the impact of uncertainty on the performance of supply chain instead of reducing the sources of uncertainty. Various analytical approaches (e.g. mathematical methods) are usually used when implementing these strategies. As early as 1993, Davis [10] suggested three strategies to cope with uncertainty: (1) TQM (total quality control), (2) new product design and (3) supply chain redesign. The first two strategies participate in reducing the process uncertainty while supply chain redesign can alleviate the demand and supply side uncertainties. In addition, van der Vorst and Beulens [85] proposed two other strategies of reducing the uncertainty: (1) collaboration with key suppliers and customers (2) reducing the role of human in supply chain processes. The first one aims to control the demand and supply uncertainties while the latter tries to reduce process uncertainty. It should be noted that the concept of collaboration has been studied broadly in supply chain management and one important prerequisite of this concept is the advanced integration of supply chain members. Furthermore, collaboration leads to uncertainty reduction in various uncertainty categories including demand, supply, process and control uncertainties. Another prerequisite of collaboration that is worthy of attention is information sharing alongside the supply chain. Other strategies such as flexibility in supply side [21, 66, 73], lead time management [66], insurance against disasters [31, 69, 79] are also suggested in the relevant literature.

1.2 Different Types of Uncertainty

Different classifications of uncertainty have ever been proposed in the literature from the different viewpoints. Tang [79] classified supply chain risks into operational and disruption risks. Operational risks are related to those uncertainties that intrinsically are involved in supply chain operations such as demand and cost uncertainty, equipment malfunction, etc. Also, disruption risks are related to those events with low-likelihood but high-severity impact, caused by natural and man-made disasters or technological threats like employee strikes. Interested readers can consult with Torabi et al. [80, 82] for more information. Another classification was expressed by Klibi et al. [32] where various types of uncertainty were categorized into three groups: (1) Randomness, (2) Hazard and (3) Deep uncertainty. They explained that randomness is characterized by random variables related to business-as-usual events while hazard is characterized by low-likelihood but high-impact extreme events. Furthermore, deep uncertainty is related to unavailability of information about likelihood of future plausible events. Mula et al. [52] classified uncertainty into two main categories: (1) flexibility in constraints and goals (2) uncertainty in data. The first category is characterized by the flexibility in satisfying the constraints and/or in target values of goals. Uncertainty in data can be also classified into two categories: (1) randomness, that stems from the random nature of some events and parameters and (2) epistemic uncertainty, which is related to lack of knowledge about the exact value of some ill-defined and imprecise

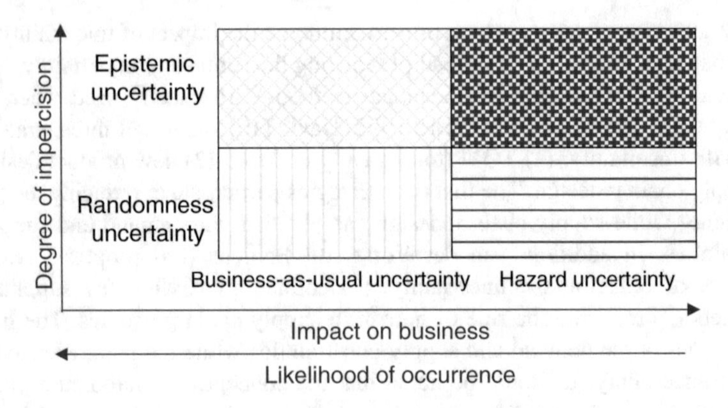

Fig. 2 Taxonomy of uncertainty

data, which are often in the form of linguistic attributes and qualitative/judgmental data extracted from field experts. According to above-mentioned classification, the intuitive taxonomy for uncertainty can be presented by considering two dimensions: (1) the impact of uncertainty and (2) the nature of imprecision. The proposed taxonomy is illustrated in Fig. 2.

1.3 Overview of Different Approaches Used to Cope with Uncertainty

Both the qualitative and quantitative approaches are applied by researchers and practitioners to handle uncertainty in SCP problems. As mentioned by Peidro et al. [57] quantitative modelling approach includes (1) analytical methods, (2) artificial intelligence methods, (3) simulation based methods, and (4) hybrid methods. This classification, along with some examples for each category is shown in Fig. 3. Since this chapter focuses on fuzzy programming methods, relevant mathematical models are elaborated in the rest of this chapter. However, it should be noted that stochastic programming (e.g. [4, 61]) and robust optimization (e.g. [62]) are the other major methods broadly used in the extant literature. Although there are many research works in the area of SCP, which focus on developing efficient methods to cope with uncertainty, there is a need for a general conceptual/methodological framework which enables the decision makers to systemically confront with uncertainty in SCP problems.

Figure 4 shows the proposed methodological/conceptual framework for handling the uncertainty in SCP area. The conceptual framework comprises of four phases: (1) SCN risk analysis: this phase is set out to select important sources of

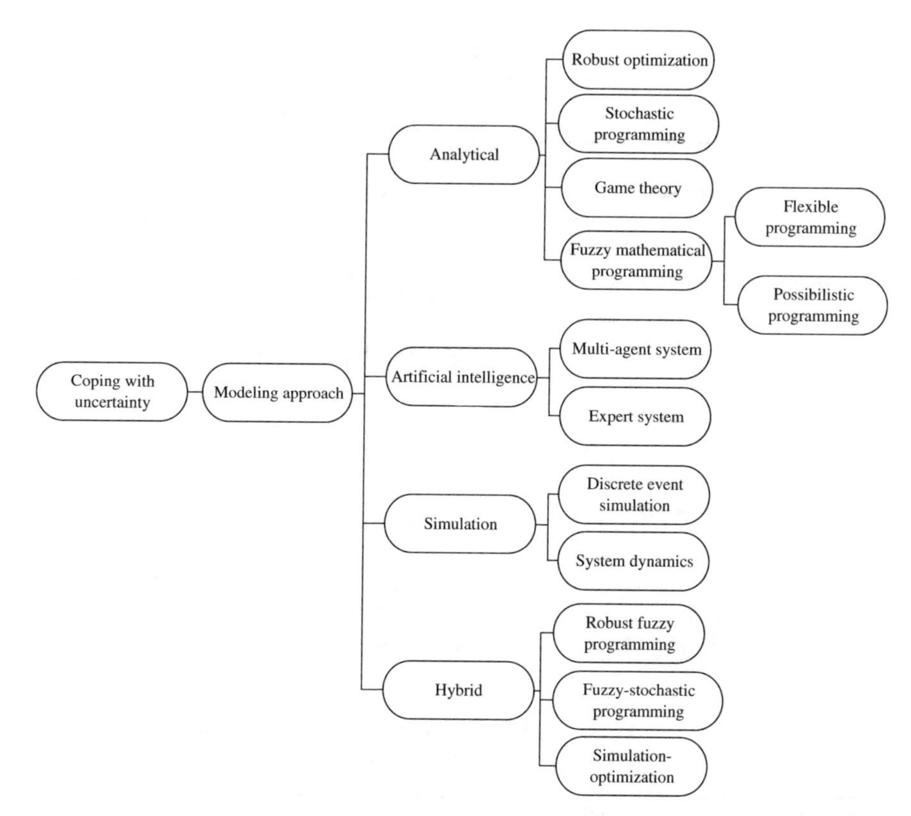

Fig. 3 Different categories of quantitative modeling approaches along with some examples (adopted from Peidro et al. [58] with some modifications)

uncertainty based on the corresponding likelihood and severity which subjectively or objectively attribute to the concerned sources; (2) Scenario development and sampling: this phase is intended to provide a plausible portfolio of scenarios or the possible states for the concerned system in the future. Scenario must be transparent, probable, relevant and compatible; (3) Solution approach: it means selecting an analytical method for modeling the uncertainty including the stochastic, fuzzy and robust programming; (4) Modeling resilience, robustness and responsiveness: decisions which are obtained from a solution method, must be remain feasible under almost all future scenarios. In addition, robustness in facing with tensions, rapidity in returning to the equilibrium state and maintaining a good level of performance in a critical condition are the measures of resilience.

To conclude this section, Fig. 5 classifies different mathematical programming methods used to deal with uncertainty in SCP problems with respect to the taxonomy of uncertainty provided in Fig. 2.

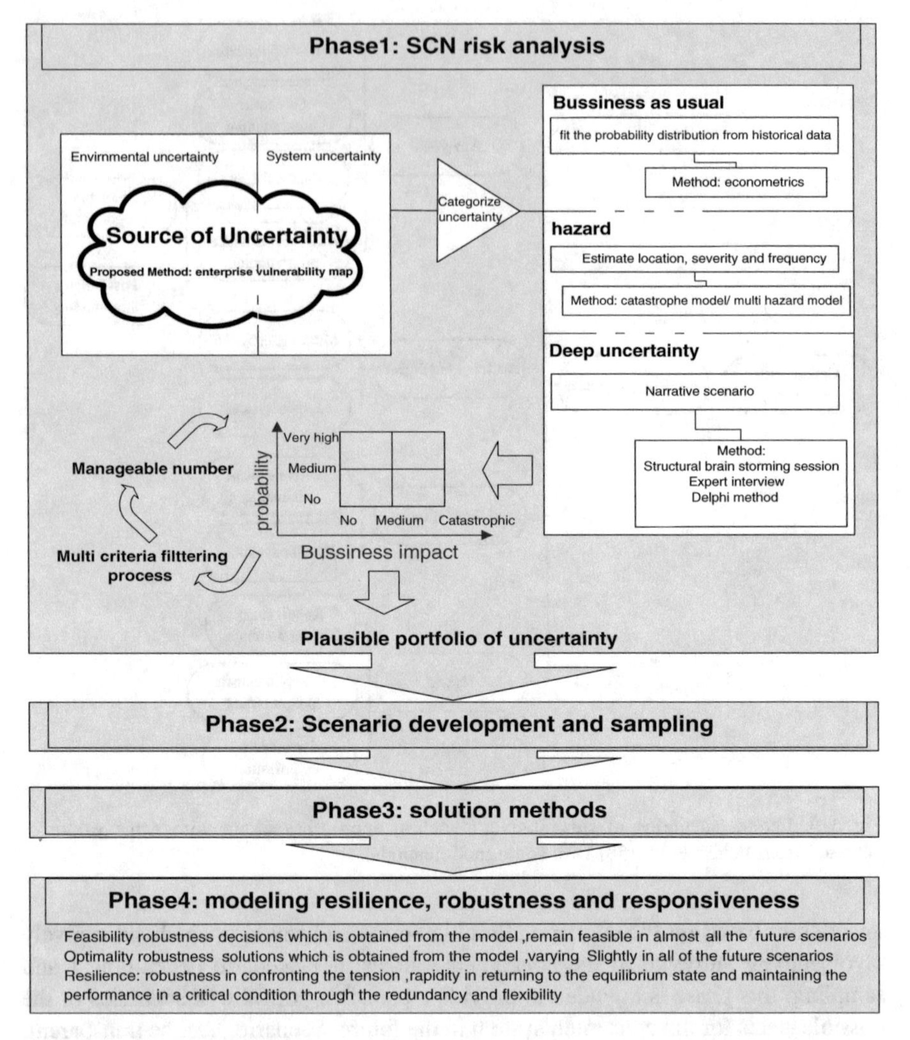

Fig. 4 Conceptual/methodological framework for modeling approach in coping with uncertainty

2 Fuzzy Mathematical Programming Models for Supply Chain Planning

As it was mentioned in the previous sections, supply chain planning problems' input data may be subject to uncertainties with different natures and sources. Imprecision associated with subjective estimations, which are based on human judgments or preferences, vagueness and flexibility in objectives and constraints, random nature of some parameters or inexactness in input data, are the examples of different types of uncertainty. Fuzzy set theory is among the main approaches that particularly, manipulate the uncertainty which arises from ambiguity in data and/or

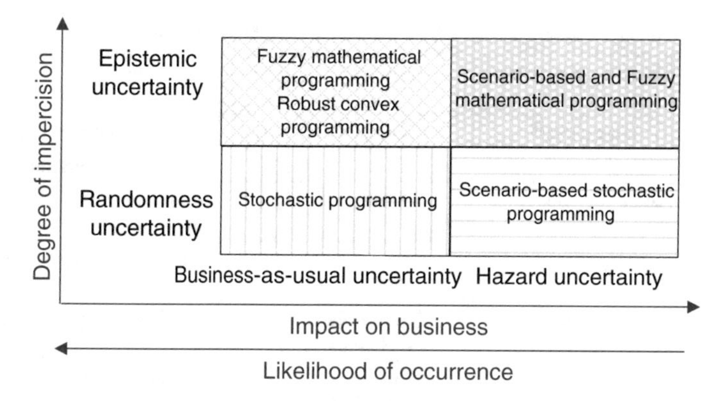

Fig. 5 Classification of analytical methods with respect to the nature of uncertainty

vagueness of target values of objectives and constraints. The imprecision can be stemmed from absence of sharp boundaries in definitions, flexibility and vagueness in objectives and constraints, subjective estimations for values of input data and lastly when the historical data does not exist sufficiently or the conditions of problem do not satisfy the presumptions of probability theory.

Fuzzy mathematical programming methods (FMPM) can be classified into two main categories ([53, 81]; Pishvaee and Torabi [60]): (1) flexible programming and (2) possibilistic programming. Flexible programming is used to deal with vagueness in target values of objectives and/or flexibility of constraint's satisfaction while possibilistic programming is applied to cope with ambiguous (i.e. imprecise) input data (i.e., epistemic uncertainty). Beside the traditional categories of FMPM (i.e. the flexible and possibilistic programming), two other types of modeling approaches called fuzzy stochastic programming and robust fuzzy programming have recently been proposed in the literature. Fuzzy stochastic programming and/or stochastic fuzzy programming, which are frequently used interchangeably (see for instance [5, 85]; Mohammadi et al. [51]) are able to handle those situations in which there are both probabilistic randomness and possibilistic imprecision in data. Also, robust fuzzy programming (see [59, 63]) enables the decision maker to be benefited from the advantages of both robust optimization and FMP approaches. To show the applicability of fuzzy mathematical programming in the context of supply chain planning, Table 2 briefly reviews and classifies a number of important FMP models developed in the literature.

2.1 Possibilistic Programming

Since the 1950s stochastic programming has been used broadly for the sake of handling randomness uncertainty in the input parameters. However, stochastic programming has some limitations in practice: (1) The low level of computational efficiency, (2) the need for sufficient and reliable historical data and (3) the inability in modelling subjective parameters and concepts ([41]; Pishvaee and Torabi [60]).

Table 2 Classification of some important FMPMs developed in the area of SCP

Problem type	Subject of planning	Authors	Modeling method
Strategic	Vendor selection	Kumar et al. [33]	Flexible goal programming
		Kumar et al. [34]	Flexible programming
	Supplier selection	Amid et al. [1]	Multi objective flexible programming
		Torabi and Baghersad [80]	Possibilistic programming
	Network design	Selim and Ozkarahan [74]	Flexible goal programming
		Pishvaee and Torabi [60]	Multi-objective possibilistic programming
		Qin and Ji [67]	Possibilistic programming
		Pishvaee et al. [63]	Robust possibilistic programming
		Pishvaee et al. [64]	Robust possibilistic programming
		Paksoy et al. [56]	Multi objective possibilistic programming
		Torabi et al. [82]	Possibilistic programming
		Fallah et al. [17]	Competitive possibilistic programming
		Pishvaee and Khalaf [59]	Robust flexible programming
Tactical	Master planning	Torabi and Hassini [81]	Possibilistic programming
		Vafa Arani and Torabi [84]	Stochastic fuzzy programming
	Production planning	Hsu and Wang [27]	Possibilistic programming
		Mula et al. [53]	Flexible programming
		Mula et al. [54]	Possibilistic programming
		Torabi et al. [84]	
	Aggregate production planning	Wang and Liang [88]	Possibilistic linear programming
	Profit management	Sakawa et al. [72]	Possibilistic programming
		Chen et al. [8]	Flexible programming
	Production-transportation	Liang [40]	Flexible programming
		Selim et al. [75]	Flexible goal programming
		Díaz-Madroñero et al. [11]	Flexible programming
Operational	Transportation problem	Liu and Kao [41]	Possibilistic programming
		Liang [39]	Multi-objective flexible programming
	Inventory management	Giannoccaro et al. [20]	Possibilistic programming
		Hsu and Wang [27]	Possibilistic programming

It is obvious that in most of real-life problems the aforementioned limitations become significant drawbacks in the application and usefulness of stochastic programming approach.

As a powerful alternative, possibilistic programming (PP) provides a framework which can appropriately handle the epistemic uncertainty in input parameters. In PP, a possibilistic distribution is attributed to an uncertain parameter, which

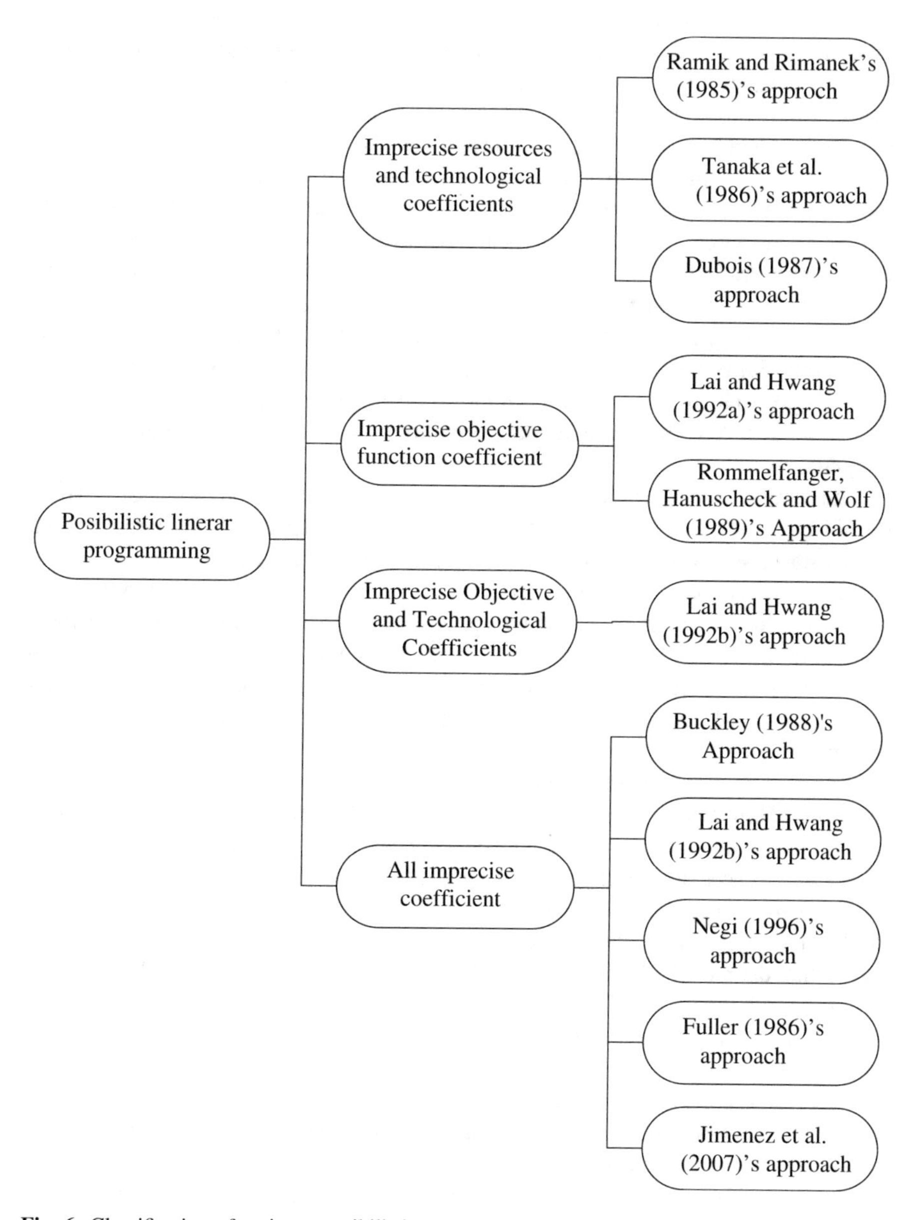

Fig. 6 Classification of various possibilistic programming methods

shows the degree of possibility of occurrence for each possible value in the related support. Figure 6 summarizes some of the important and popular PP methods which are used broadly in the literature.

For confronting the imprecise resources and technological coefficients in optimization models, Ramik [68] supposed that all the technological and resource coefficients are L-R fuzzy numbers and by the use of self-inclusion concept, they concluded that the inequalities with imprecise resources and technological coefficients are fuzzy inequalities. Accordingly, they proposed the use of fuzzy ranking methods to handle the imprecise parameters in such mathematical programming problems. Tanaka et al. [78] formulated the uncertain input parameters as symmetric triangular fuzzy numbers and based on extension principle they defined satisfaction level of α for each constraint including imprecise parameters. They defined α as a priori parameter which is specified by the decision maker. While most of developed approaches in the literature considered the constraints with imprecise resources and technological coefficients in the form of inequalities, Dubois and Prade [14] proposed their approach for different forms of constraints, especially equality constraints, and introduced two viewpoints in confronting a possibilistic constraint: (1) soft and (2) hard approach. In the first approach, by the use of possibility index, they defined the concept of α-weak feasibility and in the second approach, by the use of necessity index, the concept of α-hard feasibility was proposed. Notably, the uncertain equality constraints are substituted by two inequalities.

The imprecision of objective function parameters is a common issue which should be handled in optimization problems. Lai and Hwang [37] supposed that these imprecise coefficients in the objective function have triangular possibly distributions. They proposed the substitution of the main objective function by three functions which consist of the most possible, the most pessimistic and the most optimistic value of the main objective function. They stated that for solving the equivalent multi-objective model it is possible to use every multi criterion decision making (MODM) techniques. Rommelfanger et al. [70] assumed that the vector of parameters in objective function is an interval-valued coefficient vector. They substituted the main imprecise objective function by two deterministic objective functions. They also proposed a new approach for the models which their objective function's coefficients have general convex possibility distribution (i.e., it is not necessary to have only triangular or trapezoidal shapes).

Using the concept of exceedance and strict exceedance defined by Dubois and Prade [16], Negi [55] developed an approach which maximizes the minimum possibility degree of objective function and constraints. Jiménez et al. [29] proposed a new PP approach based on the Jiménez [28] ranking method and the concepts regarding the expected interval (EI) and expected value (EV) of imprecise parameters.

According to the approach used for dealing with possibilistic chance constraints, possibilistic programming approaches can be classified into *measure-based possibilistic programming* and *non-measure-based possibilistic programming* methods. In *measure-based possibilistic programming,* the crisp counterpart of the original problem is constructed based on a fuzzy measure (e.g. necessity measure). However, a *Non-measure based possibilistic programming* applies other techniques such as fuzzy ranking methods to form the crisp equivalent model. According to Pishvaee et al. [64], there are three main categories in *measure-based possibilistic programming* as follows: (1) expected value models (2) chance constrained programming (3) depended chance constrained programming. The first approach uses an expected value operator for every imprecise parameter in the objective function and constraints. It can be applied more conveniently without increasing the computational complexity of the original model compared to the two other methods but at the same time it has no control on the confidence level of chance constraints. The second approach is able to control the satisfaction level of constraints by the use of α-level concept; but also increases the computational complexity since it adds a new constraint for each objective function of the original model. The third approach is somehow similar to the second one but it provides more conservative decision for the decision maker since it gives more importance to maximization of satisfaction levels.

Dubois and Prade [16] defined two basic fuzzy measures including the possibility (*Poss*) and necessity (*Nes*) measures. While the first is extremely optimistic, the second is extremely pessimistic. The authors provide definitions for possibility and necessity measure based expected value operators as well as proposing methods for measuring the possibility and necessity of a chance constraint. By the use of these operators different possibilistic programming methods are extended in the literature. Another measure which is proposed by Liu and Liu [42, 43] is credibility measure. Credibility measure can be defined as the average of possibility and necessity measures. Also, Liu and Liu [42, 43] extended an expected value operator based on credibility measure. They developed both expected value and fuzzy chance constrained programming frameworks based on credibility measure. More recently, Xu and Zhou [91] defined *Me* measure which is an extension to credibility measure. By means of a parameter, this measure is able to adjust the degree of optimism or pessimism in measuring the fuzziness. They also proposed a possibilistic chance constrained programming approach based on *Me* measure, which was then extended by Torabi et al. [82].

2.1.1 Possibilistic Programming Models for Supply Chain Planning

In this section, firstly, a possibilistic programming model for supply chain network design problem is introduced. This model is adapted from Pishvaee et al. [63] work. Secondly, the method applied in this paper to achieve the crisp counterpart is described. The supply chain under consideration is of a single-product, three-echelon type, which comprises multiple production centers, distribution centers and costumer zones. Aiming at completely fulfill the demands of costumer zones, products are produced in production centers and delivered to customers through one or more distribution centers. The locations and amounts of demands are predefined. There are various candidate positions for establishing capacitated production and distribution centers. In addition, there are various candidate production technologies when a new production center is established in a location. The model must make several decisions regarding the locations and numbers of production and distribution centers as well as the amount of products which are transported between every two nodes of the network. Also the production technology must be specified for every established production center. The indices, parameters and variables that are used in mathematical formulation are as follows.

Indices:

i index of potential locations for plants $i = 1, \ldots, I$
j index of potential locations for distribution centers $j = 1, \ldots, J$
k index of fixed locations of customers $k = 1, \ldots, K$
m index of different production technologies available for plants $m = 1, \ldots, M$

Parameters:

d_k demand of customer zone k
f_i^m fixed cost of opening plant i with production technology m
g_j fixed cost of opening distribution center j
c_{ij} shipping cost per product unit from plant i to distribution center j
a_{jk} shipping cost per product unit from distribution center j to customer k
ρ_i^m production cost per unit of product at plant i with production technology m
τ_i maximum capacity of plant i
φ_j maximum capacity of distribution center j

Variables:

u_{ij}^m quantity of products produced at plant i with technology m and shipped to distribution center j
q_{jk} quantity of products shipped from distribution center j to customer k

$$x_i^m = \begin{cases} 1 & \text{if a plant with technology } m \text{ is opened at location } i \\ 0 & \text{otherwise} \end{cases}$$

$$y_j = \begin{cases} 1 & \text{if a distribution center is opened at location } j, \\ 0 & \text{otherwise} \end{cases}$$

By the use of above-mentioned notations, the formulation of the concerned supply chain network design problem is shown in below.

$$\text{Min W} = \sum_i \sum_m f_i^m x_i^m + \sum_j g_j y_j + \sum_i \sum_j \sum_m (\rho_i^m + c_{ij}) u_{ij}^m + \sum_j \sum_k a_{jk} q_{jk} \tag{1}$$

$$\text{s.t.} \quad \sum_j q_{jk} \geq d_k, \quad \forall\, k, \tag{2}$$

$$\sum_{i}\sum_{m} u_{ij}^{m} = \sum_{k} q_{jk}, \quad \forall\, j, \tag{3}$$

$$\sum_{j} u_{ij}^{m} \leq x_{i}^{m}\tau_{i}, \ \forall\, i,\, m, \tag{4}$$

$$\sum_{k} q_{jk} \leq y_{j}\varphi_{j}, \quad \forall\, j, \tag{5}$$

$$\sum_{m} x_{i}^{m} \leq 1 \quad \forall\, i, \tag{6}$$

$$x_{i}^{m}, y_{j} \in \{0,\, 1\}, \quad \forall\, i, j, m, \tag{7}$$

$$u_{ij}^{m}, q_{jk} \geq 0, \quad \forall\, i,\, j,\, m. \tag{8}$$

Equation (1) is the objective function of the problem which minimizes the total cost and includes summation of fixed opening costs, variable production costs and transportation costs. Constraint (2) is related to ensuring the full satisfaction of demands of all customers. Equation (3) shows material flow balance at each distribution center. Constraints (4) and (5) enforce capacity restrictions on plants and distribution centers and additionally ensure that if a facility is not opened, there will be no shipping from the related location. Equation (6) states that at most one technology can be attributed to each potential plant. Finally, Eqs. (7) and (8) are related to binary and non-negatively decision variables respectively.

As it is mentioned in the relevant literature [58, 62] most of the parameters of supply chain network design problems (e.g. capacities, demand of customers and transportation costs) contaminated by uncertainty in real-world situations because of fluctuation of parameters over the long-term horizon. To cope with this issue, a possibilistic chance constrained programming model is developed here.

For convenience, the supply chain network design model can be summarized in a compact form as follows:

$$\begin{aligned} \text{Min} \quad & z = fy + cx \\ s.t. \quad & Ax \geq d, \\ & Bx = 0, \\ & Sx \leq Ny, \\ & Tx \leq 1, \\ & y \in \{0, 1\}, \quad x \geq 0. \end{aligned} \tag{9}$$

where the vectors f, c, and d are related to fixed-opening costs, variable transportation and production costs and demands, respectively. The matrices A, B, C, N, S and T are technological coefficient matrices of the constraints. Also, y and x are the vectors of binary decision variables and continuous decision variables, respectively.

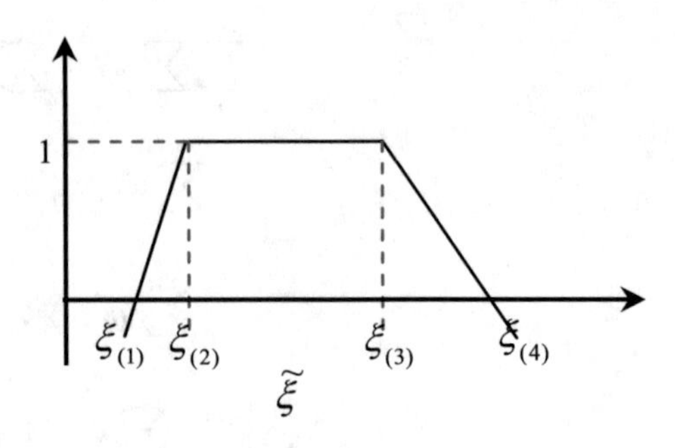

Fig. 7 The trapezoidal possibility distribution of fuzzy parameter $\tilde{\xi}$

The vectors *f, c,* and *d* and coefficient matrix *N* which respectively are related to fixed opening costs, variable transportation costs, demands and capacity of facilities are the imprecise parameters in the model. Here, trapezoidal possibility distribution (see Fig. 7) is attributed to uncertain parameters. Each trapezoidal possibility distribution can be defined by four prominent points, e.g. $\tilde{\xi} = (\xi_{(1)}, \xi_{(2)}, \xi_{(3)}, \xi_{(4)})$. It is obvious that if $\xi_{(2)} = \xi_{(3)}$, the attributed trapezoidal possibility distribution is reduced into a triangular form. The expected value operator is used to defuzzify the objective function and the necessity measure is applied to cope with chance constraints including imprecise parameters.

Accordingly, the equivalent possibilistic chance constrained programming model can be formulated as follows:

$$\begin{aligned} \text{Min} \quad & \mathrm{E}[z] = \mathrm{E}[\tilde{f}]y + \mathrm{E}[\tilde{c}]x \\ s.t. \quad & \mathrm{Nec}\{Ax \geq \tilde{d}\} \geq \alpha, \\ & Bx = 0, \\ & \mathrm{Nec}\{Sx \leq \tilde{N}y\} \geq \beta, \\ & Tx \leq 1, \\ & y \in \{0,1\}, \quad x \geq 0. \end{aligned} \tag{10}$$

Notably, the expected value used for defuzzification of the objective function was firstly developed by Yager [92]. Assume that $\tilde{\xi}$ is a fuzzy number and its membership function is defined as follow:

$$\mu_{\tilde{\xi}}(x) = \begin{cases} f_{\tilde{\xi}}(x), & \text{if } \xi_{(1)} \leq x < \xi_{(2)} \\ 1, & \text{if } \xi_{(2)} \leq x \leq \xi_{(3)} \\ g_{\tilde{\xi}}(x), & \text{if } \xi_{(3)} < x \leq \xi_{(4)} \\ 0, & \text{if } x < \xi_{(1)} \text{ or } x > \xi_{(4)} \end{cases}$$

$E^*(\tilde{\xi})$ and $E_*(\tilde{\xi})$ are the upper and lower expected (mean) values of $\tilde{\xi}$ that can be defined as follow [15, 24]:

$$E^*(\tilde{\xi}) = \xi_{(3)} + \int_{\xi_{(3)}}^{\xi_{(4)}} g_{\tilde{\xi}}(x)\,\mathrm{d}x \tag{11}$$

$$E_*(\tilde{\xi}) = \xi_{(2)} - \int_{\xi_{(1)}}^{\xi_{(2)}} f_{\tilde{\xi}}(x)\,\mathrm{d}x \tag{12}$$

The expected interval (*EI*) and expected value (*EV*) of $\tilde{\xi}$ based on Eqs. (11) and (12), can be defined as follow [15, 24]:

$$EI[\tilde{\xi}] = [E_*(\tilde{\xi}),\, E^*(\tilde{\xi})] \tag{13}$$

$$EV[\tilde{\xi}] = \frac{E_*(\tilde{\xi}) + E^*(\tilde{\xi})}{2} \tag{14}$$

If $\tilde{\xi}$ is a trapezoidal fuzzy number (see Fig. 7), then we have:

$$EI[\tilde{\xi}] = \left[\frac{\xi_{(1)} + \xi_{(2)}}{2}, \frac{\xi_{(3)} + \xi_{(4)}}{2}\right] \tag{15}$$

$$EV[\tilde{\xi}] = \frac{\xi_{(1)} + \xi_{(2)} + \xi_{(3)} + \xi_{(4)}}{4} \tag{16}$$

Let r be a real number. According to Dubois [13], Dubois and Prade [16] necessity (*Nec*) of $\tilde{\xi} \le r$ can be defined as follows:

$$\mathrm{Nec}\left\{\tilde{\xi} \le r\right\} = 1 - \sup_{x > r} \mu_{\tilde{\xi}}(x) = 1 - \mathrm{Pos}\left\{\tilde{\xi} > r\right\} \tag{17}$$

Therefore, the necessity of $\tilde{\xi} \le r$ can be stated as follows:

$$Nec\left\{\tilde{\xi} \le r\right\} = \begin{cases} 1, & \xi_{(4)} \le r, \\ \frac{r - \xi_{(3)}}{\xi_{(4)} - \xi_{(3)}}, & \xi_{(3)} \le r \le \xi_{(4)}, \\ 0, & \xi_{(3)} \ge r. \end{cases} \tag{18}$$

Based on Eqs. (17) and (18) and with respect to Dubois and Prade [18] and Luhandjula [45], it can be proved that:

$$Nec\left\{\tilde{\xi} \le r\right\} \ge \alpha \;\Leftrightarrow\; r \ge (1-\alpha)\xi_{(3)} + \alpha\xi_{(4)}. \tag{19}$$

As there is an assumption that all of uncertain parameters in formulation (11) have trapezoidal possibility distribution, the equivalent crisp model for formulation (11) can be given as follows:

$$
\begin{aligned}
\text{Min} \quad & \mathrm{E}[z] = \left(\frac{f_{(1)} + f_{(2)} + f_{(3)} + f_{(4)}}{4}\right) y + \left(\frac{c_{(1)} + c_{(2)} + c_{(3)} + c_{(4)}}{4}\right) x \\
s.t. \quad & Ax \geq (1-\alpha) d_{(3)} + \alpha d_{(4)}, \\
& Bx = 0, \\
& Sx \leq \left[(1-\beta) N_{(2)} + \beta N_{(1)}\right] y, \\
& Tx \leq 1, \\
& y \in \{0, 1\}, \quad x \geq 0.
\end{aligned}
\tag{20}
$$

2.2 Flexible Programming

In a fuzzy environment the meaning of objective function and constraints are somehow different from their classical hard meanings. For example, if a decision maker decides to "decrease the cost considerably", he/she may prefer to reach some aspiration levels (i.e., flexible targets) rather than strictly minimize the cost. Additionally, "≤", "=" and "≥" signs in a constraint might not be meant in the strict mathematical sense and slight violation is allowed. In these cases, the fuzzy distributions are attributed to these soft constraints and objective functions based on the preferences of decision makers (DMs). In this framework, the value of attributed membership function shows the degree of DMs' satisfaction. Luhandjula [45] named this sort of fuzzy mathematical programing as 'Flexible programming'.

Flexible programming models can be categorized into two groups [97]: (1) symmetric and (2) non-symmetric. The first one describes the models which both objective function and constraints have flexibility in reaching the target values. In the second

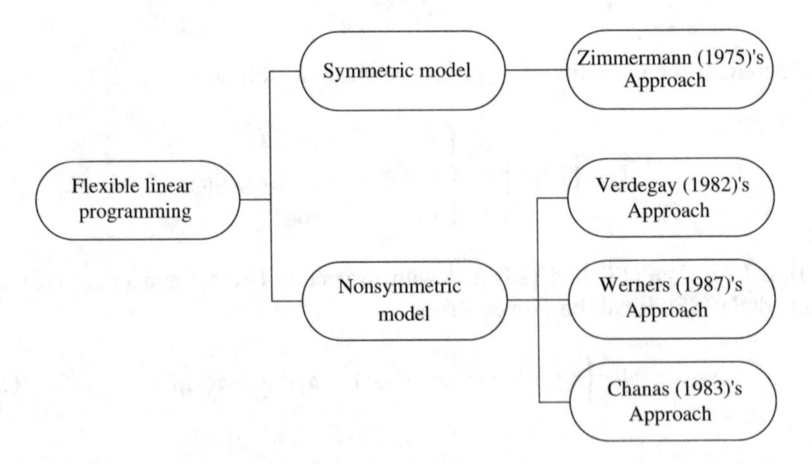

Fig. 8 Classification of different methods of flexible programming

category only constraints have fuzzy nature. Figure 8 illustrates some well-known flexible programming methods according to above-mentioned classification.

Verdegay [87] provided the crisp equivalent parametric programming model for the flexible programming problem with fuzzy constraints. Werners [89] studied the non-symmetric flexible programming problem and proposed a method in which two extreme values of objective function were used to construct a fuzzy membership function in order to attain the satisfaction degree of objective function. Zimmermann [96] applied the Zadeh's max-min operator to solve the symmetric flexible programming problem model. Chanas [7] argued that due to lack of knowledge about the fuzzy feasible region, the fuzzy membership function of objective function cannot be determined from the beginning. He proposed a framework in which the attributed fuzzy membership is specified using an interactive process.

2.2.1 Flexible Programming Models for Supply Chain Planning

Consider the supply chain network design problem described in Sect. 2.1.1. In that problem it was assumed that capacities, demand of customers, fixed opening costs and variable transportation costs are imprecise parameters. In this section, it is assumed that the demand and capacity constraints are flexible (i.e., soft constraints) and slight violation is allowed. For example, in demand constraints symbol $\tilde{\geq}$ is regarded as fuzzy version of $\geq$ which implies that demand of a customer zone must be essentially less than or *somehow similar* to the number of products that are delivered to the customer zone. According to Pishvaee and Khalaf [59] and Torabi et al. [83], the model can be represented as follows:

$$\begin{array}{ll} \text{Min} & z = cx + fy \\ \text{s.t} & Ax \tilde{\geq} d \\ & Bx = 0 \\ & Sx \tilde{\geq} Ny \\ & Ty \leq 1 \\ & y \in \{0,1\}, \quad x \geq 0 \end{array} \tag{21}$$

Based on Cadenas and Verdegay [3], the violation of soft constraints can be modeled by the use of fuzzy numbers (in this case by $\tilde{t}$ and $\tilde{r}$ denoting the maximum allowable tolerances). Therefore, model (21) is rewritten as follows:

$$\begin{array}{ll} \text{Min} & z = cx + fy \\ s.t. & Ax \geq d - \tilde{t}(1-\alpha) \\ & Bx = 0 \\ & Sx \leq Ny + [\tilde{r}(1-\beta)]y \\ & Ty \leq 1 \\ & y \in \{0,1\}, \quad x \geq 0 \end{array} \tag{22}$$

where minimum satisfaction level of flexible constraints are controlled by α and β. Now, we intended to defuzzify formulation (22) based on the fuzzy ranking method suggested by Yager [92, 93]. For this, assume that $\tilde{t}$ and $\tilde{r}$ are triangular fuzzy numbers and they are represented by their pessimistic, most possible and optimistic values (i.e., $\tilde{t} = (t^p, t^m, t^o)$ and $\tilde{r} = (r^p, r^m, r^o)$). These fuzzy numbers can be defuzzified as follows:

$$\left(t^m + \frac{\varphi_t - \varphi'_t}{3}\right) \tag{23}$$

$$\left(r^m + \frac{h_r - h'_r}{3}\right) \tag{24}$$

where parameters φ_t and φ'_t (h_r and h'_r) are lateral margins of the triangular fuzzy number $\tilde{t}$ ($\tilde{r}$) and are defined in Eqs. (25) and (24).

$$\varphi_t = t^o - t^m \tag{25}$$

$$\varphi'_t = t^m - t^p \tag{26}$$

By the use of Eqs. (23) and (24), the crisp counterpart of model (22) is as follows.

$$\begin{aligned} \text{Min} \quad & z = cx + fy \\ s.t. \quad & Ax \geq d - \left(t^m + \tfrac{\varphi_t - \varphi'_t}{3}\right)(1-\alpha) \\ & Bx = 0 \\ & Sx \leq Ny + \left[\left(r^m + \tfrac{h_r - h'_r}{3}\right)(1-\beta)\right] y \\ & Ty \leq 1 \\ & y \in \{0,1\}, \quad x \geq 0 \end{aligned} \tag{27}$$

Notably, the applied method is capable of supporting different kinds of fuzzy numbers as well as various fuzzy ranking methods when converting the soft constraints to their flexible constraints. Also, the minimum satisfaction level for constraints should be determined in the range of $0 \leq \alpha$, $\beta \leq 1$, subjectively by the decision maker.

2.3 *Robust Fuzzy Mathematical Programming*

Robust optimization provides a set of risk-averse approaches in confronting with uncertainty, which attempts to obtain a *robust solution*. A solution is *robust*, if it has *feasibility robustness* and *optimality robustness*, simultaneously [59, 63]. A solution has *feasibility robustness* if it is feasible under almost all possible realization of uncertain parameters and has *optimality robustness* if the value of objective function for the obtained solution would remain close to optimal value or have minimum (undesirable) deviation from the optimal value for (almost) all possible values of uncertain parameters. Both of *feasibility robustness* and *optimality robustness* could be used as two performance indicators, which examine the validity of a solution for an uncertain problem irrespective of the type of applied uncertainty programming approach.

Pishvaee et al. [63] classify robust optimization approaches into three main categories: (1) *hard worst case robust optimization* (2) *soft worst case robust optimization*, and (3) *realistic robust optimization*. In the first approach, the solution remains feasible under all possible values of uncertain parameters while the worst case value of objective function is optimized. This approach is applicable for those situations in which maximum immunization against uncertainty is needed (e.g. military cases). The second approach argues that it is very unlikely that all of uncertain parameters get their extreme worst case values at the same time. In this approach, similar to the worst case approach, the worst case value of objective function is optimized; however, the feasibility robustness level is not planned for the worst case value. Finally, the third approach seeks to attain a trade-off among the average performance of the objective function, worst case value of the objective function and *feasibility robustness*. As mentioned before, *feasibility robustness* and *optimality robustness* can be used in evaluation of a solution for the problems tainted with uncertainty regardless of the type of uncertainty. Therefore, we can also look for *robustness* in fuzzy environment and control the *feasibility* and *optimality robustness* of a solution. As a pioneer work in this field, Pishvaee et al. [63] proposed three robust possibilistic programming approaches, which are able to control the *feasibility* and *optimality robustness* in different degrees. These include: (1) *hard worst-case robust possibilistic programming (HWRPP)*, (2) *soft worst-case robust possibilistic programming (SWRPP)* and (3) *realistic robust possibilistic programming*. The first one optimizes the worst possible value of objective function as well as immunizing the solution against the worst-case values of uncertain parameters. This method is indifferent to the forms of uncertain parameters and only considers their extreme worst values. Therefore, there is no need to determine the possibility distributions of uncertain parameters and it is sufficient to only know the support range of each possibility distributions. The second method optimizes the worst value of objective function while considering the penalty cost for constraints' violations. Thus, this approach is more flexible than *HWRPP*, because the violation of constraints is permissible to some extent. The third approach attempts to achieve a reasonable trade-off between the average performance, feasibility robustness and optimality robustness. As was mentioned in Sect. 2, the fuzzy mathematical programming consists of *possibilistic programming* and *flexible programming*. Until now we introduced some approaches that unified *possibilistic programming* and *robust optimization* in the literature. Interested readers can consult with Pishvaee and Khalaf [59] in which new models called robust flexible programming and robust mixed flexible-possibilistic programming have been introduced.

2.3.1 Robust Fuzzy Programming Models for Supply Chain Planning

Consider the supply chain network design problem discussed in Sect. 2.1.1 where capacities, demand of customers, fixed opening costs and variable transportation costs were imprecise parameters formulated by some possibility distributions. Now, based on Pishvaee et al. [63] and for the sake of considering and controlling *feasibility* and *optimality robustness* in the mentioned problem, it is reformulated by realistic, hard worst case and soft worst case robust possibilistic programming methods.

- *Realistic robust possibilistic programming model*

At first, consider the formulation (20), where vectors f, c, d and coefficient matrix N denote the imprecise parameters regarding the fixed opening costs, variable transportation costs, demands and capacities, respectively. Based on formulation (20), the realistic robust possibilistic formulation of the concerned problem is as follows:

$$\begin{aligned} \text{Min} \quad & \mathrm{E}[z] + \gamma(z_{\max} - z_{\min}) + \delta[d_{(4)} - (1-\alpha)d_{(3)} - \alpha d_{(4)}] \\ & + \pi[\beta N_{(1)} + (1-\beta)N_{(2)} - N_1]y \\ s.t. \quad & Ax \geq (1-\alpha)d_{(3)} + \alpha d_{(4)}, \\ & Bx = 0, \\ & Sx \leq \left[\beta N_{(1)} + (1-\beta)N_{(2)}\right]y, \\ & Tx \leq 1, \\ & y \in \{0,1\}, \quad x \geq 0, \ 0.5 < \alpha, \ \beta \leq 1. \end{aligned} \tag{28}$$

The first term of objective function is the expected value of z. By minimization of the expected value of total cost, the average performance is improved. The second term, $\gamma(z_{\max} - z_{\min})$, indicates the difference between the best (z_{min}) and worst (z_{max}) possible values of z. Equation (29) defines these two values.

$$\begin{aligned} z_{\max} &= f_{(4)}y + c_{(4)}x, \\ z_{\min} &= f_{(1)}y + c_{(1)}x, \end{aligned} \tag{29}$$

Also, γ indicates the relative importance of $(z_{\max} - z_{\min})$ compared to the other terms of objective function. Therefore, $\gamma(z_{\max} - z_{\min})$ controls the maximum deviation from the expected optimal value of objective function (i.e. it controls *optimality robustness* of the solution). Furthermore, $[d_{(4)} - (1-\alpha)d_{(3)} - \alpha d_{(4)}]$ represents the difference between the worst possible value of imprecise demand and the value that is used in right hand side of the chance constraint. Also, δ is the unit penalty cost for unsatisfied demand. Therefore, the third term, $\delta[d_{(4)} - (1-\alpha)d_{(3)} - \alpha d_{(4)}]$, indicates the possible violation cost of demand constraint and it is used to control the *feasibility robustness* of the solution. Similar discussion can be performed for the fourth term in the objective function, i.e., $\pi[\beta N_{(1)} + (1-\beta)N_{(2)} - N_1]y$.

Based on the above-mentioned definitions and descriptions, we can say that formulation (22) seeks to reach a reasonable trade-off between the average performance, *optimality robustness* and *feasibility robustness*. It should be noted that when the technological coefficients (i.e., matrix N) is tainted with uncertainty the model becomes non-linear (because of multiplication of N and β). To escape from the complexity of such non-linear model, a linearization method is also proposed in Pishvaee et al. [63].

- *Hard worst case robust possibilistic programming model (HWRPP)*

In the hard worst case robust programming approach, the solution must be remain feasible under all possible values of imprecise parameters while the worst case value of objective function is optimized. The worst case approach provides the maximum conservation against uncertainty and therefore it is a fully risk-averse approach. Based on above-mentioned descriptions, we can represent the hard worst case robust possibilistic version of the concerned supply chain network design problem (1)–(8) as follows:

$$\begin{aligned} \text{Min} \quad & \sup(z) \\ s.t. \quad & Ax \geq \sup(\tilde{d}), \\ & Bx = 0, \\ & Sx \leq \inf(\tilde{N})y, \\ & Tx \leq 1, \\ & y \in \{0,1\}, \quad x \geq 0. \end{aligned} \tag{30}$$

By assuming that the possibility distributions of imprecise parameters are of trapezoidal shape, model (30) can be rewritten as follows:

$$\begin{aligned} \text{Min} \quad & z_{\max} \\ s.t. \quad & Ax \geq d_{(4)}, \\ & Bx = 0, \\ & Sx \leq N_{(1)}y, \\ & Tx \leq 1, \\ & y \in \{0,1\}, \quad x \geq 0. \end{aligned} \tag{31}$$

Notably, HWRPP model does not rely on the form of possibility distribution and it is sufficient to only know the worst possible value of imprecise parameters. It is noteworthy that some other robust possibilistic programming models are also developed by Pishvaee et al. [63] and applied in the literature for coping with uncertainty in different supply chain planning problems (e.g. [51, 95]).

- *Robust flexible programming model*

Based on model (27), the robust flexible programming model proposed by Pishvaee and Khalaf [59] can be formulated as follows:

$$\begin{aligned}
\text{Min} \quad & E = cx + fy + \gamma\left[\left(t^m + \tfrac{\varphi_t - \varphi_t'}{3}\right)(1-\alpha)\right] + \theta\left[\left(r^m + \tfrac{h_r - h_r'}{3}\right)(1-\beta)\right]y \\
s.t. \quad & Ax \geq d - \left(t^m + \tfrac{\varphi_t - \varphi_t'}{3}\right)(1-\alpha) \\
& Bx = 0 \\
& Sx \leq Ny + \left[\left(r^m + \tfrac{h_r - h_r'}{3}\right)(1-\beta)\right]y \\
& Ty \leq 1 \\
& y \in \{0,1\}, \quad x \geq 0, \quad 0 \leq \alpha, \beta \leq 1
\end{aligned} \tag{32}$$

The last two terms of objective function calculates the total penalty cost of possible violation on soft constraints (i.e., feasibility robustness) while the first two terms are the original objective function of model (27). Indeed, the third and fourth terms represent the difference between the two extreme values of the right hand side of flexible constraints which are defined as follows:

$$\left(t^m + \frac{\varphi_t - \varphi_t'}{3}\right)(1-\alpha) = d - \left[d - \left(t^m + \frac{\varphi_t - \varphi_t'}{3}\right)(1-\alpha)\right] \tag{33}$$

$$\left[\left(r^m + \frac{h_r - h_r'}{3}\right)(1-\beta)\right]y = \left[Ny + \left[\left(r^m + \frac{h_r - h_r'}{3}\right)(1-\beta)\right]y\right] - Ny \tag{34}$$

Also, penalty costs are considered for each unit of violation on soft constraints which are presented via parameters γ and θ in the proposed model. It should be noted that the model optimizes the values of satisfaction levels (i.e., α and β) as two variables. As a result, there is no need for iterative subjective experiments by setting different values of α and β. The proposed model belongs to the category of realistic robust programming since it tries to make a reasonable balance between the cost of robustness (i.e., the third and fourth terms of objective function) and the overall performance of the concerned problem (i.e., the first and second terms of objective function). Similar to model (28), the multiplication of N and β results in non-linearity of model (32) and therefore a linearization method is provided in Pishvaee and Khalaf [59]. Moreover, some other robust flexible and possibilistic-flexible programming methods are also proposed in Pishvaee and Khalaf [59].

2.4 Fuzzy-Stochastic Programming

Generally there are two main types of uncertainty including *fuzziness* and *randomness*. However, in some real world situations, *fuzziness* and *randomness* maybe appear simultaneously. We may encounter a random situation whose outcomes are not crisp numbers but fuzzy elements, i.e., in the form of linguistic descriptions. As an example, a situation can be considered in which a group of people are randomly selected and asked about the future weather in a particular day and they response in the form of linguistic descriptions such as "hot", "very hot" "more or less hot". Another example of facing with hybrid uncertainty is a model in which a constraint has a random RHS while technological coefficients of the constraint are imprecise parameters extracted from expert opinion in the form of fuzzy numbers.

To cope with an uncertain situation in which the *fuzziness* and *randomness* appear simultaneously (i.e., mixed fuzzy-stochastic data), various concepts are proposed in the literature: (1) *fuzzy random variable* which is introduced by Kwakernaak [35, 36] and refers to a random element whose prominent values are fuzzy; (2) *random fuzzy variable* which is developed by Liu and Liu [45] and refers to a fuzzy element taking random values as its prominent values; (3) *probability of a fuzzy event* [94]; (4) *probabilistic set* [25]; (5) *linguistic probabilities* [12].

As mathematical programming approaches in the form of linear and mixed integer linear are broadly used in the area of supply chain planning, now we present a classification of fuzzy-stochastic linear programming problems according to Luhandjula [47] and thereafter a number of them are introduced. Luhandjula [47] classified fuzzy stochastic linear programming models as follows:

- Flexible stochastic linear problems
- Inclusive-constrained linear programming with fuzzy random variable coefficients
- Inequality-constrained linear programming with fuzzy random variable coefficients
- Linear programming with random variables and fuzzy numbers

In flexible stochastic linear problems, since coefficients in constraints and objective functions have random distributions, the objective functions and constraints might not be meant in the strict mathematical sense and slight violations for constraints are permitted. To cope with these problems some symmetric (e.g. [48]; Luhandjula [44] and non-symmetric (e.g. [5, 6] approaches can be found in the literature. In some problems, constraints state that the possible region of LHS occurrence including some ill-defined parameters must be involved in the satisfactory region. In these cases, the problem is in the form of Inclusive-constrained and if it includes fuzzy random variables, the methodology proposed by Luhandjula [47] might be useful.

Inequality-constrained model with fuzzy random variable coefficients is a mathematical programming model in which constraints are in the form of inequality and all of coefficients are in the form of fuzzy random variables. Katagiri and Ishii [30] developed an approach for dealing with these problems in the linear form. Some problems include both random coefficients and fuzzy coefficients simultaneously. To cope with these problems, Luhandjula [46] proposed an approach and applied it in a problem in which the coefficients of the technological matrix are fuzzy numbers while the coefficients of RHS include gamma random variables.

Last but not the least, a supply chain mater planning problem with random fuzzy data is addressed by Vafa Arani and Torabi [84]. In the proposed model, market demand of final products and technical coefficients of the marketable securities are supposed to be random fuzzy variables.

2.4.1 Fuzzy Stochastic Programming Model for Supply Chain Planning

If sufficient and reliable historical data is available for a parameter, a probability distribution can be plausibly attributed to the parameter. For example, consider the demand constraint of formulation (21). It is not a far-fetched situation that the company has recorded its customer orders for a long time and based on this historical data we can attribute a probability distribution to the demands.

$$\begin{aligned} \text{Min} \quad & z = fy + cx \\ s.t. \quad & Ax \tilde{\leq} d, \\ & Bx = 0, \\ & Sx \leq Ny, \\ & Tx \leq 1, \\ & y \in \{0,1\}, \quad x \geq 0. \end{aligned} \tag{35}$$

where d is a random variable with cumulative distribution function F and "$\tilde{\leq}$" is the flexible version of "$\leq$". Now, in a bid to achieve the crisp counterpart of demand constraint we refer to Chakraborty [5] who attributed a membership function to the probability of fuzzy event $(\tilde{P}(Z \tilde{\leq} z))$ as Eq. (36), where Z is a random variable and we are about to calculate the probability of $Z \tilde{\leq} z$ which "$\tilde{\leq}$" denotes the fuzzy nature of the inequality:

$$\mu_{\tilde{P}(Z \tilde{\leq} z)}(r) = \begin{cases} 1 & \textit{if} \;\; F(z) = r \\ \frac{F(z+\Delta z) - r}{F(z+\Delta z) - F(z)}, & \textit{if} \;\; F(z) \leq r \leq F(z+\Delta z) \\ 0 & \textit{if} \;\; r > F(z+\Delta z) \end{cases} \tag{36}$$

Therefore, firstly we should obtain the membership function of $\tilde{P}(d \tilde{\leq} Ax)$ where d (demand) is a random variable with cumulative function F and r is the target value which determines the satisfaction chance of the concerned constraint. Accordingly, the membership function is formulated as follows:

$$\mu_{\tilde{P}(d \tilde{\leq} Ax)}(r) = \begin{cases} 1 & \textit{if} \;\; F(Ax) = r \\ \frac{F(Ax+\Delta Ax) - r}{F(Ax+\Delta Ax) - F(Ax)}, & \textit{if} \;\; F(Ax) \leq r \leq F(Ax+\Delta Ax) \\ 0 & \textit{if} \;\; r > F(Ax+\Delta Ax) \end{cases} \tag{37}$$

ΔAx is determined exogenously by decision maker and it is equal to the maximum allowable amount for violation of the constraint. Now, we can replace the demand constraint by $\mu_{\tilde{P}(d \tilde{\leq} Ax)}(r) \geq \alpha$ and convert the original model into an equivalent α-parametric model in which different solutions are obtained for different values of α. It should be noted that the linearity of the replaced equation depends on the linearity of cumulative distribution function of the corresponding random variable.

3 Case Study

As a sample application of fuzzy programming approaches in the context of supply chain planning, a real medical device industrial case is presented in this section. The case study is adopted from Pishvaee et al. [65]. The studied case is an Iranian single-use medical needle and syringe (SMNS) producer entitled AVAPezeshk (AVAP) (www.avapezeshk.com). AVAP has approximately 70 % of the market in Iran and has a production plant with about 600 million production capacity per year for satisfying the customers' demand. SMNS is a broadly used medical device as WHO [90] stated, 16 billion injections are carried out annually. On the other hand, reusing unsterilized needles and syringes leads to 8–16 million hepatitis B, 2.3–4.7 million hepatitis C and 80,000–160,000 human immunodeficiency virus (HIV) infections per year [90]. Therefore, the efficient management of end-of-life (EOL) SMNS can diminish its undesirable social and environmental effects. There are three main options for EOL SMNS as follows: (1) incineration, (2) safe landfill and (3) recycling by steel and plastic recycling centers. The first method is widely used due to its cost efficiency despite its significant environmental burden like air pollution. The second method also has some adverse environmental impacts and the third is employed rarely due to possibility of infection of end-of-life medical needle. The underling structure of the concerned supply chain is depicted in Fig. 9. Both forward and reverse streams involved in this network. Through the forward flow, new products are transported to customer zones to satisfy demands (unmet demand is not permitted) while in reverse flow, the EOL products which are returned after the consumption, are collected and disassembled by collection centers and then are shipped to recycling, landfill and Incineration centers.

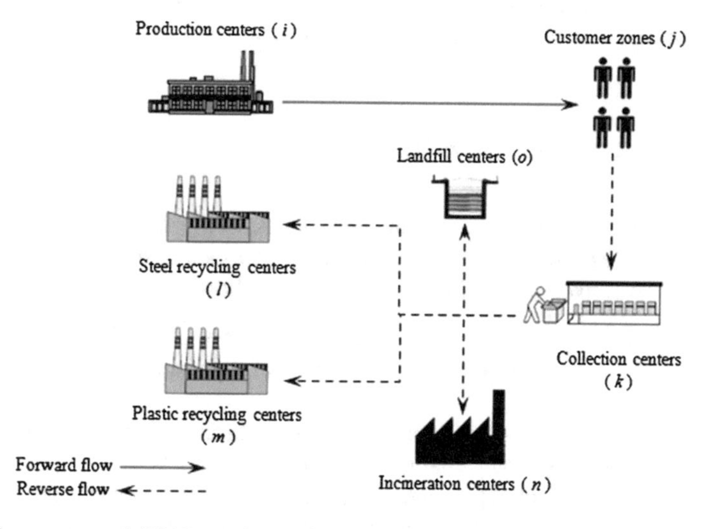

Fig. 9 The structure of AVAP supply chain network

In Pishvaee et al. [65], a mathematical model is elaborated to address the problem of AVAP supply chain network redesign. The proposed mathematical model determines the number, location and technology of production centers, the number, location and capacity of collection centers as well as the best strategies for returned EOL products aiming to achieve a reasonable trade-off between three dimensions of sustainability, i.e., economical, environmental and social aspects.

The input parameters of the problem are tainted with epistemic uncertainty due to incompleteness and unreliable data. Therefore, it is necessary to refer to an expert for estimating uncertain parameters. In the view of this fact, a group of experts and managers has been formed to attribute a trapezoidal fuzzy number to each uncertain parameter according to their knowledge. The credibility-based chance constrained programming approach which is proposed by Liu and Liu [45, 46] (an briefly described in Sect. 2.1.1) is applied to deal with the concerned possibilistic model. To cope with this multi-objective model a posteriori approach is selected in which the weights of economic, environment and social objectives must be assigned before solving the model. By referring to AVAP managers' opinions, higher weight is given to economic objective. Particularly, the range 0.8–1 is attributed to the weight of economic objective while the range of 0–0.15 is selected for environment and social objectives. The step size is set to 0.05 and the minimum confidence level of chance constraints is set as 0.9. The above-mentioned model is solved by GAMS 22.9.2 optimization software. As Fig. 10 shows, in the network with objective function weight of (1, 0, 0), in comparison to the network formed for weight vector (0.8, 0.15, 0.05) less collection centers are established and their locations are more centralized. While production centers are located in the same positions for both of instances, the more cost-oriented network, i.e., (1, 0, 0) prefers using the cost-efficient production technology type and the other uses the environmental friendly production technologies. Also, the network with the weight of (0.8, 0.15, 0.05) suggests to open some recycling centers while the network with the weight of (1, 0, 0) selects safe landfill method as the EOL strategy.

4 Future Research Directions

Given the current state-of-the-art literature in assessing the uncertainty and the application of fuzzy mathematical programming methods in supply chain planning, there are various avenues for further research as follows:

- Considering hazard and deep uncertainty in supply chain risk assessment and imbedding these sorts of uncertainty in decision support models in the context of supply chain planning.
- Assessing and fostering resilience and robustness in different types of decision support models, e.g. mathematical programming models.
- Lack of mathematical approaches for hybrid uncertain environments in spite of the fact that most of uncertain models involve multiple types of uncertainties.